Sustainable Green Synthesised Nano-Dimensional Materials for Energy and Environmental Applications

Editors

Sathish-Kumar Kamaraj
IPN-Centro de Investigación en Ciencia Aplicada y Tecnología Avanzada
Unidad Altamira, (CICATA Altamira)
Tamps, México

Arun Thirumurugan
Universidad de Atacama
Vallenar, Atacama, Chile

Shanmuga Sundar Dhanabalan
RMIT University
Melbourne, Victoria, Australia

Suresh K. Verma
School of Biotechnology, KIIT University
Bhubaneswar, Odisha, India

Shanavas Shajahan
Department of Chemistry
Khalifa University
Abu Dhabi, UAE

CRC Press
Taylor & Francis Group
Boca Raton London New York

CRC Press is an imprint of the
Taylor & Francis Group, an **informa** business

A SCIENCE PUBLISHERS BOOK

First edition published 2025
by CRC Press
2385 NW Executive Center Drive, Suite 320, Boca Raton FL 33431

and by CRC Press
4 Park Square, Milton Park, Abingdon, Oxon, OX14 4RN

© 2025 Sathish-Kumar Kamaraj, Arun Thirumurugan, Shanmuga Sundar Dhanabalan, Suresh K. Verma and Shanavas Shajahan

CRC Press is an imprint of Taylor & Francis Group, LLC

Library of Congress Cataloging-in-Publication Data (applied for)

ISBN: 978-1-032-42318-0 (hbk)
ISBN: 978-1-032-42321-0 (pbk)
ISBN: 978-1-003-36224-1 (ebk)

DOI: 10.1201/9781003362241

Typeset in Times New Roman
by Prime Publishing Services

Preface

The goals of sustainable development are the primary focus of research and development for forthcoming technology. Nanoscience and nanotechnology, in which the properties of materials change according to the nanoscale dimensions, are some of the innovative forms of technology that will be used in the future. The manufacturing of nanomaterials by greener techniques has gained a larger significance in comparison to the creation of nanomaterials through physical and chemical processes. This is because greener methods are better for the environment, have lower toxicity levels, are cheaper, have more biocompatibility, and offer better size control. In order to prepare these nano-dimensional materials, hazardous chemical compounds are frequently used. These chemicals have the potential to have an impact on the surrounding environment either directly or indirectly. As a result, there is an immediate requirement to switch to the green synthesis approach, which is favourable to the environment and focuses on the sustainable development goals set by the United Nations. Over the past few years, green synthesis has become increasingly popular in the context of the ongoing research and development of nanomaterials. This may be attributed to the fact that it addresses the problems associated with the sustainability of conventional methods. Green synthesis is a subset of green chemistry, which is a science that evolved after it was recognised that chemical businesses need sustainable procedures. The goal of green chemistry is to design safer chemical items and processes that reduce or eliminate the manufacture and use of hazardous components. Green synthesis is a subset of green chemistry. Additionally, green synthesis offers the benefits of low-cost manufacturing, efficient use of energy, safe processes and products, and decreased generation of waste.

Green synthesis of nanomaterials is a topic that is continually growing, with applications across a wide range of industries, most notably energy and the environment. However, in order to completely optimise and capitalise on the benefits that biogenic nanoparticles provide, certain fields call for a more in-depth understanding as well as further technological innovation. Concerns have been raised regarding the overuse and exhaustion of the world's resources, in particular, due to the expansion of the world's population and the ever-increasing demand for energy worldwide, both of which have pushed attention towards the viability of the evolution model of humans. In this scenario, the development of sustainable, innovative materials for next-generation energy devices is of the utmost importance, and the creation of nanostructured materials has a lot of promise in this regard.

Emerging as a major problem on a global scale is environmental contamination. The existence of persistent organic pollutants in the aquatic environment is one of the primary causes for concern across the globe, even though all of the natural resources are subjected to the potentially harmful impacts of various contaminants in the same measure. For example, the UNESCO development report for 2020 states that 68% of the world's population is drinking water that is not appropriate from a sanitary standpoint, and it estimates that water pollution costs the United States of America 2.2 billion dollars in losses.

In this context, our book summarised the numerous experimental research that focused on the sustainable green chemistry approach to address the various problems associated with sustainable energy generation systems and energy storage systems. And then expand it towards the many environmental issues that need to be remedied.

Sathish-Kumar Kamaraj
Arun Thirumurugan
Shanmuga Sundar Dhanabalan
Suresh K. Verma
Shanavas Shajahan

Acknowledgements

First and foremost, we give thanks to the Almighty for giving us good health and the crucial chance to finish this book. We want to express our sincere gratitude to the series editor and advisory board for accepting our book and for their encouragement and assistance during the writing process. We would like to express our gratitude to all the authors and reviewers for their significant contributions and unwavering support in making this book possible. We take great delight in thanking the several publishers and writers who kindly permitted us to use their tables and figures.

Sathish-Kumar Kamaraj would like to thank the director of the Centro de Investigación en Ciencia Aplicada y Tecnología Avanzada, Unidad Altamira (CICATA Altamira) and the director general of the Instituto Politécnico Nacional (IPN) for their ongoing support and facilities that enable the research activities. I appreciate the work of Secretara de Investigación y Posgrado (SIP) – IPN on project SIP:20231443. Additional extensions have been granted to the Secretary of Public Education (SEP-Mexico) and the National Council of Humanities, Sciences, and Technologies (CONAHCyT-Mexico). He expresses his gratitude to Mrs. Mounika Kamaraj and Bbg Aarudhra for providing family assistance.

Arun Thirumurugan would like to express his gratitude to Dr. Justin Joseyphus (NIT-T, India), Prof. P.V Satyam (IOP, India), Dr. Ali Akbari-Fakhrabadi (FCFM, University of Chile, Chile), and Prof. RV Mangala Raja (University of Adolfo Ibanez, Santiago, Chile) for their kindness and guidance. Arun Thirumurugan would like to thank Dr. R. Udaya Bhaskar and Mauricio J. Morel (University of ATACAMA, Chile), Carolina Venegas, Yerko Reyes, and Juan Campos, Sede Vallenar, University of ATACAMA, Chile for their support. Arun Thirumurugan acknowledges Agencia Nacional de Investigación y Desarrollo de Chile (ANID) for the financial support through PAI project SA 77210070.

Shanmuga Sundar Dhanabalan would like to express his sincere thanks to Prof. Sivanantha Raja Avaninathan (Alagappa Chettiar Government College of Engineering & Technology, Karaikudi, Tamil Nadu, India), Prof. Marcos Flores Carrasco (FCFM, University of Chile, Chile), Prof. Sharath Sriram, and Prof. Madhu Bhaskaran (Functional Materials and Microsystem, RMIT University, Australia) for their continuous support, guidance, and encouragement.

Suresh K. Verma would like to thank Prof. M. Suar, Director of R&D and Innovation (KIIT University), Prof. V. Raina (KIIT University), and Mr. Krishn Verma for their continuous guidance and support. He also acknowledges his team, including Ms. Aishee Ghosh, Ms. Dibyangshee Singh, Mr. Aditya Nandi, Ms. Adrija Sinha, and Ms. Anmol Choudhary, for their support. Dr. Suresh K. Verma also acknowledges the Department of Biotechnology, Gov. of India, for infrastructure support available through the DBT-BUILDER program (BT/INF/22/SP42155/2021) at KIIT UNIVERSITY.

Contents

CHAPTER 1

Biogenic Metal Nanoparticles
Biosynthesis to Applications

Shaheen Husain,[1,]* *Lopamudra Bhadra,*[2] *Faizan Zarreen Simnani,*[2] *Aditya Nandi,*[2] *Pritam Kumar Panda*[3,]* *and Suresh K. Verma*[2,]*

1. Introduction

Nanotechnology is a branch of science that deals with nanoparticles or things that are smaller than a nanometer (NPs). Nanomaterials are tiny solid particles with a size between one and one hundred nanometers. Due to their enhanced and unique physicochemical characteristics, such as their extremely small dimensions, enormous surface area compared to their bulk, and greater reactivity, nanomaterials hold promise for many applications for human welfare (Khan et al. 2019). Adding many functional groups to the nanoparticle can improve the quality of products. Therefore, nanoproducts are widely used in different industrial, medical, and dental sectors. Nanomaterials play a vital role in different applications in medical science that commonly occur in a few basic forms. As the field of nanotechnology advanced, novel nanomaterials became apparent, having different properties as compared to their larger counterparts. This difference in the physiochemical properties of nanomaterials can be attributed to their high surface-to-volume ratio. Due to these unique properties, they make excellent candidates for biomedical applications, as a variety of biological processes occur at nanometer scales. NPs can be synthesized using a variety of chemicals and physical methods (Figure 1) (Iravani et al. 2014), photochemical reduction (Sergeev and Klabunde 2013), electrochemical reduction (Aresta et al. 2014), and heat vaporization (Che et al. 2022). These processes involve

[1] Amity Institute of Nanotechnology, Amity University, Noida, Sector 125, Uttar Pradesh, 201313.

[2] KIIT School of Biotechnology, KIIT University, Bhubaneswar 751024, Odisha, India.

[3] Condensed Matter Theory Group, Materials Theory Division, Department of Physics and Astronomy, Uppsala University, Box 516, SE-751 20 Uppsala, Sweden.

* Corresponding authors: shaheen.husain11@gmail.com; pritamkp15@gmail.com; sureshverma22@gmail.com

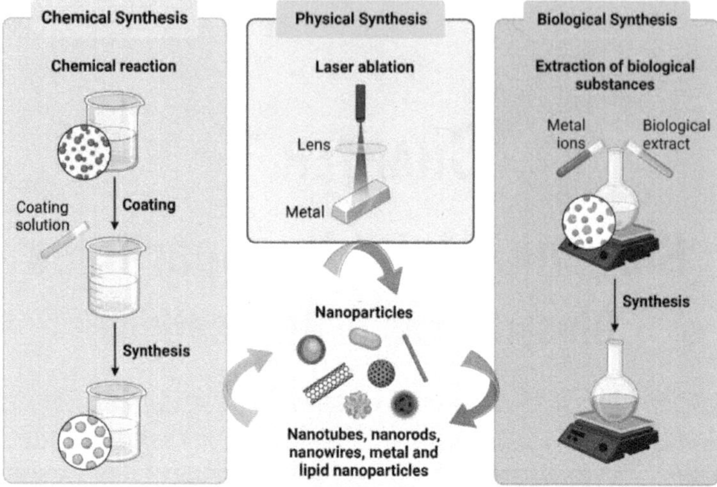

Figure 1. Nanoparticle synthesis routes via chemical, physical and biological methods.

several toxic chemicals as reducing agents. Noble metal nanoparticles are being used in places where people come into contact with them, necessitating the development of eco-friendly biosynthesis techniques that reduce the use of hazardous chemicals (Husain et al. 2021, 2015). The biological methods for the synthesis of nanoparticles are better than the chemical and physical methods due to slow kinetics, which offers better control over crystal growth (Husain et al. 2019). Chemical methods are not eco-friendly, and physical methods have limitations of time and money (Carmona et al. 2017). In the biological approach, microorganisms, enzymes, and plant extracts provide biological reducing agents (Anu et al. 2020, Husain et al. 2019). As a result, technologies for synthesizing "green" nanoparticles have been created. Because no hazardous compounds are utilized, this alternative uses biological systems such as yeast, fungus, bacteria, and plant extracts, making it a safer and more environmentally friendly alternative to chemical approaches. Green extracts are widely used for a variety of reasons, including their enormous and accessible reserves, global distribution, safe handling, availability of a diverse range of metabolites with high reducing potentials, and low waste and energy costs (Husain et al. 2015, Hashem and Salem 2022). Biologically synthesized nanoparticles are appealing for a variety of applications due to their distinctive features, such as drug delivery, imaging, catalysis, energy production, environmental remediation, etc. Overall, biologically synthesized nanoparticles are appealing for a variety of applications due to their special qualities, and scientists are actively looking for novel ways to maximize their potential. This chapter provides a general overview of the different types, synthesis methods, characterizations, properties, and applications of NPs.

2. Biosynthesis of Metal Nanoparticles (MNPs)

Traditional techniques have long been utilized, but studies have shown that biosynthesis is the most successful method for creating NPs because it has fewer

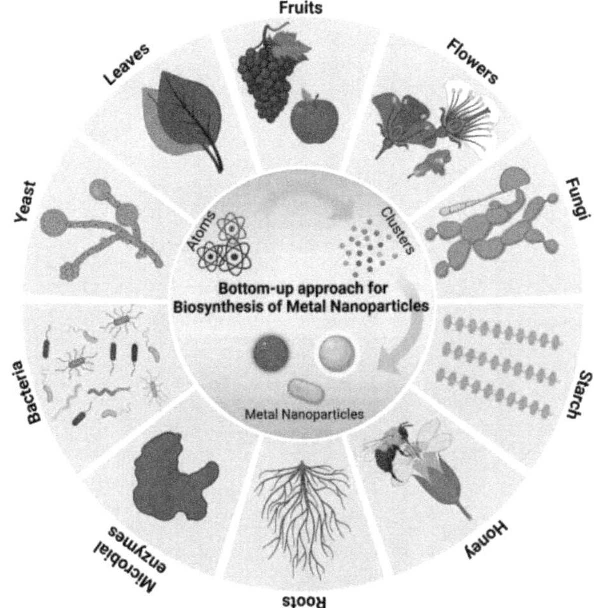

Figure 2. Schematic representation of bottom-up approach for the biosynthesis of metal nanoparticles via several biogenic compounds.

risks of failure, is less expensive, is easier to characterize, and particles are distinct from those created by physical and chemical methods. The bottom-up mechanism for creating MNPs is like the chemical approach in which biological components such as a plant extract replace a costly chemical-reducing agent (Figure 2). Biological organisms have great potential for producing NPs. Green reduction of MPs to NPs is favorable to the environment and is sustainable, chemical-free, less expensive, and scalable.

There are different methods of biosynthesis of metal nanoparticles, which are as follows:

2.1 Green Synthesis of MNPs Through Plants

Plant nanotechnology has recently opened new pathways to produce nanoparticles and is an environmentally benign, simple, quick, and stable technique. Using water as a reducing solvent to synthesize nanoparticles has several benefits, including biocompatibility, scalability, and medicinal application (Liaqat et al. 2022). This means that plant-derived nanoparticles may meet the rising demand for nanoparticles with applications in biomedicine and the environment since they are made from easily accessible plant components and are not hazardous (Simon et al. 2022). Gold and silver nanoparticle synthesis utilizing *Panax ginseng* leaf and root extract has recently been shown to be possible using medicinal plants as sources of raw materials (Singh et al. 2015). The use of plants or plant extracts for metal and metal-based hybrid nanoparticle synthesis represents a novel green synthesis approach

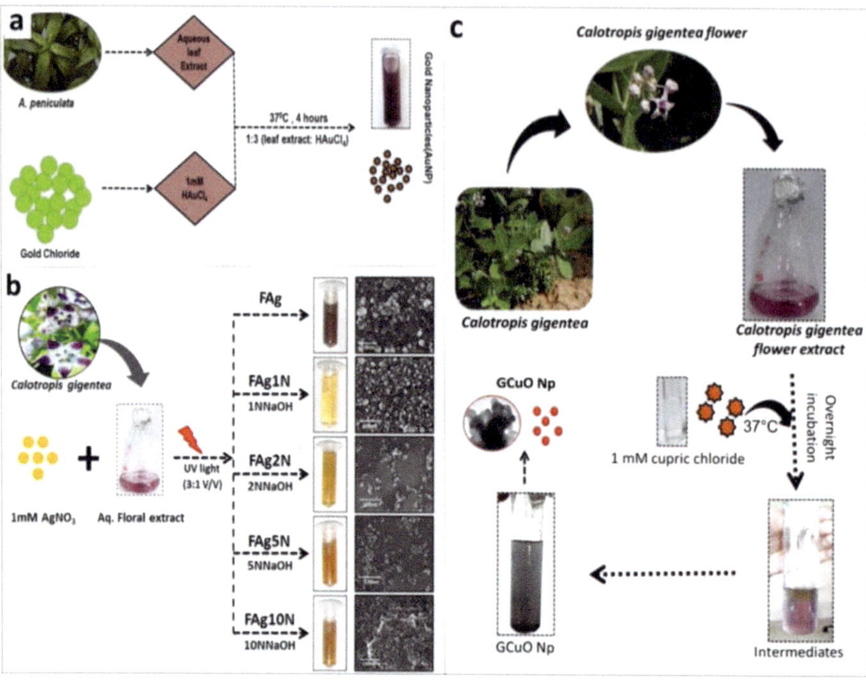

Figure 3. a. Green synthesis of gold nanoparticles (AuNPs) using aqueous leaf extract of *Andrographis peniculata* (figure adapted and modified from Kumari et al. 2019). b. Schematic diagram showing the green synthesis of silver nanoparticles (AgNPs) using *Calotropis gigantea* floral extract (figure adapted and modified from Verma et al. 2018). c. Green synthesis of copper oxide nanoparticles from the floral extract of *Calotropis gigantea* using Cupric chloride as a precursor. GCuO: Green prepared CuO; Np: Nanoparticle (figure adapted and modified from Kumari et al. 2018).

that has attracted significant research interest. In addition, metal nanoparticles have been synthesized using different plant components, such as leaves, fruits, and stems, and their extracts (Figure 3). Furthermore, it has been suggested that secondary metabolites such as flavonoids and alkaloids have important functions in metal salt reduction and capping and stabilizing agents for nanoparticles generated from proteins and amino acids (Duan et al. 2015).

Metallic NPs produced from plant extracts are stable and mono-dispersible when the pH, incubation time, mixing ratio, and temperature are accurately regulated. Curry, mango, neem, turmeric, and guava have all been utilized to create Gold NPs. Plant extracts are rich in polyphenols, which hasten the breakdown of organic materials (*Green Metallic Nanoparticles: Biosynthesis to Applications*, n.d.). According to the metal bioaccumulation study, nanoparticles (NPs) are the most frequent metal deposits (Chopra et al. 2022). If you look at extracts from plants and look for compounds like terpenoids and phenolic acids (Md Mominur Rahman et al. 2022) as well as proteins in spectroscopic measurements, you will see that metal ions may be reduced to nanostructured forms. Satoaki Onitsuka and others synthesized gold and silver nanoparticles (NPs) from the *Camellia sinensis* plant extracts. In the production of silver NPs, the leaves of *Azadirachta indica* and *Triphala* were employed. Neem leaves yielded Ag NPs of 43 nm in diameter, whereas *Triphala* leaves yielded 59 nm in diameter.

2.2 Fungi-Assisted MNPs

Synthesis fungal biomass and associated metabolites are used to synthesize NPs in a relatively young field of nanotechnology known as "myco nanotechnology" (Gade et al. 2010). Micro- and macrofungi alike have several reducing enzymes and proteins, which provide a significant advantage in the production of NP. As opposed to bacteria, fungi generate a diverse spectrum of enzymes, which allows the transformation of metal salts into NPs to occur very quickly in contrast. The bio-potential of the fungal cell wall is also assumed to be significant in the absorption and reduction of metal ions and generation of metal NPs (Khan et al. 2018). It is still unclear precisely how NPs are generated or what biological components are involved in the process. It is suggested that fungus-mediated NPs arise either *in vivo* or *in vitro*. Most of the harmful transition metal ions are converted to a non-toxic form in the mycelia of the fungus during the *in vivo* method, which exploits this process to make NPs intracellularly. It is possible to directly use washed mushroom mycelia to produce NPs intracellularly in this method. Mycelia must undergo additional treatment to remove the NPs from the mycelia before being used again (Molnár et al. 2018). In contrast, there are three approaches to making NPs from fungal cell-free extracts using the *in vitro* methodology. The first technique to produce NPs is to use the fermented fungus's supernatant, which contains extracellular proteins and enzymes (Vágó et al. 2016), according to the second approach (Siddiqi and Husen 2016). Bioengineered nanoparticles (e.g., AuNPs) may be generated by the intracellular components released into the medium due to the breakdown of cell walls. The use of fungal mycelia's aqueous extract to produce NPs is also an option. NPs may be produced by the autolysis of fungal cells, followed by the dissolution of membrane proteins and surface carbohydrates in the solution. The washing and re-suspension of fungal mycelia in a pathogen-free environment is a difficult and not always possible operation using this technique, as previously indicated (Kitching et al. 2016). Most well-known NP syntheses use Basidiomycota to produce edible mushrooms (Aygün et al. 2020). Oyster and *ganoderma* species have been the subject of several studies in the last few years to learn more about the synthesis of NPs. The mushrooms produced through pure culture are non-pathogenic and non-toxic and can create a wide variety of physiologically active proteins (Chopra et al. 2021). *Ganoderma* sp. is one of the most investigated mushrooms to produce NPs. Over 250 different varieties of this fungus have been identified. This includes *G. lucidum*, *G. applanatum*, *G. capense*, and *G. tsugae*. Pharmacological evaluation of *Ganoderma* sp. has demonstrated its antibacterial, anti-HIV, anti-inflammatory, anti-proliferative, anti-diabetic, anti-cancer, hypocholesterolemic, and hepatoprotective potential (Sudheer et al. 2019, Mominur Rahman et al. 2021, Mohammad Mominur Rahman et al. 2022). These mushrooms have also been proven to be helpful in the synthesis of NPs, particularly AgNPs. There are certain drawbacks when using mushrooms to synthesize metallic NPs, such as the need to maintain aseptic growing conditions, the possibility of contamination in samples, and the variability in the NP size when using these mushrooms (Li et al. 2011). AgNPs were produced using *G. sessiliforme* and showed significant antibacterial and antioxidant activity (Mohanta et al. 2018). There are two prominent mushrooms, *G. lucidum*, and *G. applanatum*, recognized for their bioactive components and antibacterial qualities in the culinary and medical worlds. To synthesize AgNPs, the mushrooms' extracts were employed (Poudel et al. 2017, Al-Ansari et al. 2020). Most AuNPs were synthesized from *P. eous* and *P. florida*, which had the most remarkable ability to reduce ferric oxide. A human colon cancer cell line, HCT-116, was tested with the AuNPs produced using Lentinus sajor-caju (Fr.) extract as the reducing agent (Chaturvedi et al. 2020).

2.3 Bacterial Mediated Synthesis of MNPs

Critical metals must permeate the cell wall into the cytoplasm and then return through the meshwork of the cell wall to be discharged into the environment. Since the cell wall's peptidoglycans provide polyanions for the metal-to-chemical reactive group stoichiometric interaction, metal is deposited on the cell wall in an inorganic form. Many metal-binding sites on the wall may be changed using chemical methods to convert positive charge to negative charge. Bacteria may benefit from dipole moments created by metal deposition by aligning themselves with the geomagnetic field. A cell's internal and exterior environments and bacteria species with different morphologies frequently influence the crystalline and non-crystalline phases of particle formation when particles are created. In the silver-resistant bacterial strain, *Pseudomonas stutzeri AG259*, which was isolated from a silver mine, internal accumulation of silver NPs, as well as some silver sulfide, with sizes ranging from 35 nm to 46 nm, was observed (Slawson et al. 1992). When *P. stutzeri AG259* was exposed to high concentrations of silver ions during growth, larger particles were produced, resulting in the intracellular synthesis of silver NPs ranging in size from a few nm to 200 nm (Klaus-Joerger et al. 2001). The thickness of the crystals was regulated by the periplasmic space but not their width, which might be rather large (100–200 nm) due to the presence of the periplasmic space. Psychrophilic bacteria *Phaeocystis antarctica, Pseudomonas meridiana, Arthrobacter kerguelensis, Arthrobacter gangotriensis*, and two mesophilic bacteria, *Bacillus indicus* and *Bacillus cecembensis*, were employed to biosynthesize silver nanoparticles bacteria break down Ag^+ to its elemental form (Ag^0) outside the cell. Several shapes and sizes of AgNPs may be found in extracellularly produced AgNPs. These include hexagonal, spherical, triangular, circular, and cuboidal, depending on the culture medium utilized for the growth of bacteria. The reducing agent for the biogenic reduction of Ag^+ to Ag^0 is the proteins on the bacterial cell wall or tiny soluble secretory enzymes. Extracellular synthesis of AgNPs by many bacterial taxa has been functionally described in the natural environment (Islam et al. 2021). Extracellular and intracellular synthesis of AgNPs by *Proteus mirablis* and *Vibrio alginolyticus*, respectively, have been observed in different media and growth conditions (Samadi et al. 2009, Rajeshkumar and Malarkodi 2014).

2.4 Algae-Assisted Synthesis of MNPs

Pterocladia capillacae, Jania rubins, Ulva faciata, and *Colpmenia sinus* (El-Rafie et al. 2013) algae species have been used to synthesize silver nanoparticles. The NPs were 7–20 nm in diameter and spherical. Researchers believe their antibacterial action is caused by a blockage of bacterial cell processes caused by their adhesion to the cell wall. *Pithophora oedogonia,* a freshwater green alga, has been used to synthesize silver nanoparticles with a diameter of 25–44 nm. Carbohydrates, saponins, steroids, and proteins were shown to decrease $AgNO_3$ to silver nanoparticles by IR spectroscopy and quantitative analysis of the extract. Compared with Gram-positive bacteria, they were shown to be more efficacious (Sinha et al. 2015). The production of silver nanoparticles from the marine alga *Caulerpa racemosa* and their antibacterial efficacy against human diseases have also been reported by Kathiraven et al. (2015). *Staphylococcus aureus* and *Proteus mirabilis bacteria* were killed at a low 5–15 L (5–25 nm) silver nanoparticles having face-centered cubic shape. Using a combination of 14 bacteria and microalgae, silver nanoparticles were created. Even in the dark, extracellular polysaccharides were producing nanoparticles. Silver nanoparticles of varying sizes and morphologies were found, ranging from species to species (Patel et al. 2015). Six harmful microorganisms were used to assess the antibacterial activity. The cell membrane is damaged due to the production of free radicals. Living cells of the *Euglena gracilis* microalgae produce gold nanoparticles by Dahoumane and others (2016). Like other marine algae, the biomaterial in the alga acts as a reducing agent, capping agent, and catalyst. The yield of nanoparticles is influenced by several variables, including pH, reaction time, temperature, and concentration. *Padina tetrastromatica,* a macroalga, produced crystalline spherical nanoparticles with a size range of 5–35 nm (Gopinath 2015). However the antifungal, antibacterial, and anticancer effects of Ag nanoparticles generated by seaweeds have also been discovered. Biosynthesis of both Ag and Au nanoparticles has been achieved by Ramkumar Vijayan et al., using an aqueous solution containing an extract from *Turbinaria conoides.* The nanoparticles were also tested for their ability to inhibit the development of biofilms (Nag et al. 2021). Antimicrobial nanoparticles derived from a *Sargassum plagiophyllum* aqueous extract have also been demonstrated to be effective against bacterial pathogens such as *Escherichia coli* (Stalin Dhas et al. 2014). Brown seaweed (*Fucus vesiculosus*) was used to bio-sorb and reduce Au, resulting in nanoparticles of varied sizes and morphologies (Mata et al. 2009). *Sargassum wightii Greville,* a marine alga, has also been used to produce Au nanoparticles by Singaravelu et al. (2007). Alga generated stable nanoparticles that ranged from 8 nm to 15 nm in diameter and were spherical. Researchers have been able to produce a wide range of stable nanoparticle sizes in similar studies by Luangpipat et al. (2011) using *Chlorella vulgaris,* while Senapati et al. (2012) reported the biosynthesis of Au nanoparticles. Green alga spirogyra insignis and red alga *Chondrus crispus* were used by Castro et al. (2013) to synthesize Au nanoparticles. A brown alga, *Stoechospermum marginatum,* was used to biosynthesize gold nanoparticles by Arockiya Aarthi Rajathi et al. (2012a). Their analysis found that the nanoparticles were crystallized and varied in size from 18.7 nm to 93.7 nm, with a limited number of hexagonal and triangular platelets in the mix. Diterpenoids in brown seaweed were discovered to be directly engaged in reducing Au by the hydroxyl groups. The nanoparticles also showed antibiotic efficacy against various bacterial pathogens (Arockiya Aarthi Rajathi et al. 2012b). For example, brown seaweed (*Turbinaria ornate* and *Padina pavonica*) exhibit biosynthesized Au nanoparticles ranging in size from 7 nm to 11 nm (Jegadeeswaran et al. 2016) and from 30 to 70 nm (*Padina pavonica*). The biosynthesis of gold nanoparticles by two freshwater algae species has also been shown (Sharma et al. 2014). These species are the green alga *Prasiola crispa* and the red alga *Lemanea fluviatilis*. Abboud et al. (2014b) reported a biosynthesis of copper oxide nanoparticles utilizing a brown alga extract (*Bifurcaria bifurcata*). Nanoparticles of cuprous oxide (Cu_2O) and cupric oxide (CuO) were generated in a simple technique. A few nanoparticles were elongated, but most of the particles were spherical. An average particle size of 22.6 nm was discovered for the samples, ranging from 5 nm to 45 nm. Copper oxide nanoparticles were shown to be effective against both *Enterobacter aerogenes* and *Staphylococcus aureus* in subsequent antibacterial experiments (Abboud et al. 2014a). The manufacture of copper-cored copper oxide nanoparticles utilizing red seaweed extracts (*Kappaphycus alvarezii*) was also described in recent work by Khanehzaei et al. (2014); it showed how stabilized copper-cored cuprous oxide nanoparticles with a mean particle size of 53 nm were synthesized in the presence of seaweed. Nanoparticle surfaces were also discovered to be capped by pairs of electrons, some hydroxide and sulfur groups from the water-soluble sulfated polysaccharides present in seaweed cell walls (Khanehzaei et al. 2014). One-step green biogenic synthesis of ferric oxide (Fe_3O_4) nanoparticles using brown seaweed was recently shown in work by Mahdavi et al. 2013 (*Sargassum muticum*). To make Fe_3O_4 nanoparticles, an aqueous seaweed extract was combined with an aqueous ferric chloride solution. Reduction and capping are accomplished by the amino, carboxy, and hydroxyl functional groups produced from the water-soluble polysaccharide cell walls (Mahdavi et al. 2013). The average size of the particles formed was 18 nm, crystalline, and cubic in shape. It was shown that the Fe_3O_4 nanoparticles generated by Namvar et al. (2014) have anticancer efficacy against human cancer cell lines, including leukemia, breast cancer, cervical cancer, and liver cancer when used *in vitro* tests. The buildup of Fe_3O_4 nanoparticles in treated cells was shown to increase cell death *in vitro* tests and proved their potential utility in cancer therapy (Namvar et al. 2014). Another work used the marine green alga *Caulerpa serrulate* to bio-fabricate stable colloidal crystalline AgNPs. The manufactured NPs were found to be between 10 nm and 2 nm in diameter and spherical in form by TEM. The photocatalytic activity was shown, with 99% of Congo red dye degraded after only 6 minutes of incubation. They also showed antibacterial action against Gram-negative and Gram-positive bacteria, including *Staphylococcus aureus*, *Shigella* sp., *Salmonella typhi*, and *Escherichia coli* (Aboelfetoh et al. 2017). Another complex dye, methylene blue (MB), is hazardous to living beings, making its breakdown a critical concern for both the environment and biology. After 30 minutes of exposure to light and NaBH4, Edison et al. (2016)

could bio-generate AgNPs that could totally break down MB in the presence of the marine green alga *Caulerpa racemosa*. Bioactive molecules to produce AgNPs have been discovered by combining a sulfated polysaccharide from the marine red alga *Porphyra vietnamensis* with silver nitrate. Antibacterial activity against Gram-negative and Gram-positive bacteria was shown by the production of NPs with an average diameter of 13 nm (Venkatpurwar and Pokharkar 2011). A chloroauric acid (HAuCl4) solution and an aqueous extract of marine microalgae (*Tetraselmis suecica*) were used to synthesize and characterize gold nanoparticles (AuNPs). There was a distinct band in the UV–Vis spectrum that corresponded to the formation of AuNPs. However, the most common is 79 nm diameter with a polydisperse and crystalline structure (Shakibaie et al. 2010). Their diameter ranged from 51 nm to 120 nm. Polysaccharide hydroxyl groups from the algal polysaccharides were shown to have an essential role in the biosynthesis of AuNPs from Padina gymnospora. The generated NPs' crystalline nature was verified by X-ray diffraction (XRD), and an AFM study showed that they were between 53 nm and 67 nm in size (M. Singh et al. 2013). Ramakrishna and his colleagues used *Sargassum tenerrimum* and *Turbinaria conoides* as reducing agents for gold ions. Two extracts of the gold nanoparticles showed photocatalytic activity by degrading 4-nitrophenol and p-nitroaniline into their corresponding amino arenes (4-aminophenol and p-phenylenediamine) and rendering naturally colored solutions (Rhodamine B and Sulforhodamine) into colorless solution in the presence of NaBH4 as a catalyst (Ramakrishna et al. 2016). Because of the additive action of nanosilver and the wide range of phytoconstituents with intrinsic antimicrobial capabilities, silver nanoparticles with exceptional stability and environmental friendliness can be easily manufactured from plant extracts and demonstrate a broad spectrum of antimicrobial activities, anticancer activities, and catalytic reduction of 4-nitrophenol (Bharadwaj et al. 2021).

3. Metal Nanoparticles Characterization Techniques

Nanoparticles have been studied using a variety of approaches to determine their size, crystal structure, elemental content, and a range of other physical features. Physical attributes can be examined using more than one technique in numerous instances. The different strengths and limits of each methodology make selecting the best method difficult, and combinatorial characterization is frequently required. Size and shape are two essential criteria addressed in the characterization of NPs. We may also assess the surface chemistry and estimate the size distribution, degree of aggregation, surface charge, and surface area. Other features and applications of NPs may be influenced by their size, size distribution, and organic ligands on their surfaces (Husain et al. 2019). There are microscopy-based techniques, e.g., scanning electron microscopy (SEM), transmission electron microscopy (TEM), and atomic force microscopy (AFM), which provide information on the nanomaterials' size, shape, and crystal structure. Other approaches, such as magnetic procedures, are tailored to certain families of materials (Titus et al. 2019). Many more techniques give further information on the nanoparticle samples' structure, elemental content, optical characteristics, and other common and more particular physical qualities. X-ray, spectroscopy, and scattering techniques are examples of these techniques. The microstructure and dispersion of NPs must be described as a function of different process parameters to optimize the material qualities of MNPs (Campbell et al. 2020). UV/visible spectroscopy is a method for determining how much light is absorbed and dispersed by a substance. UV/Vis spectroscopy is a valuable method for identifying, characterizing, and investigating gold and silver plasmonic nanoparticles because their optical properties are sensitive to size, shape, concentration, agglomeration state, and refractive index near the nanoparticle surface. Transmission electron microscopy is a high-magnification imaging technique that records the transmission of an electron beam through a sample. The preferred way for directly measuring the particle size, grain size, size distribution, and morphology of nanoparticles is to use TEM imaging. Sizing precision is usually within 3% of the actual value. DLS (dynamic light scattering) is a valuable technology for determining the properties of nanoparticles and other colloidal solutions. Because it offers information on the aggregation state of nanoparticle solutions, the hydrodynamic diameter is a valuable complement to other size studies (Campbell et al. 2020).

4. Factors Affecting Biosynthesis of Nanoparticles

Several elements influence the formation and shape of nanoparticles that have been developed. Researchers have linked these variances to the synthetic process's choice of adsorbate and catalyst (Patra and Baek 2015). Nanoparticle creation from biological extracts may also be affected by reaction conditions. Studies have shown that a reaction solution's pH significantly impacts the production of the resulting nanoparticles. The form and size of the generated nanoparticles may be affected by changes in the reaction pH. When comparing lower acidic pH values to higher acidic pH values, bigger particles are produced. The bigger particles (25–85 nm) were generated at pH two, whereas the smaller particles (5–20 nm) were created at pH three and four in research using *Avena sativa* biomass (Armendariz et al. 2004). According to the researchers, particle aggregation may have been caused by the lack of functional groups at pH 2. The bacteria *Rhodopseudomonas* capsulate was shown to produce gold nanoparticles similarly. It was discovered that, with a pH rise of 7, spherical particles measuring 10–20 nm were present. Nanoplates were formed when the reaction pH was lowered to 4 (He et al. 2007).

Another researcher demonstrated that the pH of Saudi Dates extract had an impact on the shape, reaction rate, and size of biosynthesized Pt NPs (Al-Radadi 2019). The reaction rate was found to be quicker when the dispersive medium's hydroxyl content rose. The acidified media, on the other hand, created a variety of different-sized particles. The shape and size of synthesized Pt NPs are expected to be rod-shaped at pHs 1.5, 3.5, 5, and 7, with a diameter of 700.5 nm, spherical at sizes 5.0–5.4 nm, 2.5–13.8 nm, and rod-shaped at pHs 1.5–5.5 with 700.5 nm diameter. Another key part of any synthesis is temperature. The temperature increase has shown catalytic behavior by boosting the reaction rate and efficiency of nanoparticle synthesis while using biological entities to formulate nanoparticles. According to research on neem leaf extracts and the production of AgNPs, temperature elevation (10–50°C) was linked to an increase in the reduction of Ag+ (Verma and Mehata 2016). Smaller AgNPs were formed at 50°C in the same way as Kaviya et al. found in the generation of AgNPs from citrus peel extract using different temperatures (Kaviya et al. 2011). AgNPs were also produced in this manner from Escherichia coli wasted culture supernatants (Gurunathan et al. 2009). The scientists speculated that a critical enzyme involved in the creation of nanoparticles may have been affected by elevated temperatures. But the study's findings showed that temperatures over 60°C favored the creation of larger-sized particles, which was surprising. Molecular kinetics at high temperatures causes a fast reduction of Ag^+ (which aids reduction and nucleation) at the expense of secondary reduction on nascent particle surfaces, which is why this finding was made. At higher incubation temperatures, Saudi's date extract was used by Al-Radadi to study the effect of temperature on the biogenesis of Pt NPs. According to microscopy measurements, the average particle size was 3.4 nm at 20°C and 2.6 nm at 30°C (Al-Radadi 2019). It has also been shown that temperature affects the structure of nanoparticles as well. While AgNPs were generated at ambient temperature using Cassia fistula extracts, spherical AgNPs were created at higher temperatures (over 60°C) (Lin et al. 2010). Plant macromolecules' interactions with Ag faces were assumed to be altered by high temperatures in the research, preventing the coalescence of nearby nanoparticles.

5. Applications of Metal Nanoparticles

5.1 Metal-Based Nanoparticles in Medicine

The ability to control the properties of nanoparticles makes these nanomaterials very interesting for medicine and pharmacology. The application of nanoparticles in medicine is associated with the design of specific nanostructures, which can be used as novel diagnostic and therapeutic modalities. There are a lot of applications for nanoparticles, e.g., drug delivery systems, radiosensitizers in radiation or proton therapy, bioimaging, or bactericides/fungicides (Figure 4a).

5.2 Metal Nanoparticles for Drug Delivery

Drug targeting specific organs and tissues has become a key challenge in recent years. This is because, in the case of conventional drug administration routes, there are difficulties in reaching the target with the desired dose during a defined period. Currently, thanks to the use of modern forms of drugs, so-called drug delivery systems (DDS), it is possible to overcome the problem of biochemical barriers in the body (e.g., the brain-blood barrier). The novel systems can solve difficulties related to drug solubility and protect the drug from photodegradation and pH changes (Wilczewska et al. 2012). The most popular forms of drug delivery include liposomes, liquid crystals, dendrimers, cyclodextrins, micelles, polymersomes, hydrogels and nanoparticles (nanospheres and nanocapsules) (Husain et al. 2023). Among all types of noble metal-based nanoparticles, gold nanoparticles (Au NPs) have shown a great capacity for use as potential drug delivery carriers

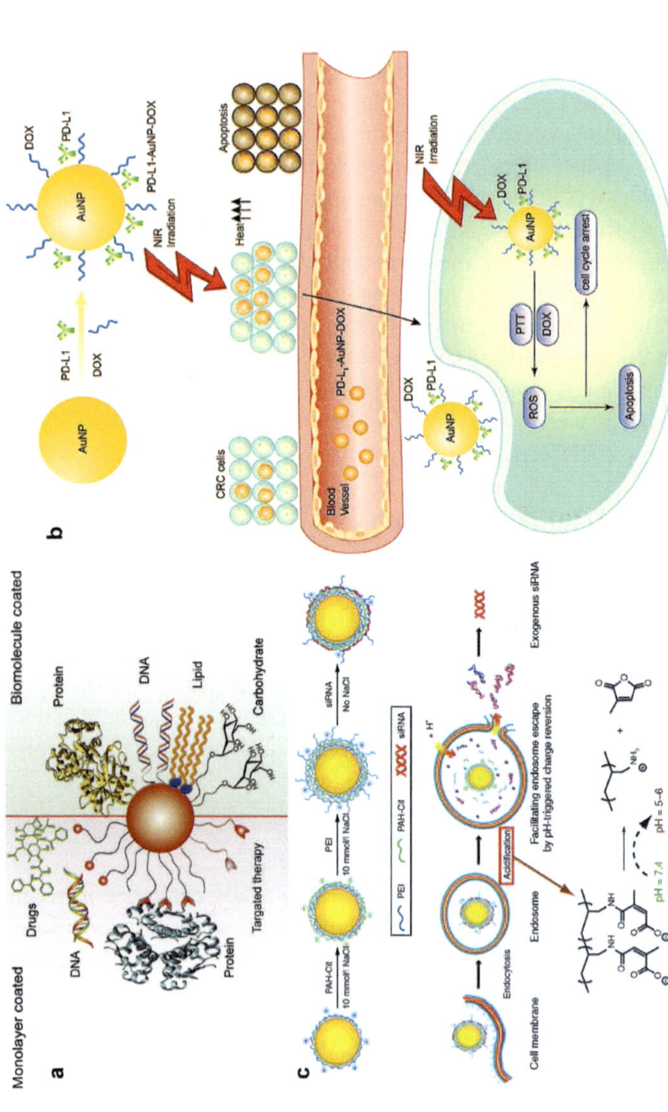

Figure 4. (a) Schematic presentation of gold nanoparticle surface structure commonly employed in the application of delivery of various biomolecules (figure adapted and modified from Rana et al. 2012). (b) Model for Colorectal cancer using anti-PD-L1-targeting gold nanoparticles conjugated with doxorubicin (PD-L1-AuNPs-DOX). Under NIR radiation, the PD-L1-AuNPs-DOX model may enhance the effectiveness of targeted delivery and facilitate the buildup of PD-L1 and DOX in cancer cells. The approach then efficiently induces ROS to promote apoptosis and cell cycle arrest to prevent Colorectal cancer cells from proliferating in vitro (Figure adapted and modified from He et al. 2021). (c) Layered-by-layered gold nanoparticle in siRNA delivery. Charge-reversal layer (cis-aconitic anhydride functionalized poly(allylamine) contained layered-by-layered gold nanoparticle synthesis. A pH-dependent layer on AuNPs facilitates the release of intracellular payload at the endosome (figure adapted and modified from Ding et al. 2014b). Applications of biosynthesized gold nanoparticles (figure adopted and modified from Ramalingam (2019).

(Figure 4b). It is relatively easy to obtain Au NPs with various sizes (1–100 nm) and shapes (spherical, rod-like, cage-like, etc.). Moreover, Au NPs can be easily functionalized with different types of molecules. Biocompatibility, stability, macroscopic quantum tunneling effect, and the presence of surface plasmon resonance (SPR) bands are also important features of Au NPs. Unfortunately, Au NPs are not biodegradable, and their surface modification can change the toxicity, biodistribution, and pharmacokinetics of the transported drugs (Klębowski et al. 2018).

Bhattacharya and co-authors discovered that Au NPs interact with folic acid PEG-amine, which enabled the delivery of medicines to tumors via ligation to the folate receptor found on the surface of some cancer cells (Klębowski et al. 2018). Tran and co-authors have obtained, using a one-pot synthesis, 3 nm and 20 nm methotrexate-conjugated gold nanoparticles (MTX-Au NPs). Methotrexate (MTX) is a cytotoxic agent, which is an antagonist of folic acid. Researchers have proven that 3 nm MTX-Au NPs show a stronger cytotoxic effect on human choriocarcinoma cell lines compared to free MTX (Klębowski et al. 2018). In other studies, novel doxorubicin-conjugated Au NPs shielded by PEGylation on the surface of Au NPs have been synthesized. This strategy made it possible to obtain better drug solubility as well as drug release in an acidic environment (Au NPs-Dox-PEG NPs). Using this approach, the intratumoral drug concentration of Au NPs-Dox-PEG NPs was twice as high as free doxorubicin (Cui et al. 2017). A similar effect has been achieved in the case of the synthesis of Au NPs conjugated with paclitaxel (PTX). This strategy has enabled more efficient anticancer therapy in a murine liver cancer model (Ding et al. 2013) (Schematic representation of siRNA delivery is depicted in Figure 4c). Bergen et al. (2006) have conjugated Au NPs with galactose-targeting ligands, allowing delivery of a ligand to the asialoglycoprotein receptor. This strategy could be used to treat hepatocellular carcinoma (Bergen et al. 2006). Recently, Farooq et al. (2018) have obtained in one-step synthesis the Au NPs loaded with a combination of two anticancer drugs having different mechanisms of action: bleomycin and doxorubicin. Using such nanohybrids has reduced systemic drug toxicity and decreased the possibility of the development of cancer drug resistance (Farooq et al. 2018). Silver nanoparticles (Ag NPs) obtained using *Morus alba* leaf extract can be successfully used to treat hepatocellular ailments. The results have shown that Ag NPs exhibit cytotoxic hepatoprotective properties in the rat model (Singh et al. 2018). Palladium-based nanoparticles (Pd NPs) may also have potential importance in delivering anticancer drug agents such as cisplatin against A549, SKOV-3, and HepG2 cell lines (Schmidt et al. 2016).

5.3 *Metal Nanoparticles for Gene Delivery*

Gene therapy is a method of delivering exogenous DNA or RNA to treat or prevent diseases. Popular viral vectors frequently activate host immune systems, reducing the efficiency of gene therapy. The abovementioned troubles can be solved using non-viral systems such as metallic nanoparticles. Recent studies showed that Au NPs with different shapes (e.g., nanospheres or nanorods) protect nucleic acid by preventing

DNA or RNA from degradation by nuclease. Au NPs conjugated to oligonucleotides show unique properties that can make them potential gene regulatory agents. These carriers can be divided into covalent (Au NPs may be functionalized with thiolated oligonucleotides) and noncovalent (Ding et al. 2014a). For example, covalent Au NPs can activate immune-related genes in peripheral blood mononuclear cells but not an immortalized, lineage-restricted cell line. This finding is promising in the application of such conjugates in the development of gene delivery systems (Kim et al. 2011). Son and co-authors have attached three fragments of nucleic acids to the surface of Au NPs. In this way, a nanomachine that silenced the polo-like kinase 1 via siRNA was obtained (Son et al. 2014). In turn, Peng and co-authors have synthesized lactoferrin-derived peptides coated Au NPs. The obtained conjugates can efficiently deliver genes encoding vascular endothelial growth factor (VEGF), inducing blood vessel formation (Peng et al. 2016).

5.4 Metal Nanoparticles for Protein Delivery

There is a growing list of evidence documenting the application of nanoparticles as protein carriers. Organothiol, a molecular probe, could be used to study the structure and the morphology of proteins attached to Au NPs (Kong et al. 2017). Joshi et al. (2006) have obtained insulin-functionalized Au NPs, which have been found useful in the transmucosal delivery of drugs for the treatment of diabetes in rat models (Joshi et al. 2006). Enhancement of insulin delivery efficiency can be achieved by covering Au NPs with a non-toxic biopolymer, such as chitosan, which strongly adsorbs insulin on its surface. Schäffler and co-authors have dealt with the conjugation of Au NPs, either with human serum albumin or apolipoprotein E. The results of these experiments showed that the attachment of proteins reduces liver retention compared to traditional citrate-stabilized Au NPs (Schäffler et al. 2014). Rathinaraj et al. (2015) obtained Herceptin (anti-HER-2/neu monoclonal antibody) immobilized on 29 nm Au NPs, improving the interaction of this drug with the suitable receptors on the surface of the breast cancer cells (SK-BR3). Ag NPs have also been used as protein carriers. For example, Farkhani et al. have combined Ag NPs with cell-penetrating peptides (CPP). CPP increases the penetration of Ag NPs across the cell membrane, causing a reduction of surviving breast cancer cells (MCF-7 cell line) (Mussa Farkhani et al. 2016). Di Pietro and co-authors have functionalized Ag NPs with a specific peptide sequence consisting of arginine, glycine, and aspartic acid (RGD), allowing for the effective entry of Ag NPs into leukemia and neuroblastoma cells (Di Pietro et al. 2016). Numerous applications of noble metal-based nanoparticles as biologically active compound carriers give hope for more effective treatment of cancer and other civilization diseases. However, the main difficulty in using these nanoparticles *in vivo* is problems with their degradation and elimination from the body. Therefore, the improvement of the pharmacokinetics of such nanoparticles should be the main goal of scientists considering this issue.

5.5 Metal Nanoparticles in Radiation-Based Anticancer Therapy

Radiation therapy (RT) is, next to chemotherapy and surgery, a commonly used method in the treatment of various types of cancer. To improve the efficiency of radiotherapy, researchers can use some small molecules (oxygen and its mimics, gemcitabine, and capecitabine), macromolecules (microRNAs, some proteins, and peptides), and nanoparticles (Wang et al. 2018). The main goal of radiation or proton therapy is the delivery of a destructive dose of radiation to cancer cells with simultaneous protection of the surrounding healthy tissue. It is possible to achieve this goal via two methods: conforming the amount to the tumor volume or enhancing the sensitivity of the cancer cells to radiation. Gamma and X-rays are characterized by exponential dose deposition with tissue depth, so part of the radiation dose is delivered in front of or behind the tumor, which is a disadvantage of RT (Retif et al. 2015). High-energy ionizing radiation such as X-rays is mainly used to ionize cellular organelles or water. Water, the main component of the cell, is the main target of ionizing radiation. As a result of this interaction, lysis of the water molecules occurs. This process, called radiolysis, causes the formation of free radicals, such as hydroxyl and hydrogen radicals, as well as charged water species. The interaction of free radicals with DNA and cellular structures induces apoptosis. It has been reported that hydroxyl ions are major sources of cellular damage by lipid peroxidation (*E Nanoparticles in Radiation Therapy: A Summary of Various Approaches to Enhance Radiosensitization in Cancer*, n.d.). High atomic number nanoparticles, such as Au NPs, can increase the production of secondary electrons or reactive oxygen species (ROS), improving the efficiency of RT (Retif et al. 2015). Nanoparticles can reduce the total dose of radiation and increase the dose administrated locally to the tumor by using this approach. Moreover, the side effects can be reduced as well (Jeynes et al. 2014). The photoelectric and Compton effects are the main physical phenomena between X-ray beams and metal-based nanoparticles. A photon is fully or partially absorbed by the nanoparticles, which causes the removal of an electron from the surface of the nanoparticles. The ionizing radiation has enough energy to separate at least one electron from the nanoparticle, resulting in ion production. Charged particles are directly ionized because they can interact directly with atomic electrons through Coulomb forces and transfer part of their kinetic energy (*Radiosensitization by Gold Nanoparticles: Effective at Megavoltage Energies and Potential Role of Oxidative Stress*, n.d.). In turn, photons are not charged and hence more penetrating. They are not directly ionized but have enough energy to free an orbital electron, the Compton electron, which is directly ionized. Photoelectric absorptions lead to an increase in the absorbed dose. Auger electrons are produced using energy released from electrons, which fall from higher orbits during a process of replacement of ejected electrons. Auger electrons can produce a higher density of ionization at a localized area so that they can deposit their energy within the vicinity of the Au NPs. This can lead to a high non-homogenous dose distribution in the nanoscale. It may suggest that the combination of RT and Au NPs enhances radiosensitization (Saberi et al. 2016). The interaction of X-rays with high-Z nanoparticles has been summarized.

The size of nanoparticles used for radiosensitization affects their interactions with biological systems and radiation. Thus, it is necessary to avoid the accumulation of Au NPs in organs such as the heart or liver. Otherwise, severe side effects may occur. Akhter et al., have shown that the toxicity of Au NPs is minimal if their sizes vary between 5 nm and 50 nm (Akhter et al. 2012). The size of nanoparticles is important from the point of view of interaction with radiation. When Au NPs become larger, more ionizing events from interaction with secondary electrons and radiation occur in the bulk of the NPs. As a result, the dose deposited in the surroundings of the Au NPs is reduced (Haume et al. 2016).

Concerning the charge of nanoparticles, there are few papers indicating that positively charged nanoparticles significantly improve cell uptake by interacting with the negatively charged cell membrane. Moreover, positively charged nanoparticles can selectively target cancer cells because of the presence of highly negatively charged glycocalyx on the surface of some cancer cells (Beddoes et al. 2015). Silver nanoparticles also exhibit radiosensitizing properties. Their antitumor mechanism is based on the induction of apoptosis, activating oxidative stress, and increasing the fluidity of the cell membrane. The radiosensitization mechanism of Ag NPs is probably related to the release of an Ag^+ cation from the silver-based structures inside the cells, which penetrated the cells. This cation can capture an electron and act as an oxidizing agent, causing ROS production (Su et al. 2014). In corroboration, Liu and co-authors have shown an increased radiation effect using Ag NPs in the treatment of glioma, obtaining anti-proliferative and pro-apoptotic effects (Liu et al. 2016). Proton therapy (PT) is another method of radiation-based treatment. In this case, the tumor is treated with a focused proton beam, allowing the tumor to be more precisely irradiated, and at the same time, healthy tissues are spared. The main advantage of PT over RT is that the maximum dose of protons occurs at a certain depth depending on the energy of the beam (so-called Bragg Peak) but not on the patient's skin. In addition, after reaching its maximum, the dose quickly falls to zero, enabling very precise irradiation of the tumor, while saving healthy tissues. Protons have similar biological effects to X-rays used in RT because the relative biological effectiveness (RBE) of the protons is approximately 1.1. RBE is defined as the ratio of a dose of standard radiation to the dose of test radiation necessary to obtain the same biological effect. The reference radiation is usually ^{60}Co photons. Additional injection of non-toxic, high-Z nanoparticles into tumor tissue can increase the absorbed dose by up to 6% (Torrisi, n.d.).

Au NPs are especially desirable in PTT because they can give a strong heating effect upon laser irradiation. This is due to enhanced absorption induced by localized surface plasmon resonance (SPR) (Yao et al. 2016). The four most common forms of Au NPs used in PTT are silica core/gold shell nano-shells (clinical trials), gold nanorods, hollow gold nanocages, and nano-stars (all in the preclinical state) (Riley and Day 2017). Camerin et al. (2010) have carried out studies using Au NPs conjugated with phthalocyanine (PC). These conjugates were effectively taken up by melatonin melanoma cells (B78H1 cell line) and resulted in more than a double increase in cell death compared to free PC in PDT (Camerin et al. 2010). Epidermal growth factor peptide-targeted Au NPs are an innovative approach to delivering PC to cancer cells, increasing the efficiency of PTD (Meyers et al. 2015). In addition to Au NPs, it is also possible to use PTD Ag NPs or Pt NPs (El-Hussein 2016, Thapa et al. 2018). As shown above, noble metal-based nanoparticles have enormous potential in improving cancer treatment. It would be interesting to check whether simultaneous treatment of tumor cells with two (or more) types of nanoparticles would give a synergistic effect. The synthesis of noble metal-based nano complexes (e.g., consisting of gold and platinum) also seems promising. Perhaps a spectacular result would also be reached via the study of the simultaneous effect of noble metal-based nanoparticles and a radiosensitizing agent of another type (e.g., gemcitabine).

6. Conclusion

Researchers focus their research on metallic nanoparticles from biogenic sources because they are easier to process and develop these kinds of metallic nanoparticles as theranostics for various infectious and noninfectious diseases. In contrast to traditional physical and chemical techniques, the biosynthesis method provides a fundamentally new way to create nanomaterials. The most recent developments in the synthesis of metallic nanoparticles using biogenic compounds elucidate the uniqueness of such nanomaterials. The biological manufacture of metal nanoparticles employing bacteria, fungi, algae, plants, etc., has been systematically demonstrated. In addition, we promoted the use of biogenic metal nanoparticles for biomedical applications like targeted drug delivery and gene therapy. The adjustments that must be made to create metallic nanoparticles as a safe, biocompatible agent are also highlighted in this chapter. In this regard, further improvement in the biosynthesis of metal nanoparticles may offer important, eco-friendly end products with a wide range of applications.

References

Abboud, Y., Saffaj, T., Chagraoui, A., El Bouari, A., Brouzi, K., Tanane, O. and Ihssane, B. (2014a). Biosynthesis, characterization and antimicrobial activity of copper oxide nanoparticles (CONPs) produced using brown alga extract (Bifurcaria bifurcata). *Applied Nanoscience (Switzerland)*, 4(5): 571–576. https://doi.org/10.1007/S13204-013-0233-X/FIGURES/7.

Abboud, Y., Saffaj, T., Chagraoui, A., El Bouari, A., Brouzi, K., Tanane, O. and Ihssane, B. (2014b). Biosynthesis, characterization and antimicrobial activity of copper oxide nanoparticles (CONPs) produced using brown alga extract (Bifurcaria bifurcata). *Applied Nanoscience*, 4(5): 571–576. https://doi.org/10.1007/s13204-013-0233-x.

Aboelfetoh, E. F., El-Shenody, R. A. and Ghobara, M. M. (2017). Eco-friendly synthesis of silver nanoparticles using green algae (Caulerpa serrulata): reaction optimization, catalytic and antibacterial activities. *Environmental Monitoring and Assessment*, 189(7): 1–15. https://doi.org/10.1007/S10661-017-6033-0/METRICS.

Akhter, S., Ahmad, M. Z., Ahmad, F. J., Storm, G. and Kok, R. J. (2012). Gold nanoparticles in theranostic oncology: Current state-of-the-art. *Http://Dx.Doi.Org/10.1517/17425247.2012.716824*, 9(10): 1225–1243. https://doi.org/10.1517/17425247.2012.716824.

Al-Ansari, M. M., Dhasarathan, P., Ranjitsingh, A. J. A. and Al-Humaid, L. A. (2020). Ganoderma lucidum inspired silver nanoparticles and its biomedical applications with special reference to drug resistant Escherichia coli isolates from CAUTI. *Saudi Journal of Biological Sciences*, 27(11): 2993–3002. https://doi.org/10.1016/j.sjbs.2020.09.008.

Al-Radadi, N. S. (2019). Green synthesis of platinum nanoparticles using Saudi's dates extract and their usage on the cancer cell treatment. *Arabian Journal of Chemistry*, 12(3): 330–349. https://doi.org/10.1016/J.ARABJC.2018.05.008.

Anu, K., Devanesan, S., Prasanth, R., AlSalhi, M. S., Ajithkumar, S. and Singaravelu, G. (2020). Biogenesis of selenium nanoparticles and their anti-leukemia activity. *Journal of King Saud University - Science*, 32(4): 2520–2526. https://doi.org/10.1016/J.JKSUS.2020.04.018.

Aresta, M., Dibenedetto, A. and Angelini, A. (2014). Converting "exhaust" carbon into "working" carbon. *Advances in Inorganic Chemistry*, 66: 259–288. https://doi.org/10.1016/B978-0-12-420221-4.00008-1.

Armendariz, V., Herrera, I., Peralta-Videa, J. R., Jose-Yacaman, M., Troiani, H., Santiago, P. and Gardea-Torresdey, J. L. (2004). Size controlled gold nanoparticle formation by Avena sativa biomass: Use of plants in nanobiotechnology. *Journal of Nanoparticle Research*, 6(4): 377–382. https://doi.org/10.1007/S11051-004-0741-4/METRICS.

Arockiya Aarthi Rajathi, F., Parthiban, C., Ganesh Kumar, V. and Anantharaman, P. (2012a). Biosynthesis of antibacterial gold nanoparticles using brown alga, Stoechospermum marginatum (kützing). *Spectrochimica Acta Part A: Molecular and Biomolecular Spectroscopy*, 99: 166–173. https://doi.org/10.1016/j.saa.2012.08.081.

Arockiya Aarthi Rajathi, F., Parthiban, C., Ganesh Kumar, V. and Anantharaman, P. (2012b). Biosynthesis of antibacterial gold nanoparticles using brown alga, Stoechospermum marginatum (kützing). *Spectrochimica Acta Part A: Molecular and Biomolecular Spectroscopy*, 99: 166–173. https://doi.org/10.1016/J.SAA.2012.08.081.

Aygün, A., Özdemir, S., Gülcan, M., Cellat, K. and Şen, F. (2020). Synthesis and characterization of Reishi mushroom-mediated green synthesis of silver nanoparticles for the biochemical applications. *Journal of Pharmaceutical and Biomedical Analysis*, 178. https://doi.org/10.1016/j.jpba.2019.112970.

Beddoes, C. M., Case, C. P. and Briscoe, W. H. (2015). Understanding nanoparticle cellular entry: A physicochemical perspective. *Advances in Colloid and Interface Science*, 218: 48–68. https://doi.org/10.1016/J.CIS.2015.01.007.

Bergen, J. M., Von Recum, H. A., Goodman, T. T., Massey, A. P. and Pun, S. H. (2006). Gold nanoparticles as a versatile platform for optimizing physicochemical parameters for targeted drug delivery. *Macromolecular Bioscience*, 6(7): 506–516. https://doi.org/10.1002/MABI.200600075.

Bharadwaj, K. K., Rabha, B., Pati, S., Choudhury, B. K., Sarkar, T., Gogoi, S. K., Kakati, N., Baishya, D., Kari, Z. A. and Edinur, H. A. (2021). Green synthesis of silver nanoparticles using diospyros malabarica fruit extract and assessments of their antimicrobial, anticancer and catalytic reduction of 4-Nitrophenol (4-NP). *Nanomaterials*, 11(8): 1999. https://doi.org/10.3390/NANO11081999.

Camerin, M., Magaraggia, M., Soncin, M., Jori, G., Moreno, M., Chambrier, I., Cook, M. J. and Russell, D. A. (2010). The *in vivo* efficacy of phthalocyanine–nanoparticle conjugates for the photodynamic therapy of amelanotic melanoma. *European Journal of Cancer*, 46(10): 1910–1918. https://doi.org/10.1016/J.EJCA.2010.02.037.

Campbell, J., Burkitt, S., Dong, N. and Zavaleta, C. (2020). Nanoparticle characterization techniques. *Nanoparticles for Biomedical Applications: Fundamental Concepts, Biological Interactions and Clinical Applications*, 129–144. https://doi.org/10.1016/B978-0-12-816662-8.00009-6.

Carmona, E. R., Benito, N., Plaza, T. and Recio-Sánchez, G. (2017). *Green Synthesis of Silver Nanoparticles by using Leaf Extracts from the Endemic Buddleja Globosa Hope*. https://doi.org/10.1080/17518253.2017.1360400

Castro, L., Blázquez, M. L., Muñoz, J. A., González, F. and Ballester, A. (2013). Biological synthesis of metallic nanoparticles using algae. *IET Nanobiotechnology*, 7(3): 109–116. https://doi.org/10.1049/IET-NBT.2012.0041.

Chaturvedi, V. K., Yadav, N., Rai, N. K., Abd Ellah, N. H., Bohara, R. A., Rehan, I. F., Marraiki, N., Batiha, G. E. S., Hetta, H. F. and Singh, M. P. (2020). Pleurotus sajor-caju-mediated synthesis of silver and gold nanoparticles active against colon cancer cell lines: A new era of herbonanoceutics. *Molecules*, 25(13). https://doi.org/10.3390/molecules25133091.

Che, X., Guo, R., Wu, F., Ren, H. and Zhou, W. (2022). Simulations of water vaporization in novel internal-intensified spouted beds: Multiphase-flow, heat and mass transfer. *Journal of Industrial and Engineering Chemistry*, 116: 411–427. https://doi.org/10.1016/J.JIEC.2022.09.032.

Chopra, H., Bibi, S., Singh, I., Hasan, M. M., Khan, M. S., Yousafi, Q., Baig, A. A., Rahman, M. M., Islam, F., Emran T. Bin and Cavalu, S. (2022). Green metallic nanoparticles: Biosynthesis to applications. *Frontiers in Bioengineering and Biotechnology*, 10. https://doi.org/10.3389/fbioe.2022.874742.

Chopra, H., Mishra, A. K., Baig, A. A., Mohanta, T. K., Mohanta, Y. K. and Baek, K. H. (2021). Narrative review: Bioactive potential of various mushrooms as the treasure of versatile therapeutic natural product. *Journal of Fungi*, 7(9). https://doi.org/10.3390/jof7090728.

Cui, T., Liang, J. J., Chen, H., Geng, D. D., Jiao, L., Yang, J. Y., Qian, H., Zhang, C. and Ding, Y. (2017). Performance of doxorubicin-conjugated gold nanoparticles: Regulation of drug location. *ACS Applied Materials and Interfaces*, 9(10): 8569–8580. https://doi.org/10.1021/ACSAMI.6B16669/SUPPL_FILE/AM6B16669_SI_001.PDF.

Dahoumane, S. A., Yéprémian, C., Djédiat, C., Couté, A., Fiévet, F., Coradin, T. and Brayner, R. (2016). Improvement of kinetics, yield, and colloidal stability of biogenic gold nanoparticles using living cells of Euglena gracilis microalga. *Journal of Nanoparticle Research*, 18(3). https://doi.org/10.1007/s11051-016-3378-1.

Di Pietro, P., Zaccaro, L., Comegna, D., Del Gatto, A., Saviano, M., Snyders, R., Cossement, D., Satriano, C. and Rizzarelli, E. (2016). Silver nanoparticles functionalized with a fluorescent cyclic RGD peptide: a versatile integrin targeting platform for cells and bacteria. *RSC Advances*, 6(113): 112381–112392. https://doi.org/10.1039/C6RA21568H.

Ding, Y., Jiang, Z., Saha, K., Kim, C. S., Kim, S. T., Landis, R. F. and Rotello, V. M. (2014a). Gold nanoparticles for nucleic acid delivery. *Molecular Therapy*, 22(6): 1075–1083. https://doi.org/10.1038/MT.2014.30.

Ding, Y., Jiang, Z., Saha, K., Kim, C. S., Kim, S. T., Landis, R. F. and Rotello, V. M. (2014b). Gold Nanoparticles for Nucleic Acid Delivery. *Molecular Therapy*, 22(6): 1075–1083. https://doi.org/10.1038/mt.2014.30.

Ding, Y., Zhou, Y. Y., Chen, H., Geng, D. D., Wu, D. Y., Hong, J., Shen, W. Bin, Hang, T. J. and Zhang, C. (2013). The performance of thiol-terminated PEG-paclitaxel-conjugated gold nanoparticles. *Biomaterials*, 34(38): 10217–10227. https://doi.org/10.1016/J.BIOMATERIALS.2013.09.008.

Duan, H., Wang, D. and Li, Y. (2015). Green chemistry for nanoparticle synthesis. *Chemical Society Reviews*, 44(16): 5778–5792. https://doi.org/10.1039/C4CS00363B.

Edison, T. N. J. I., Atchudan, R., Kamal, C. and Lee, Y. R. (2016). Caulerpa racemosa: A marine green alga for eco-friendly synthesis of silver nanoparticles and its catalytic degradation of methylene blue. *Bioprocess and Biosystems Engineering*, 39(9): 1401–1408. https://doi.org/10.1007/S00449-016-1616-7/METRICS.

El-Hussein, A. (2016). Study DNA damage after photodynamic therapy using silver nanoparticles with A549 cell line. *Research Article Journal of Molecular Nanotechnology and Nanomedicine*, 7(2).

El-Rafie, H. M., El-Rafie, M. H. and Zahran, M. K. (2013). Green synthesis of silver nanoparticles using polysaccharides extracted from marine macro algae. *Carbohydrate Polymers*, 96(2): 403–410. https://doi.org/10.1016/j.carbpol.2013.03.071.

Farooq, M. U., Novosad, V., Rozhkova, E. A., Wali, H., Ali, A., Fateh, A. A., Neogi, P. B., Neogi, A. and Wang, Z. (2018). Retracted article: Gold nanoparticles-enabled efficient dual delivery of anticancer therapeutics to HeLa cells. *Scientific Reports*, 8(1): 1–12. https://doi.org/10.1038/s41598-018-21331-y.

Gade, A., Ingle, A., Whiteley, C. and Rai, M. (2010). Mycogenic metal nanoparticles: Progress and applications. *Biotechnology Letters*, 32(5): 593–600. https://doi.org/10.1007/s10529-009-0197-9.

Gopinath, K. F. P. A. (2015). Eco-Friendly synthesis and characterization of silver nanoparticles using marine macroalga padina tetrastromatica. *International Journal of Science and Research (IJSR)*, 4(6): 1050–1054.

Gurunathan, S., Kalishwaralal, K., Vaidyanathan, R., Venkataraman, D., Pandian, S. R. K., Muniyandi, J., Hariharan, N. and Eom, S. H. (2009). Biosynthesis, purification and characterization of silver nanoparticles using Escherichia coli. *Colloids and Surfaces B: Biointerfaces*, 74(1): 328–335. https://doi.org/10.1016/J.COLSURFB.2009.07.048.

Hashem, A. H. and Salem, S. S. (2022). Green and ecofriendly biosynthesis of selenium nanoparticles using Urtica dioica (stinging nettle) leaf extract: Antimicrobial and anticancer activity. *Biotechnology Journal*, 17(2): 2100432. https://doi.org/10.1002/BIOT.202100432.

Haume, K., Rosa, S., Grellet, S., Śmiałek, M. A., Butterworth, K. T., Solov'yov, A. V., Prise, K. M., Golding, J. and Mason, N. J. (2016). Gold nanoparticles for cancer radiotherapy: A review. *Cancer Nanotechnology*, 7(1): 1–20. https://doi.org/10.1186/S12645-016-0021-X.

He, J., Liu, S., Zhang, Y., Chu, X., Lin, Z., Zhao, Z., Qiu, S., Guo, Y., Ding, H., Pan, Y. and Pan, J. (2021). The application of and strategy for gold nanoparticles in cancer immunotherapy. *Frontiers in Pharmacology*, 12. https://doi.org/10.3389/fphar.2021.687399.

He, S., Guo, Z., Zhang, Y., Zhang, S., Wang, J. and Gu, N. (2007). Biosynthesis of gold nanoparticles using the bacteria Rhodopseudomonas capsulata. *Materials Letters*, 61(18): 3984–3987. https://doi.org/10.1016/J.MATLET.2007.01.018.

Husain, S., Afreen, S., Hemlata, Yasin, D., Afzal, B. and Fatma, T. (2019). Cyanobacteria as a bioreactor for synthesis of silver nanoparticles-an effect of different reaction conditions on the size of nanoparticles and their dye decolorization ability. *Journal of Microbiological Methods*, 162: 77–82. https://doi.org/10.1016/j.mimet.2019.05.011.

Husain, S., Nandi, A., Simnani, F. Z., Saha, U., Ghosh, A., Sinha, A., Sahay, A., Samal, S. K., Panda, P. K. and Verma, S. K. (2023). Emerging trends in advanced translational applications of silver nanoparticles: A progressing dawn of nanotechnology. *Journal of Functional Biomaterials*, 14(1): 47. https://doi.org/10.3390/JFB14010047.

Husain, S., Sardar, M. and Fatma, T. (2015). Screening of cyanobacterial extracts for synthesis of silver nanoparticles. *World Journal of Microbiology and Biotechnology*, 31(8): 1279–1283. https://doi.org/10.1007/s11274-015-1869-3.

Husain, S., Verma, S. K., Yasin, D., Hemlata, A. Rizvi, M. M. and Fatma, T. (2021). Facile green bio-fabricated silver nanoparticles from Microchaete infer dose-dependent antioxidant and anti-proliferative activity to mediate cellular apoptosis. *Bioorganic Chemistry*, 107: 104535. https://doi.org/10.1016/j.bioorg.2020.104535.

Iravani, S., Korbekandi, H., Mirmohammadi, S. V. and Zolfaghari, B. (2014). Synthesis of silver nanoparticles: chemical, physical and biological methods. *Research in Pharmaceutical Sciences*, 9(6): 385.

Islam, F., Bibi, S., Meem, A. F. K., Islam, M. M., Rahaman, M. S., Bepary, S., Rahman, M. M., Rahman, M. M., Elzaki, A., Kajoak, S., Osman, H., Elsamani, M., Khandaker, M. U., Idris, A. M. and Emran T. Bin. (2021). Natural bioactive molecules: An alternative approach to the treatment and control of covid-19. *International Journal of Molecular Sciences*, 22(23). https://doi.org/10.3390/ijms222312638.

Jegadeeswaran, P., Rajiv, P., Vanathi, P., Rajeshwari, S. and Venckatesh, R. (2016). A novel green technology: Synthesis and characterization of Ag/TiO2 nanocomposites using Padina tetrastromatica (seaweed) extract. *Materials Letters*, 166: 137–139. https://doi.org/10.1016/J.MATLET.2015.12.058.

Jeynes, J. C. G., Merchant, M. J., Spindler, A., Wera, A. C. and Kirkby, K. J. (2014). Investigation of gold nanoparticle radiosensitization mechanisms using a free radical scavenger and protons of different energies. *Physics in Medicine and Biology*, 59(21): 6431. https://doi.org/10.1088/0031-9155/59/21/6431.

Joshi, H. M., Bhumkar, D. R., Joshi, K., Pokharkar, V. and Sastry, M. (2006). Gold nanoparticles as carriers for efficient transmucosal insulin delivery. *Langmuir*, 22(1): 300–305. https://doi.org/10.1021/LA051982U/ASSET/IMAGES/MEDIUM/LA051982UN00001.GIF.

Kathiraven, T., Sundaramanickam, A., Shanmugam, N. and Balasubramanian, T. (2015). Green synthesis of silver nanoparticles using marine algae Caulerpa racemosa and their antibacterial activity against some human pathogens. *Applied Nanoscience (Switzerland)*, 5(4): 499–504. https://doi.org/10.1007/s13204-014-0341-2.

Kaviya, S., Santhanalakshmi, J., Viswanathan, B., Muthumary, J. and Srinivasan, K. (2011). Biosynthesis of silver nanoparticles using citrus sinensis peel extract and its antibacterial activity. *Spectrochimica Acta Part A: Molecular and Biomolecular Spectroscopy*, 79(3): 594–598. https://doi.org/10.1016/J.SAA.2011.03.040.

Khan, A. U., Malik, N., Khan, M., Cho, M. H. and Khan, M. M. (2018). Fungi-assisted silver nanoparticle synthesis and their applications. *Bioprocess and Biosystems Engineering*, 41(1). https://doi.org/10.1007/s00449-017-1846-3.

Khan, I., Saeed, K. and Khan, I. (2019). Nanoparticles: Properties, applications and toxicities. *Arabian Journal of Chemistry*, 12(7): 908–931. https://doi.org/10.1016/j.arabjc.2017.05.011.

Khanehzaei, H., Ahmad, M. B., Shameli, K., Ajdari, Z., Kebangsaan Malaysia, U. and Darul Ehsan, S. (2014). Synthesis and characterization of $Cu@Cu_2O$ core shell nanoparticles prepared in seaweed Kappaphycus alvarezii Media. *Int. J. Electrochem. Sci.*, 9: 8189–8198.

Kim, E. Y., Schulz, R., Swantek, P., Kunstman, K., Malim, M. H. and Wolinsky, S. M. (2011). Gold nanoparticle-mediated gene delivery induces widespread changes in the expression of innate immunity genes. *Gene Therapy*, 19(3): 347–353. https://doi.org/10.1038/gt.2011.95.

Kitching, M., Choudhary, P., Inguva, S., Guo, Y., Ramani, M., Das, S. K. and Marsili, E. (2016). Fungal surface protein mediated one-pot synthesis of stable and hemocompatible gold nanoparticles. *Enzyme and Microbial Technology*, 95: 76–84. https://doi.org/10.1016/j.enzmictec.2016.08.007.

Klaus-Joerger, T., Joerger, R., Olsson, E. and Granqvist, C. G. (2001). Bacteria as workers in the living factory: Metal-accumulating bacteria and their potential for materials science. *Trends in Biotechnology*, 19(1): 15–20. https://doi.org/10.1016/S0167-7799(00)01514-6.

Klębowski, B., Depciuch, J., Parlińska-Wojtan, M. and Baran, J. (2018). Applications of noble metal-based nanoparticles in medicine. *International Journal of Molecular Sciences*, 19(12): 4031. https://doi.org/10.3390/IJMS19124031.

Kong, F. Y., Zhang, J. W., Li, R. F., Wang, Z. X., Wang, W. J. and Wang, W. (2017). Unique roles of gold nanoparticles in drug delivery, targeting and imaging applications. *Molecules*, 22(9), 1445. https://doi.org/10.3390/MOLECULES22091445.

Kumari, P., Panda, P. K., Jha, E., Pramanik, N., Nisha, K., Kumari, K., Soni, N., Mallick, M. A. and Verma, S. K. (2018). Molecular insight to in vitro biocompatibility of phytofabricated copper oxide nanoparticles with human embryonic kidney cells. *Nanomedicine*, 13(19): 2415–2433. https://doi.org/10.2217/nnm-2018-0175.

Kumari, S., Kumari, P., Panda, P. K., Pramanik, N., Verma, S. K. and Mallick, M. A. (2019). Molecular aspect of phytofabrication of gold nanoparticle from Andrographis peniculata photosystem II and their in vivo biological effect on embryonic zebrafish (Danio rerio). *Environmental Nanotechnology, Monitoring and Management*, 11: 100201. https://doi.org/10.1016/j.enmm.2018.100201.

Li, X., Xu, H., Chen, Z. S. and Chen, G. (2011). Biosynthesis of nanoparticles by microorganisms and their applications. *Journal of Nanomaterials*, 2011. https://doi.org/10.1155/2011/270974.

Liaqat, N., Jahan, N., Khalil-ur-Rahman, Anwar, T. and Qureshi, H. (2022). Green synthesized silver nanoparticles: Optimization, characterization, antimicrobial activity, and cytotoxicity study by hemolysis assay. *Frontiers in Chemistry*, 10: 995. https://doi.org/10.3389/FCHEM.2022.952006/BIBTEX.

Lin, L., Wang, W., Huang, J., Li, Q., Sun, D., Yang, X., Wang, H., He, N. and Wang, Y. (2010). Nature factory of silver nanowires: Plant-mediated synthesis using broth of Cassia fistula leaf. *Chemical Engineering Journal*, 162(2): 852–858. https://doi.org/10.1016/J.CEJ.2010.06.023.

Liu, P., Jin, H., Guo, Z., Ma, J., Zhao, J., Li, D., Wu, H. and Gu, N. (2016). Silver nanoparticles outperform gold nanoparticles in radiosensitizing U251 cells *in vitro* and in an intracranial mouse model of glioma. *International Journal of Nanomedicine*, 11: 5003. https://doi.org/10.2147/IJN.S115473.

Luangpipat, T., Beattie, I. R., Chisti, Y. and Haverkamp, R. G. (2011). Gold nanoparticles produced in a microalga. *Journal of Nanoparticle Research*, 13(12): 6439–6445. https://doi.org/10.1007/S11051-011-0397-9/METRICS.

Mahdavi, M., Namvar, F., Ahmad, M. Bin and Mohamad, R. (2013). Green biosynthesis and characterization of magnetic iron oxide (Fe_3O_4) nanoparticles using seaweed (Sargassum muticum) aqueous extract. *Molecules*, 18(5): 5954–5964. https://doi.org/10.3390/MOLECULES18055954.

Mata, Y. N., Torres, E., Blázquez, M. L., Ballester, A., González, F. and Muñoz, J. A. (2009). Gold(III) biosorption and bioreduction with the brown alga Fucus vesiculosus. *Journal of Hazardous Materials*, 166(2-3): 612–618. https://doi.org/10.1016/j.jhazmat.2008.11.064.

Meyers, J. D., Cheng, Y., Broome, A. M., Agnes, R. S., Schluchter, M. D., Margevicius, S., Wang, X., Kenney, M. E., Burda, C. and Basilion, J. P. (2015). Peptide-targeted gold nanoparticles for photodynamic therapy of brain cancer. *Particle and Particle Systems Characterization*, 32(4): 448–457. https://doi.org/10.1002/PPSC.201400119.

Mohanta, Y. K., Nayak, D., Biswas, K., Singdevsachan, S. K., Abd_Allah, E. F., Hashem, A., Alqarawi, A. A., Yadav, D. and Mohanta, T. K. (2018). Silver nanoparticles synthesized using wild mushroom show potential antimicrobial activities against food borne pathogens. *Molecules*, 23(3). https://doi.org/10.3390/molecules23030655.

Molnár, Z., Bódai, V., Szakacs, G., Erdélyi, B., Fogarassy, Z., Sáfrán, G., Varga, T., Kónya, Z., Tóth-Szeles, E., Szucs, R. and Lagzi, I. (2018). Green synthesis of gold nanoparticles by thermophilic filamentous fungi. *Scientific Reports*, 8(1). https://doi.org/10.1038/s41598-018-22112-3.

Mominur Rahman, M., Islam, F., Saidur Rahaman, M., Sultana, N. A., Fahim, N. F. and Ahmed, M. (2021). Studies on the prevalence of HIV/AIDS in Bangladesh including other developing countries. *Advances in Traditional Medicine*. https://doi.org/10.1007/s13596-021-00610-6.

Mussa Farkhani, S., Asoudeh Fard, A., Zakeri-Milani, P., Shahbazi Mojarrad, J. and Valizadeh, H. (2016). Enhancing antitumor activity of silver nanoparticles by modification with cell-penetrating peptides. *Http://Dx.Doi.Org/10.1080/21691401.2016.1200059*, 45(5): 1029–1035. https://doi.org/10.1080/21 691401.2016.1200059.

Nag, M., Lahiri, D., Sarkar, T., Ghosh, S., Dey, A., Edinur, H. A., Pati, S. and Ray, R. R. (2021). Microbial fabrication of nanomaterial and its role in disintegration of exopolymeric matrices of biofilm. *Frontiers in Chemistry*, 9. https://doi.org/10.3389/fchem.2021.690590.

Namvar, F., Rahman, H. S., Mohamad, R., Baharara, J., Mahdavi, M., Amini, E., Chartrand, M. S. and Yeap, S. K. (2014). Cytotoxic effect of magnetic iron oxide nanoparticles synthesized via seaweed aqueous extract. *International Journal of Nanomedicine*, 9(1): 2479. https://doi.org/10.2147/IJN. S59661.

Patel, V., Berthold, D., Puranik, P. and Gantar, M. (2015). Screening of cyanobacteria and microalgae for their ability to synthesize silver nanoparticles with antibacterial activity. *Biotechnology Reports*, 5: 112–119. https://doi.org/10.1016/j.btre.2014.12.001.

Patra, J. K. and Baek, K. H. (2015). Green nanobiotechnology. *Journal of Nanomaterials*, 2014. https://doi.org/10.1155/2014/417305.

Peng, L. H., Huang, Y. F., Zhang, C. Z., Niu, J., Chen, Y., Chu, Y., Jiang, Z. H., Gao, J. Q. and Mao, Z. W. (2016). Integration of antimicrobial peptides with gold nanoparticles as unique non-viral vectors for gene delivery to mesenchymal stem cells with antibacterial activity. *Biomaterials*, 103: 137–149. https://doi.org/10.1016/J.BIOMATERIALS.2016.06.057.

Poudel, M., Pokharel, R., K.C., S., Awal, S. C. and Pradhananga, R. (2017). Biosynthesis of silver nanoparticles using ganoderma lucidum and assessment of antioxidant and antibacterial activity. *International Journal of Applied Sciences and Biotechnology*, 5(4): 523–531. https://doi.org/10.3126/ijasbt.v5i4.18776.

Radiosensitization by gold nanoparticles: Effective at megavoltage energies and potential role of oxidative stress. (n.d.).

Rahman, Md Mominur, Rahaman, M. S., Islam, M. R., Rahman, F., Mithi, F. M., Alqahtani, T., Almikhlafi, M. A., Alghamdi, S. Q., Alruwaili, A. S., Hossain, M. S., Ahmed, M., Das, R., Emran T. Bin and Uddin, M. S. (2022). Role of phenolic compounds in human disease: Current knowledge and future prospects. *Molecules*, 27(1). https://doi.org/10.3390/molecules27010233.

Rahman, Mohammad Mominur, Islam, M. R., Islam, M. T., Harun-Or-rashid, M., Islam, M., Abdullah, S., Uddin, M. B., Das, S., Rahaman, M. S., Ahmed, M., Alhumaydhi, F. A., Emran T. Bin, Mohamed, A. A. R., Faruque, M. R. I., Khandaker, M. U. and Mostafa-Hedeab, G. (2022). Stem cell transplantation therapy and neurological disorders: Current status and future perspectives. *Biology*, 11(1). https://doi.org/10.3390/biology11010147.

Rajeshkumar, S. and Malarkodi, C. (2014). *In vitro* antibacterial activity and mechanism of silver nanoparticles against foodborne pathogens. *Bioinorganic Chemistry and Applications*, 2014. https://doi.org/10.1155/2014/581890.

Rajeshkumar, Shanmugam, Malarkodi, C., Paulkumar, K., Vanaja, M., Gnanajobitha, G. and Annadurai, G. (2014). Algae mediated green fabrication of silver nanoparticles and examination of its antifungal activity against clinical pathogens. *International Journal of Metals*, 2014: 1–8. https://doi.org/10.1155/2014/692643.

Ramakrishna, M., Rajesh Babu, D., Gengan, R. M., Chandra, S. and Nageswara Rao, G. (2016). Green synthesis of gold nanoparticles using marine algae and evaluation of their catalytic activity. *Journal of Nanostructure in Chemistry*, 6(1): 1–13. https://doi.org/10.1007/S40097-015-0173-Y/TABLES/1.

Rana, S., Bajaj, A., Mout, R. and Rotello, V. M. (2012). Monolayer coated gold nanoparticles for delivery applications. *Advanced Drug Delivery Reviews*, 64(2): 200–216. https://doi.org/10.1016/j.addr.2011.08.006.

Ramalingam, V. (2019). Multifunctionality of gold nanoparticles: Plausible and convincing properties. *Advances in Colloid and Interface Science*, 271: 101989. https://doi.org/10.1016/j.cis.2019.101989.

Rathinaraj, P., Al-Jumaily, A. M. and Huh, D. S. (2015). Internalization: acute apoptosis of breast cancer cells using herceptin-immobilized gold nanoparticles. *Breast Cancer: Targets and Therapy*, 7: 51. https://doi.org/10.2147/BCTT.S69834.

Retif, P., Pinel, S., Toussaint, M., Frochot, C., Chouikrat, R., Bastogne, T. and Barberi-Heyob, M. (2015). Nanoparticles for radiation therapy enhancement: The key parameters. *Theranostics*, 5(9): 1030. https://doi.org/10.7150/THNO.11642.

Riley, R. S. and Day, E. S. (2017). Gold nanoparticle-mediated photothermal therapy: Applications and opportunities for multimodal cancer treatment. *Wiley Interdisciplinary Reviews: Nanomedicine and Nanobiotechnology*, 9(4): e1449. https://doi.org/10.1002/WNAN.1449.

Saberi, A., Shahbazi-Gahrouei, D., Abbasian, M., Fesharaki, M., Baharlouei, A. and Arab-Bafrani, Z. (2016). Gold nanoparticles in combination with megavoltage radiation energy increased radiosensitization and apoptosis in colon cancer HT-29 cells. *Http://Dx.Doi.Org/10.1080/09553002.2017.1242816*, 93(3): 315–323. https://doi.org/10.1080/09553002.2017.1242816.

Samadi, N., Golkaran, D., Eslamifar, A., Jamalifar, H., Fazeli, M. R. and Mohseni, F. A. (2009). Intra/extracellular biosynthesis of silver nanoparticles by an autochthonous strain of Proteus mirabilis isolated from photographic waste. *Journal of Biomedical Nanotechnology*, 5(3): 247–253. https://doi.org/10.1166/jbn.2009.1029.

Schäffler, M., Sousa, F., Wenk, A., Sitia, L., Hirn, S., Schleh, C., Haberl, N., Violatto, M., Canovi, M., Andreozzi, P., Salmona, M., Bigini, P., Kreyling, W. G. and Krol, S. (2014). Blood protein coating of gold nanoparticles as potential tool for organ targeting. *Biomaterials*, 35(10): 3455–3466. https://doi.org/10.1016/J.BIOMATERIALS.2013.12.100.

Schmidt, A., Molano, V., Hollering, M., Pöthig, A., Casini, A. and Kühn, F. E. (2016). Evaluation of new palladium cages as potential delivery systems for the anticancer drug cisplatin. *Chemistry—A European Journal*, 22(7): 2253–2256. https://doi.org/10.1002/CHEM.201504930.

Senapati, S., Syed, A., Moeez, S., Kumar, A. and Ahmad, A. (2012). Intracellular synthesis of gold nanoparticles using alga Tetraselmis kochinensis. *Materials Letters*, 79: 116–118. https://doi.org/10.1016/J.MATLET.2012.04.009.

Sergeev, G. B. and Klabunde, K. J. (2013). Synthesis and stabilization of nanoparticles. *Nanochemistry*, 11–54. https://doi.org/10.1016/B978-0-444-59397-9.00002-5.

Shakibaie, M., Forootanfar, H., Mollazadeh-Moghaddam, K., Bagherzadeh, Z., Nafissi-Varcheh, N., Shahverdi, A. R. and Faramarzi, M. A. (2010). Green synthesis of gold nanoparticles by the marine microalga Tetraselmis suecica. *Biotechnology and Applied Biochemistry*, 57(2): 71–75. https://doi.org/10.1042/BA20100196.

Sharma, B., Purkayastha, D. D., Hazra, S., Gogoi, L., Bhattacharjee, C. R., Ghosh, N. N. and Rout, J. (2014). Biosynthesis of gold nanoparticles using a freshwater green alga, Prasiola crispa. *Materials Letters*, 116: 94–97. https://doi.org/10.1016/J.MATLET.2013.10.107.

Siddiqi, K. S. and Husen, A. (2016). Fabrication of metal nanoparticles from fungi and metal salts: Scope and application. *Nanoscale Research Letters*, 11(1): 1–15. https://doi.org/10.1186/s11671-016-1311-2.

Simon, S., Remaliah, N., Sibuyi, S., Fadaka, A. O., Meyer, S., Josephs, J., Onani, M. O., Meyer, M. and Madiehe, A. M. (2022). Biomedical applications of plant extract-synthesized silver nanoparticles. *Biomedicines*, 10(11): 2792. https://doi.org/10.3390/BIOMEDICINES10112792.

Singaravelu, G., Arockiamary, J. S., Kumar, V. G. and Govindaraju, K. (2007). A novel extracellular synthesis of monodisperse gold nanoparticles using marine alga, Sargassum wightii Greville. *Colloids and Surfaces B: Biointerfaces*, 57(1): 97–101. https://doi.org/10.1016/j.colsurfb.2007.01.010.

Singh, A., Dar, M. Y., Joshi, B., Sharma, B., Shrivastava, S. and Shukla, S. (2018). Phytofabrication of silver nanoparticles: Novel drug to overcome hepatocellular ailments. *Toxicology Reports*, 5: 333–342. https://doi.org/10.1016/J.TOXREP.2018.02.013.

Singh, M., Kalaivani, R., Manikandan, S., Sangeetha, N. and Kumaraguru, A. K. (2013). Facile green synthesis of variable metallic gold nanoparticle using Padina gymnospora, a brown marine macroalga. *Applied Nanoscience (Switzerland)*, 3(2): 145–151. https://doi.org/10.1007/S13204-012-0115-7/FIGURES/6.

Singh, P., Kim, Y. J. and Yang, D. C. (2015). A strategic approach for rapid synthesis of gold and silver nanoparticles by Panax ginseng leaves. *Http://Dx.Doi.Org/10.3109/21691401.2015.1115410*, 44(8): 1949–1957. https://doi.org/10.3109/21691401.2015.1115410.

Sinha, S. N., Paul, D., Halder, N., Sengupta, D. and Patra, S. K. (2015). Green synthesis of silver nanoparticles using fresh water green alga Pithophora oedogonia (Mont.) Wittrock and evaluation of their antibacterial activity. *Applied Nanoscience (Switzerland)*, 5(6): 703–709. https://doi.org/10.1007/s13204-014-0366-6.

Slawson, R. M., Van Dyke, M. I., Lee, H. and Trevors, J. T. (1992). Germanium and silver resistance, accumulation, and toxicity in microorganisms. *Plasmid*, 27(1): 72–79. https://doi.org/10.1016/0147-619X(92)90008-X.

Son, S., Nam, J., Kim, J., Kim, S. and Kim, W. J. (2014). I-motif-driven au nanomachines in programmed siRNA delivery for gene-silencing and photothermal ablation. *ACS Nano*, 8(6): 5574–5584. https://doi.org/10.1021/NN5022567/SUPPL_FILE/NN5022567_SI_001.PDF.

Stalin Dhas, T., Ganesh Kumar, V., Karthick, V., Jini Angel, K. and Govindaraju, K. (2014). Facile synthesis of silver chloride nanoparticles using marine alga and its antibacterial efficacy. *Spectrochimica Acta - Part A: Molecular and Biomolecular Spectroscopy*, 120: 416–420. https://doi.org/10.1016/j.saa.2013.10.044.

Su, X. Y., Liu, P. D., Wu, H. and Gu, N. (2014). Enhancement of radiosensitization by metal-based nanoparticles in cancer radiation therapy. *Cancer Biology and Medicine*, 11(2): 86. https://doi.org/10.7497/J.ISSN.2095-3941.2014.02.003.

Sudheer, S., Alzorqi, I., Manickam, S. and Ali, A. (2019). Bioactive compounds of the wonder medicinal mushroom "Ganoderma lucidum." *Reference Series in Phytochemistry*, 1863–1893. https://doi.org/10.1007/978-3-319-78030-6_45.

Thapa, R. K., Soe, Z. C., Ou, W., Poudel, K., Jeong, J. H., Jin, S. G., Ku, S. K., Choi, H. G., Lee, Y. M., Yong, C. S. and Kim, J. O. (2018). Palladium nanoparticle-decorated 2-D graphene oxide for effective photodynamic and photothermal therapy of prostate solid tumors. *Colloids and Surfaces B: Biointerfaces*, 169: 429–437. https://doi.org/10.1016/J.COLSURFB.2018.05.051.

Titus, D., James Jebaseelan Samuel, E. and Roopan, S. M. (2019). Nanoparticle characterization techniques. *Green Synthesis, Characterization and Applications of Nanoparticles*, 303–319. https://doi.org/10.1016/B978-0-08-102579-6.00012-5.

Torrisi, L. (n.d.). *Gold Nanoparticles Enhancing Protontherapy Efficiency.*

Vágó, A., Szakacs, G., Sáfrán, G., Horvath, R., Pécz, B. and Lagzi, I. (2016). One-step green synthesis of gold nanoparticles by mesophilic filamentous fungi. *Chemical Physics Letters*, 645: 1–4. https://doi.org/10.1016/j.cplett.2015.12.019.

Venkatpurwar, V. and Pokharkar, V. (2011). Green synthesis of silver nanoparticles using marine polysaccharide: Study of in-vitro antibacterial activity. *Materials Letters*, 65(6): 999–1002. https://doi.org/10.1016/J.MATLET.2010.12.057.

Verma, A. and Mehata, M. S. (2016). Controllable synthesis of silver nanoparticles using Neem leaves and their antimicrobial activity. *Journal of Radiation Research and Applied Sciences*, 9(1): 109–115. https://doi.org/10.1016/J.JRRAS.2015.11.001.

Verma, S. K., Jha, E., Panda, P. K., Thirumurugan, A., Patro, S., Parashar, S. K. S. and Suar, M. (2018). Molecular insights to alkaline based bio-fabrication of silver nanoparticles for inverse cytotoxicity and enhanced antibacterial activity. *Materials Science and Engineering: C*, 92: 807–818. https://doi.org/10.1016/j.msec.2018.07.037.

Wang, H., Mu, X., He, H. and Zhang, X. D. (2018). Cancer Radiosensitizers. *Trends in Pharmacological Sciences*, 39(1): 24–48. https://doi.org/10.1016/J.TIPS.2017.11.003.

Wilczewska, A. Z., Niemirowicz, K., Markiewicz, K. H. and Car, H. (2012). Nanoparticles as drug delivery systems. *Pharmacological Reports: PR*, 64(5): 1020–1037. https://doi.org/10.1016/S1734-1140(12)70901-5.

Yao, C., Zhang, L., Wang, J., He, Y., Xin, J., Wang, S., Xu, H. and Zhang, Z. (2016). Gold nanoparticle mediated phototherapy for cancer. *Journal of Nanomaterials*, 2016. https://doi.org/10.1155/2016/5497136.

CHAPTER 2

Role of Green Reducing Agents in Synthesis of Nanomaterials

Simran Singh Deo,[1] *Sudipta K. Jyotish,*[2] *Swagat K. Mohapatra*[3]
and *Jagnyaseni Tripathy*[4,*]

1. Introduction

Recent advances in nanoscience and nanotechnology led to the development of novel nanomaterials, greatly impacting energy harvesting, devices and the biomedical industry. Many early nanomaterials required chemical processing routes, ultimately posing hazards to the biohealth and environment. Developing environmentally benign procedures for the synthesis of metallic and non-metallic nanoparticles is the need of the hour. Green reducing agents make up for a large fraction of the environmental approaches amid the advances in eco-friendly synthesis methods. These green reducers mostly comprise phytochemicals or various forms of plant extracts, among others. All of these natural reducers cut down on the usage of harsh chemicals and replace those with greener modes, such as heat, pressure, phytochemicals or bioderived products, viz., microorganism extracts. With no hazardous or excess chemicals, these methods are not only environmentally friendly but also efficient and cost-effective. The phytoconstituents in plant extracts are rich in polyols, and anti-oxidants often act as reducing and stabilizing agents in the synthesis of nanomaterials. The desired nanostructure can be obtained by reducing the starting materials using leaf, flower, root, seed, starch, extracts, juice, etc. These

[1] Dept. of Physics, School of Applied Sciences, KIIT University 754021 India; 0000-0003-0475-2237.

[2] Dept. of Physics, Utkal University, 751004, India; 0000-0002-7266-1858.

[3] Dept. of Chemistry, ICT-IOCB, 751013, India; 0000-0002-4485-3967.

[4] Dept. of Physics, School of Applied Sciences, KIIT University 754021 India; 0000-0002-4685-3610.
Emails: simransinghdeo7894@gmail.com; sudiptakumarjyotish214@gmail.com; sk.mohapatra@iocb.ictmumbai.edu.in

* Corresponding author: jtripathyfpy@kiit.ac.in

have been successfully used in the synthesis of nanomaterial in various shapes and sizes, such as nanodot, nanorod, flowered shaped or hexagonal nanoparticles, etc. Among the many nature-friendly reducing methods, the ones based on plant extracts are more common and have been successfully used for obtaining both metallic as well as non-metallic nanoparticles in various shapes, sizes and physico-chemical properties aimed at various applications.

2. Plant Extracts as Reducing Agents

Plant extracts with several rich phytoconstituents make the most common green reducers in the synthesis of nanomaterials. These are easily obtained from living or nonliving plants and are often used as reduction and capping agents. As shown schematically in Figure 1, plant extracts can be classified as leaf, root, seed, bark, fruit extract, juice/pulp, etc.

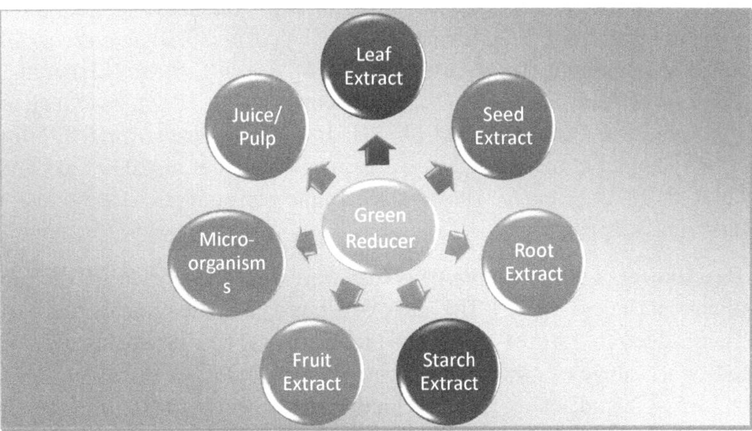

Figure 1. Schematic representation of various plant-based green reducing agents for the synthesis of metallic and non-metallic nanoparticles.

2.1 Leaf Extract

Leaf extracts of many plants, e.g., bay, acacia, fenugreek, holy basil, spinach, aloe vera, etc., have been successfully used as reducing agents in the synthesis of various types of metallic and non-metallic nanoparticles. Leaf extracts are easy to obtain following simple extraction procedures. Once the required leaves are collected and cleaned with fresh water, the extracts can be prepared in two slightly different ways. In the first method, the leaves are dried post-rinsing with water; the dried leaves are then finely powdered and dissolved in deionized water to obtain an aqueous extract. Alternately, finely chopped, freshly cleaned leaves are boiled thoroughly in water. Figure 2 shows a schematic representation of the synthesis process using leaf extract. The filtered extract in either method is then added to an aqueous solution of metal or non-metal salt. The addition of leaf extract results in a change of color with time, indicating the formation of nanoparticles further confirmed and characterized by various experimental and analytical methods.

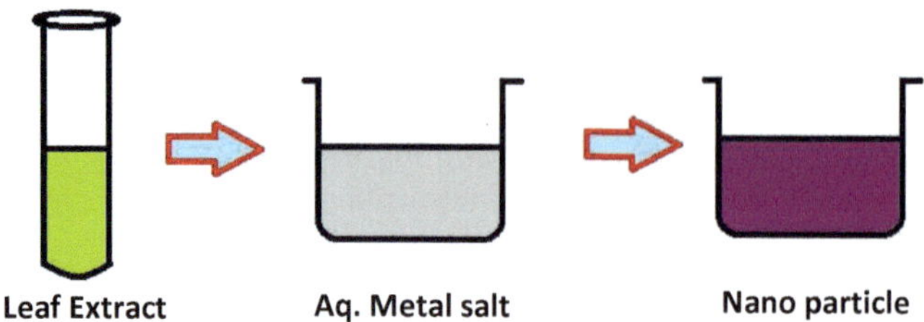

Figure 2. A schematic representation of the leaf extract in synthesizing nanoparticles.

As summarized in Table 1 Few leaf extracts, viz., *Bamboo* (Sharma et al. 2023), *Rhus Coriaria L* (Rahimzadeh et al. 2022) are used in the synthesis of non-metallic nanoparticles, while most are used for the synthesis of metallic nanoparticles, e.g., *bay laurel* (Al-Ghamdi et al. 2019), *black tea* (Uddin et al. 2012, Kamath et al. 2020), *Mangifera indica* (Philip et al. 2011), *lemon* (Shukla et al. 2012), *curry* (Shukla et al. 2012), *olive* (Khalil et al. 2014), *Eucalyptus globulus* (Kamath et al. 2020), *Acalypha indica* (Shanthi et al. 2016), *Dalbergia sissoo* (Khan et al. 2020), *fenugreek* (Ghoshal and Singh 2022), *common sage* (Elia et al. 2014), *lemon verbena* (Elia et al. 2014), *rose geranium* (Elia et al. 2014), *shikakai* (Kumar et al. 2018), *reetha* (Kumar et al. 2018), *aloe vera* (Kandregula et al. 2015), *hibiscus rosa sinensis* (Philip et al. 2010), *oak* (Kamath et al. 2020), *pomegranate* (Kamath et al. 2020) and *green tea* (Kamath et al. 2020). Figure 2 shows a schematic representation of the leaf extract in synthesizing nanoparticles.

2.1.1 Bay Laurel (*Laurus nobilis*): Its leaf extract has been used for the synthesis of silver nanoparticles (AgNP). The leaf extract, prepared the same way as explained above, was added to 20 mM aqueous solution of $AgNO_3$, producing the AgNP as indicated by a change of color of the solution from colorless to yellow to reddish-brown (Al-Al-Ghamdi et al. 2019). Notably, the concentration of extract in salt solution determined the speed of reduction. A peak absorbance at 437 nm in the UV-visible absorption spectrum confirmed the formation of the metal NP, and the TEM studies show the shape of AgNP as spherical and oval.

2.1.2 Black Tea Leaf: Its extract has been used as a reducing agent in synthesizing polyvinyl alcohol (PVA)-silver (Ag) nanofilm and iron oxide nanoparticles. As reported by (Jamal et al. 2012), black tea leaf extract was prepared in deionized water (0.02 g/ml), and the infusion was filtered until all the insoluble matter disappeared. The filtrate was subsequently used for the preparation of PVA-Ag nanofilms or iron oxide nanoparticles.

2.1.3 PVA-Ag Nanofilm From Black Tea Leaf Extract: Equal parts of leaf extract and 1 mM aqueous AgNO3 solution were mixed in a beaker, forming the AgNP indicated by a change in the solution color to light brown. The mixture was stirred at 80°C for 1 hour after adding the PVA to form the nanofilm. The residual solution was then poured into a Petri dish and air-dried to form a transparent film. The nanocomposite film thus produced was characterized spectroscopically. While the AgNP formed was found to be crystalline in nature with FCC structure, the nanofilm was semi-crystalline in nature as verified by XRD, with peak absorbance in the 326 to 386 nm range (Jamal et al. 2012).

2.1.4 Iron Oxide Nanoparticle From Black Tea Leaf Extract: Black tea leaf extract in distilled water was added to 0.1 M aqueous $FeCl_3$ solution in a ratio of 2:1 dropwise manner, along with continuous stirring at 300 rpm. As soon as the pH was adjusted to 6 using NaOH, a black precipitate was obtained, indicating the formation of iron oxide nanoparticle (Fe_2O_3 NP), which was further centrifuged at 10,000 rpm for 20 minutes; the pellet collected after discarding supernatant was washed and dried overnight at 40°C, and spherical nanoparticles (Fe_2O_3 NP) of 60–70 nm size were collected (Kamath et al. 2020).

2.1.5 Mango Leaf (Mangifera Indica): Its extract can also be used for producing AgNP. Fresh mango leaves were cut into small pieces and boiled at 300 K (0.09 g/ml) for 1 minute, which was then filtered to get the extract. Also, 0.5 mM $AgNO_3$ maintained at pH 8 was added to the leaf extract in a 4:1 ratio under continuous stirring for 5 minutes at 300 K to produce AgNP. The color of the solution changed from greenish-gray to brown, indicating the formation of the NP, which was further confirmed by UV-Vis with an absorption peak at 439 nm (Philip et al. 2011).

2.1.6 Lemon Leaf (Citrus Limon): Its extracts were used for the production of gold nanoparticles (Au NPs). Finely cut lemon leaves were boiled in distilled water (0.2 g/ml) for 5 to 10 minutes and filtered. The leaf extract was then added to 2 mM $HAuCl_4$ solution in the volumetric ratio of 1:9 and kept in the dark overnight for the formation of the nanoparticle. A peak absorbance at 550 nm confirmed the formation of Au NPs (Shukla et al. 2012).

2.1.7 Curry Leaf (Murraya Koenigii Linn): Its extract was utilized for the synthesis of gold nanoparticles. Curry leaf extract is prepared in the same way, as mentioned above, by boiling the cut leaves in water and filtering. Filtrate was added to 2 mM $HAuCl_4$ in the volumetric ratio of 1:9. Formation of Au NP was realized after 30 minutes of mixing the two solutions, but the complete formation of AuNP required the solution to be kept in the dark overnight. The resulting AuNP showed a peak absorbance of 550 nm (Khalil et al. 2014).

2.1.8 Olive Leaf: Its extract was used for the synthesis of silver nanoparticles (AgNP). The extract, prepared by boiling olive leaves in deionized water (0.02 g/ml), was filtered, and the filtrate was added to 20 mM $AgNO_3$ solution, where the mixture showed a change in color from yellow-brown to deep brown, indicating the formation of AgNP. An increase in the concentration of leaf extract resulted in a peak shift from 458 nm to 441 nm in the observed UV-Vis absorption spectra (Balaji et al. 2017).

2.1.9 Reetha and Shikakai Leaf: Their extracts are also used for producing AgNP. Finely cut reetha and shikakai leaves were separately boiled in Milli-Q water (0.2 g/ml) at 100°C, cooled to ambient temperature and filtered. To a solution mixture of 1 mM $AgNO_3$ and ammonia (2:1), the extracts were added in the ratio of 10:3 to prepare the AgNP. The AgNP, with an absorption peak at 401 nm, was obtained after centrifugation, followed by washing with water and ethanol (Kumar et al. 2018).

2.1.10 Fenugreek Leaf (Trigonella Foenumgraecum): Its extract was used as a simple means for the production of AgNP. Leaf extract was prepared with sun-dried leaves boiled in distilled water at 80°C and incubated for 30 minutes. The extract was then added to 1 mM $AgNO_3$ solution in the volumetric ratio 1:10 and centrifuged at room temperature for 30 minutes at 5,000 rpm. After 30 minutes, the solution changed from colorless to brown to dark brown, suggesting the formation of silver nanoparticles. The UV-Vis absorption spectrum showed a sharp band at 432 nm, confirming the formation of Ag NP (Gargi et al. 2022).

2.1.11 Common Sage (Salvia Officinalis): Its leaf extracts were used for synthesizing AgNP. The soxhlet method was used to obtain the extract from the clean and dried leaves of Common Sage. The resulting solution was filtered to remove the unwanted particles, and the filtrate was added to the $AgNO_3$ solution. A change in color of the solution from light yellow to dark brown occurred with the addition of $AgNO_3$, indicating the formation of AgNP. The UV-Vis absorbance spectrum showed a peak absorbance at 430 nm (Elia et al. 2014).

2.1.12 Rose Mallow (Hibiscus Rosa Sinensis): Its leaf extract was used for the synthesis of both gold and silver nanoparticles. Finely cut leaf was added to deionized water (0.1g/ml) at 300 K for 1 minute under stirring. The filtrate obtained acts both as a reducing and stabilizing agent in synthesizing Au and Ag NP (Philip et al. 2010).

a. Silver Nanoparticle Synthesis: The extract of hibiscus rosa was added to an aqueous solution of 0.8 mM $AgNO_3$ in the volumetric ratio of 4:5 under vigorous stirring for 1 minute. The pH of the solution was maintained at 6.8. The color of the solution changed to golden yellow, indicating the formation of AgNP. The solution was centrifuged, and the solute was collected as the nanoparticles. The AgNP formed showed a peak absorbance of 399 nm.

b. Gold Nanoparticle Synthesis: Hibiscus extract was added to 0.5 mM aqueous solution of $HAuCl_4.3H_2O$ in the volumetric ratio 1:5 under a vigorous stirring of 1 minute. After one and a half hours, the solution showed a stable light violet color. A rapid color change was noticed with an increase in the extract concentration. A powder containing AuNP was collected upon drying the solute post-centrifugation. The UV-Vis absorbance showed a peak at 548 nm.

2.1.13 Eucalyptus Leaf (Eucalyptus Globulus): Its extract was used in synthesizing zinc oxide nanoparticles (ZnO NP). The powdered eucalyptus leaves boiled in deionized water (0.2 gm/ml) at 80°C for 1 hour formed a light black solution. Once the solution settled at room temperature, it was filtered and stored for further use. Also, 0.1 N zinc nitrate hexahydrate solution was mixed with the leaf extract in a 1:1 ratio dropwise, stirred at 600 rpm for 3 hours. A brown-colored precipitate indicated the formation of ZnO NP. Moreover, 24 hours after the settlement of the precipitation, the powder was collected and annealed in a muffle furnace at 400°C for 2 hours (Balaji et al. 2017).

2.1.14 Acalypha Indica: Its leaf extract was used for the preparation of zirconium dioxide nanoparticles (ZrO_2 NP). Powdered acalypha leaves were boiled in double distilled water for 30 minutes, filtered and stored for future use. ZrO_2 NPs were prepared by adding the leaf extract to 0.1 M aqueous solution of zirconyl nitrate octahydrate in 1:5 volumetric ratio and boiling at 80°C for 2 hours. The solution was then kept in an oven at 200°C for a few days, yielding the ZrO_2 NP powder (Shanthi et al. 2016).

2.1.15 Dalbergia Sissoo Leaf (North Indian Rosewood): Its extract was used in the formation of magnesium oxide nanoparticles (MgO_2 NPs). A mixture of the *D. Sissoo* leaf extract and double distilled deionized water in a 4:1 ratio was heated at 80°C and then cooled to 30°C. The resulting solution was added to 0.1 M magnesium nitrate hexahydrate in a dropwise manner. After 4 hours, the precipitate was filtered out and dried in an oven for 2 hours, resulting in MgO_2 NPs (Khan et al. 2020).

2.1.16 Aloe Vera: Leaf extract was used as a reducing agent to synthesize titanium oxide nanoparticles (TiO_2 NPs). Aloe vera extract was prepared by boiling cut pieces of aloe leaves in distilled water (0.25 g/ml) at 90°C for 2 hours. The filtrate was then used as a reducing agent. To 1 N titanium chloride solution, the extract was added dropwise, and pH was adjusted to 7 and stirred for 4 hours. The resulting nanoparticles were separated by filter paper and washed to remove the other product. The dried and calcined TiO_2 NPs were then characterized by various spectroscopic methods (Kandregula et al. 2015).

2.1.17 Oak (*Quercus Virginiana*), Green Tea, Pomegranate (*Punica Granatum*) and Eucalyptus (*Eucalyptus Globulus*) Leaves: They were used for the synthesis of iron oxide nanoparticles (Fe_2O_3 NPs). Finely cut leaves dried at 50°C for 24 hours were boiled in distilled water at 80°C for 20–60 minutes under constant stirring at 700 rpm. To obtain the Fe_2O_3 NPs, respective leaf extracts were added dropwise to an aqueous of 0.1M $FeCl_3$ solution in a 2:1 ratio at room temperature, and pH was adjusted to 6 and stirred at 300 rpm. The formation of Fe_2O_3 NPs was indicated by the appearance of black precipitation. The precipitate was centrifuged using a cooling centrifuge at 10,000 rpm, and the nanoparticle pellets collected after discarding the supernatant were washed and dried in a hot air oven at 40°C overnight, producing the Fe_2O_3 NPs (Kamath et al. 2020).

2.1.18 Bamboo Leaf: Silicon oxide nanoparticles (SiO_2 NPs) were prepared by using bamboo leaf extract. The extract was prepared by taking a fine powder of the soaked bamboo leaves dried under daylight. The powder was added to nitric acid (0.04 g/ml) for 24 hours, followed by the addition of deionized water until the pH of the solution became 7. The solution was dried in an oven at 90°C to 110°C for 12 hours and refluxed at 110°C with 1N NaOH solution. The sodium silicate solution thus obtained was added dropwise to a mixture of ammonia and ethanol. The final solution mixture was centrifuged, washed, and dried to produce SiO_2 NPs (~ 80 nm in size) (Sharma et al. 2023).

Table 1. Leaf extracts used as green reducing agents.

Sl. no.	Leaf	NPs synthesized	Size (nm)	Characteristic properties (UV-VIS Absorbance/Structure (XRD)/Shape (SEM/FESEM/TEM)	Indicator	Application	Reference
1.	Bay laurel	Ag	22–28	UV-Vis: 437 nm FCC structure	Colorless-yellow-brown	Anti-microbial and catalytic study	(Al-Ghamdi et al. 2019)
2.	Black tea	PVA-Ag nanocomposite		UV-Vis: 323 nm, 374 nm, and 386 nm FCC structure spherical (FESEM)	Light brown	Dielectric study	(Jamal et al. 2012)
3	Black tea	Iron oxide	60–70	SEM: Spherical			(Kamath et al. 2020)
4.	Mango	Ag	20	UV: 439 nm, FCC structure, TEM: Spherical, hexagonal, nanorods	Greenish gray	pH and temperature study	(Philip et al. 2011)
5.	Lemon	Au	30–130	UV: 550 nm, SEM: Triangular, pentagonal, octagonal nanostructures		Phytochemical study	(Shukla et al. 2012)
6.	Curry	Au	30–130	UV: 550 nm, SEM: Triangular shape, TEM: 30–130 nm		Phytochemical study	(Khalil et al. 2014)
7.	Olive	Ag	20–25	UV: 440–458 nm, FCC structure, TEM: Quasi-spherical	Yellow-brownish yellow-deep brown	Anti-bacterial study	(Balaji et al. 2017)
8.	Reetha and Shikakai	Ag	20–40	UV: 401 nm, FCC structure, TEM: 20–40 nm (Spherical)		Bacteria and dye molecule detection	(Kumar et al. 2018)
9.	Common sage	Ag	30–50	UV: 430 nm, SEM: Spherical (30–50 nm)	Colorless-Brown	Anti-cancer activity	(Elia et al. 2014)
10.	Fenugreek	Ag	25	UV: 432 nm, FCC, FESEM: 4–30 nm	Colorless-brown-deep brown	Anti-microbial properties	(Gargi et al. 2022)
11.	Hibiscus Rosa Sinesis	Ag		UV: 399 nm, TEM: 13 nm	Golden yellow	Synthesis technique	(Philip et al. 2010)

12.	Hibiscus Rosa Sinesis	Au		UV: 548 nm, FCC Structure, TEM: Triangular, hexagonal, spherical	Light violet color	Synthesis technique	(Philip et al. 2010)
13.	Eucalyptus	ZnO	11.6	UV: 361 nm Spherical, 10–20 nm	Brown	Photocatalytic and anti-oxidant activity	(Kamath et al. 2020)
14.	Acalypha Indica	ZrO		Monoclinic, SEM: 20–100 nm			(Shanthi et al. 2016)
15.	Dalbergia Sisso	MgO	42	268 nm, Cubic Structure, SEM: Spherical		Photocatalytic and anti-bacterial activity	(Khan et al. 2020)
16.	Aloe vera	TiO$_2$	30	Tetragonal, SEM: 60–80 nm, TEM: Crystalline nature, TGA-DTA: 0.98%		Green Synthesis technique	(Kandregula et al. 2015)
17.	Bamboo	SiO$_2$	80	Mono-dispersed and spherical, Zeta Potential: –30.5 mV, +31.7 mV		Removal of toxic water pollutants	(Sharma et al. 2023)

2.2 Seed Extract

Seed extracts are used in many cases of nanomaterial synthesis (Jagtap et al 2013, He et al. 2017, Baghizadeh et al. 2015, Nazeruddin et al. 2014, Hussein et al. 2018, Nawabjohn et al. 2022]. As indicated by FTIR studies, these extracts act as reducing and capping agents; they reduce the metal and non-metal salts to their nano-forms without any further use of chemicals. To prepare the extract, wash the seeds thoroughly under running water, followed by distilled water or other types of purified water as necessary. Properly cleaned seeds are then dried and powdered. The fine powder of a chosen seed is then boiled in water, and the extract is then collected as the filtrate. The solution, in some cases, is centrifuged to remove any unwanted seed residues and diluted with distilled water prior to collecting the desired seed extract for synthesis. Some of the seed extracts used for the synthesis of nanoparticles are *Artocarpus heterophyllus Lam* (Jagtap et al. 2013), *Alpinia katsumadai* (He et al. 2017), *Calendula officinalis* (Baghizadeh et al. 2015), *Coriandrum sativum* (Nazeruddin et al. 2014), *Fenugreek* (Hussein et al. 2018), *Cassia Torra* (Nawabjohn et al. 2022), *Rhus coriaria L* (Rahimzadeh et al. 2022), etc., as given in Table 2.

Figure 3 shows a schematic representation of the seed extract in synthesizing nanoparticles.

2.2.1 Artocarpus Heterophyllus Lam (Jack Fruit): Its seed extract was used to synthesize silver nanoparticles (AgNP). Extract from sun-dried Jack fruit seeds collected after centrifugation to remove any unwanted seed residues; the extract was then diluted with distilled water and filtered. The filtrate containing the seed extract was added to 6 mM aqueous $AgNO_3$ solution in a volumetric ratio of 1:4. The mixture was then placed in an autoclave at 15 psi at 121°C for 5 minutes, after which a color change was observed from yellow-brown-deep red confirming the formation of AgNPs confirmed by a UV-Visible absorbance peak at 410 nm (Jagtap et al. 2013).

2.2.2 Alpinia Katsumadai (Cardamon): They are also used for the synthesis of silver nanoparticles. The seeds, in this case, were dried, chopped, sonicated in deionized water, and filtered to get the extract. The extract was added to an aqueous solution of $AgNO_3$ in the volume ratio 1:3 at a pH of 8, 10 or 12. A color change in

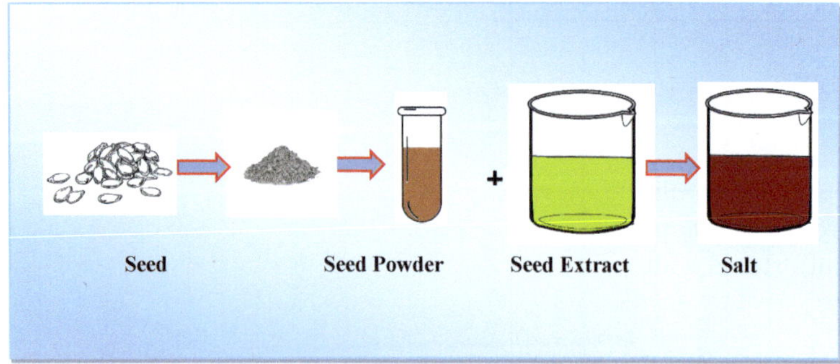

Figure 3. Schematic representation of the addition of seed extract to metal salt solution and resulting nanoparticle formation indicated by a change in color.

the solution indicated the formation of Ag NPs. At pH 8 the color of the solution changed to reddish brown, while for pH 10 and 12, it changed to dark brown. Increasing pH came with a rapid change in color, indicating the rapid formation of AgNP. The solution was then centrifuged, washed and dried to collect the AgNP. The peak absorbance in this case, as seen in UV-Vis, ranges between 416–428 nm (Yangqing He et al. 2017).

2.2.3 Calendula Officinalis (Marigold): Their seed extract was also used for synthesizing Ag NPs. The seed extract of the marigold was added to 1 mM aqueous solution of $AgNO_3$ in a 1:8 volume ratio and incubated at room temperature. The color of the solution changed to pale yellow or brown, indicating the formation of AgNP. At neutral or slightly basic conditions (around pH 9.0), Ag NPs were found to be formed and not under acidic conditions. The UV-Vis showed an absorption peak at 440 nm, confirming the formation of AgNP (Baghizadeh et al. 2015).

2.2.4 Coriandrum Sativum (Chinese Parsley): Their seed extract was used for the synthesis of AgNP. The seed extract, in this case, was prepared by soaking the washed seeds overnight in deionized water. The macerated seeds were then ground and diluted with more deionized water to collect the extract. To synthesize the AgNP, a solution of the extract was added dropwise to 0.1 N $AgNO_3$ solution in a 1:2 volume ratio. A dark brown colored precipitate appeared at the bottom, which was further centrifuged and washed with alcohol to collect the AgNP. The formation of Ag nps was confirmed by an absorbance peak at 421 nm (Nazeruddin et al. 2014).

2.2.5 Fenugreek Seed: Its extract was utilized as a reducing agent for the synthesis of AgNP. The seed extract was prepared by boiling the seeds in alkaline water at pH 9 using a magnetic stirrer with heating. The filtrate of the extract was collected for further use. The diluted extract was added dropwise to the $AgNO_3$ solution under constant stirring. After some time, the yellow color of the solution changed to dark brown, indicating the formation of AgNP. A UV-Vis absorbance peak at 430 nm confirmed the AgNP (Hussein et al. 2018).

2.2.6 Cassia Tora: The seed extract was used for the synthesis of silver nanoparticles. Seed extract prepared by dissolving Cassia Tora seed powder in distilled water and diluted with distilled water was used in the synthesis of AgNP. The diluted extract was added to an aqueous solution of $AgNO_3$ in a 1:1 volume ratio under constant stirring and incubated overnight. The color change from yellow to red indicates the formation of AgNP. The formation of the Ag NP was confirmed by a UV-Vis absorbance peak at 423 nm (Nawabjohn et al. 2022).

2.2.7 Rhus Coriaria L (Elm-Leaved Sumach): The seed extract was used for the preparation of silicon oxide nanoparticles (SiO_2 nps). The shade-dried seeds were ground to fine powder, and the powder was added to double distilled water (0.5 g/ml), stirred with heating on a magnetic stirrer and filtered. The filtrate was added to an aqueous solution of sodium silicate dropwise in a 1:5 ratio and stirred at 60°C for 12 hours, keeping the pH at 9. The mixture was finally filtered, and the precipitate was washed with water and ethanol, centrifuged at 7,000 rpm and oven-dried to collect the AgNP (Rahimzadeh et al. 2022).

Table 2. Seed extracts used as green-reducing agents.

Sl. no.	Seed extract	NPs synthesized	Size (nm)	Characteristic Properties (UV-VIS absorbance/ Structure (XRD) /Shape (SEM/ FESEM/TEM)	Indicator	Application	Reference
1.	Artocarpus Heterophyllus Lam	Ag	3–25	UV: 410 nm, TEM: Irregular structure	Yellow-brown-deep red	Anti-bacterial activity	(Jagtap et al. 2013)
2.	Alpinia katsumadai	Ag		UV: 416–428 nm, FCC, Crystalline, HRTEM: Quasi spherical	Reddish brown-deep brown	Anti-bacterial activity	(Yangqing He et al. 2017)
3.	Calendula Officinalis	Ag	7.5	UV: 440 nm, FCC Crystalline, TEM: Spherical Cubic	Pale yellow/ brown	Synthesis technique	(Baghizadeh et al. 2015)
4.	Corriandrum Sativum	Ag		UV: 421 nm, FCC, SEM: Spherical bead	Dark brown	Anti-microbial activity	(Nazeruddin et al. 2014)
5.	Cassia Tora	Ag		UV: 423 nm, FCC, SEM: 60	Yellow-red	Anti-bacterial activity	(Nawabjohn et al. 2022)
6.	Fenugreek	Ag	17	UV: 430 nm, FCC Crystalline, HRTEM: 17nm	Yellow-brown	Anti-bacterial activity	(Hussein et al. 2018)
7.	Rhus Coriaria L	SiO_2	23	Amorphus, FESEM: Spherical, TEM: 55 nm	Red-brown-white	Green vs. Chemical synthesis	(Rahimzadeh et al. 2022)

2.3 Root Extract

In the green synthesis of nanoparticles, root extract is used as an efficient method of reduction or stabilization of the nanoparticle, similar to the leaf or seed extracts. The various root extracts mentioned below contain both reducing and capping agents essential for the reduction of the ions to their respective nano-form and stabilizing it. To prepare the root extract, we need to either make a powder of dried root and boil it with water or boil the thoroughly washed root without drying. Some of the root extracts used for the synthesis of nanoparticles are *Morinda Citrifolia L* (Suman et al. 2014) for gold nanoparticles, *Berberis asiatica* (Dangi et al. 2020) for silver nanoparticles, *Sphagneticola trilobata Lin* (Shaikh et al. 2020) for zinc oxide nanoparticles, *Mimosa Pudica* (Niraimathee et al. 2016) for iron oxide nanoparticles. A schematic representation of preparing root extract for green reduction is shown below in Figure 4.

2.3.1 Morinda Citrifolia L: The root extract was used in the synthesis of gold nanoparticles (AuNP). The shade-dried root of Morinda Citrifolia L was boiled in distilled water for 15 minutes and filtered to obtain the root extract. The extract was added to a 1 mM solution of $HAuCl_4$ in a ratio of 3:34 by volume and left overnight for complete reduction. A pink-ruby-red color was observed, indicating the formation of AuNPs. The formation of AuNP was confirmed spectro-photometrically from the UV-Vis absorbance peak at 540 nm (Suman et al. 2014).

2.3.2 Berberis Asiatica: The root extract was used for synthesizing silver nanoparticles (AgNP). The root extract, in this case, was prepared by boiling the powder of root and briefly centrifuging the solution at 3,000 rpm. The filtered extract was added to 1 mM $AgNO_3$ solution in a 1:7 volume ratio. After 4 hours of mixing, the light-yellow colored solution turned reddish brown, indicating the formation of AgNP. Twenty-four hours were given for the complete reduction of AgNP indicated by a UV-visible absorption peak at 427 nm (Dangi et al. 2020).

2.3.3 The Root Extract of *Sphagneticola Trilobata Linn*: It is commonly known as Singapore Daisy and can also be used as a reducing agent. The root extract obtained by boiling the powdered root in Milli-Q water (0.1 g/ml) was used to synthesize zinc oxide nanoparticles (ZnO NP). Also, 0.1 M zinc acetate solution was added slowly into the root extract solution in a 1:4 volume ratio, stirring continuously at 60°C for 2 hours, and the solution appeared yellow in color, indicating the formation of ZnO NP. However, the precipitate was allowed to settle completely, and after 8 hours, the solution was centrifuged, washed and dried to collect the ZnO NP (Shaikh et al. 2020).

2.3.4 Mimosa Pudica (Sensitive Plant): The root extract was used as a reducing agent for the synthesis of iron oxide nanoparticles (Fe_2O_3 NP). The root extract, prepared in the usual manner of powdering and boiling, was added to 20 mM Ferrous Sulfate solution in a 3:50 volume ratio. The formation of the Fe_2O_3 NP was indicated by a change in the color of the solution from light to dark brown (Niraimathee et al. 2016).

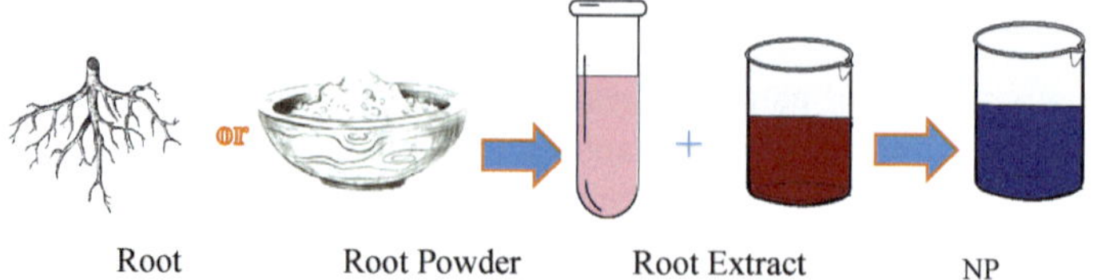

Figure 4. Schematic representation of the addition of root extract to metal salt solution and resulting nanoparticle formation indicated by a change in color.

2.4 Starch Extract

Similar to how leaf, seed or root extract can be used as a reducing agent in the synthesis of various metallic and non-metallic nanoparticles, one can also consider the use of plant starch extract as an alternative green reducing agent. To prepare the starch extract, mix the starch powder with water or soak the fruit for some time in water. Some commonly used starch extracts for the green reduction of nanoparticles are corn starch, Cassava starch, Sago starch, etc. (Ponsanti et al. 2020).

2.4.1 Corn, Cassava and Sago Starch: The extracts prepared separately from corn, cassava and sago starch powder were used as reducing agents for the synthesis of flowered-shaped silver nanoparticles (AgNP). The extract was prepared by dissolving the starch powder of corn, cassava and sago in water and boiling at 100°C for an hour under continuous stirring. The solution was then centrifuged, and the filtered extract was collected. To prepare the AgNP, 15 mM silver nitrate solution was mixed with 0.1 M solution of cetyltrimethylammonium bromide (CTAB) solution. Considering that CTAB was added, this may not be considered a fully green reduced synthesis process. The color of the solution in each changed to brown-yellow, indicating the formation of AgNP. The UV-Vis characterization showed the absorbance peaks at 422 nm, 426 nm and 435 nm up on reduction with corn, cassava, and sago extract, respectively (Ponsanti et al. 2020).

2.4.2 Starch Powder: This was used as a reducing agent to synthesize nickel oxide nanoparticles. A 0.5 M aqueous solution of $Ni(NO_3)_2 \cdot 6H_2O$ at pH 11 was stirred at 80°C until the color of the solution turned dark blue, following which the soluble starch extract was added slowly under vigorous stirring. Notably, the soluble starch extract used in this method was purchased commercially from Aldrich. The completion of the reaction was marked with the appearance of a light green precipitate. The precipitate was filtered, washed, and oven-dried for 12 hours to get a grayish-black color powder, which was the desired nickel oxide nanoparticles. The nanoparticles showed a UV-Vis absorbance peak at 319 nm, confirming the formation of the nickel oxide nanoparticle (Sabouri et al. 2018).

2.5 Bark Extract

One can use bark extracts as an alternative green-reducing agent in synthesizing nanoparticles similar to the leaf, seed and root extract. Bark extracts are prepared by taking powder from the bark of a specific plant and boiled in distilled or deionized

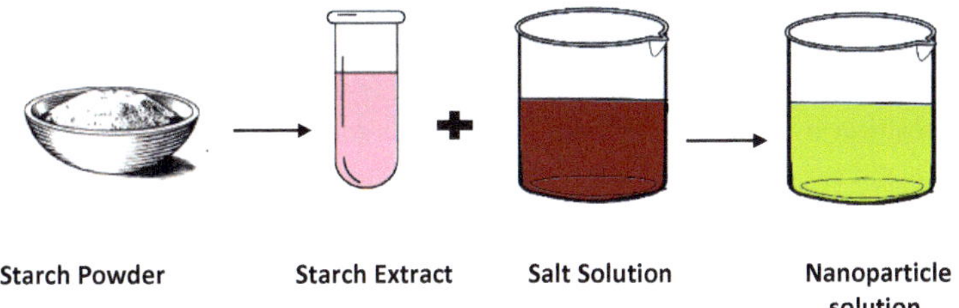

Figure 5. Schematic representation of the addition of starch extract to metal salt solution and resulting nanoparticle formation indicated by a change in color.

water. The prepared solution is then filtered and used as the extract. Some examples of nanoparticles prepared with bark extract are silver nanoparticles from bark extract of *pine* (*Punica eldarica*) (Iravani et al. 2013) and gold nanoparticles from *cinnamon extract* (Mitwalli et al. 2020). (Bark extracts are also used for the synthesis of carbon quantum dots, which is not covered here.)

2.5.1 The Bark Extract of Pine (Punica Eldarica): It was used as a reducing agent to synthesize Ag NP. The bark extract was prepared by drying the bark at room temperature to make a fine powder, which was then boiled with deionized water (0.2 gm/ml) for 15 minutes and filtered to separate the insoluble substances. The pine bark extract was added to an aqueous solution of silver nitrate; a color change in the solution was seen. The changed color from yellowish brown to dark brown indicated the formation of AgNP. The UV-Vis absorbance peak was observed at 430 nm for the synthesized Ag NP (Iravani et al. 2013).

2.5.2 The Cinnamon Bark Extract: This was used as a reducing or capping agent. The bark extract was prepared by taking the finely powdered dry cinnamon mixed with double distilled water, boiling it for 5 minutes and filtering it twice. The filtered extract was added to the $HAuCl_4$ solution and heated for 15 minutes in the microwave, following which a change of color was observed from light yellow to purple red. The formation of Au NP was further confirmed by a UV-Vis absorbance peak at 535 nm (Mitwalli et al. 2020).

2.6 Juice Extract

Nanoparticles can also be synthesized using juice/pulp extracts as the reducing agents. The use of juice extract is one of the easiest methods, as collecting the extract requires only filtering in most cases. Some examples of juice extract as a reducing agent are grape and tomato juice (Zia et al. 2017) for synthesizing silver nanoparticles, lemon juice (Sujitha et al. 2017) for synthesizing gold nanoparticles, etc.

2.6.1 Grape or Tomato Juice for the Synthesis of Silver Nanoparticles: Ag NPs were synthesized by using tomato or grape juice extract as a reducing agent (Zia et al. 2017). The extract, prepared by squeezing the thoroughly washed fruit, was added to the aqueous solution of $AgNO_3$ at neutral pH and kept in the dark.

The color of the solution changed from yellowish brown to dark reddish-dark brown, indicating the formation of Ag NP in the solution. In the case of grapes juice extract, the UV-Vis absorbance peak shifted toward the longer wavelengths with an increase in the concentration of $AgNO_3$ in the solution. The peak absorbance for 1 mM and 50 mM silver nitrate solution (with grape juice) were observed at 396 nm and 419 nm, respectively. For tomato juice extract, for lower concentrations of $AgNO_3$ (1 mM to 10 mM) in the initial solution, the peak was observed at 396 nm; however, for more concentrated solutions, a pattern similar to the nanoparticle synthesized from the grape extract was observed.

2.6.2 Lemon Juice (Citrus Limon) Extract: It was used as a reducing agent in a one-pot synthesis method for synthesizing gold nanoparticles (Au NP). Freshly extracted lemon juice was centrifuged at 10,000 rpm for 10 minutes. The pot was cleaned with distilled water and aqua regia prior to use. Juice extract was added to 1 mM solution of $HAuCl_4.3H_2O$ under vigorous stirring. A change in solution color from colorless to purple-ruby red within a mere 10 minutes of the addition of the extract indicated the formation of Au NP (Sujitha et al. 2017).

2.7 Vegetable and Fruit Extract

Fruit extracts are one of the simplest methods to reduce salt solutions and synthesize NPs. These are prepared by washing the fruits under running water to remove any kind of dirt, cutting them into small pieces, and thoroughly grinding them after boiling them in water. The solution is then filtered to collect the extract and used for the synthesis of nanoparticles. Some of the commonly used fruit extracts are apple extract (Ali et al. 2016), used in the synthesis of Ag NP, pomegranate (Basavegowda et al. 2013) for Au NP and potato (Buazar et al. 2016) for the synthesis of ZnO NP.

2.7.1 Apple Fruit Extract for Silver Nanoparticles: Red apple extract prepared the same way as explained above was used for the synthesis of silver nanoparticles (AgNP). The extract was added to 0.1 M aqueous $AgNO_3$ solution in a volume ratio 1:9 and stirred at 80°C for 1 hour. Within a few minutes, AgNP was formed, indicated by the change in solution color. The solution was centrifuged, and AgNP pellets were obtained with a broad characteristic peak between 420 and 450 nm (Ali et al. 2016).

2.7.2 Pomegranate (Punica granatum) Extract for Gold Nanoparticles: Pomegranate was used as a reducing agent in synthesizing Au NP. The fruit extract in double distilled water was added to 2 mM $HAuCl_4$ aqueous solution prepared and incubated at 90°C. Au NP was obtained within 3–4 hours of mixing with a UV-Vis peak absorption at 536 nm (Basavegowda et al. 2013).

2.7.3 Zinc Oxide Nanoparticles From Potato Extract: Potato extract was easily obtained by boiling small pieces of potatoes in deionized water at 85°C for 30 minutes (Buazar et al. 2016). The milky solution was cooled to room temperature and centrifuged to remove the insoluble part. To the extract solution at 85°C, zinc nitrate powder was added under constant stirring. The pH of the solution was maintained between 6.5–7.5. A milky white solution was obtained, indicating the formation of ZnO NP. The solution was centrifuged, washed, and dried at 50°C to obtain the ZnO powder.

2.8 Microorganism Extract

Besides plant-based reducing agents, metallic nanoparticles can also be prepared using micro-organisms, yeast, and algae as they are environmentally friendly, non-toxic, and less hazardous. (Nithya et al. 2009, Gupta et al. 2018). To prepare the extract, collect the micro-organisms and then prepare an aqueous extract using ultrahigh pure water. The mixture would have to be heated for several minutes to denature the enzymes in it. The solution has to be filtered twice to remove any unwanted residues. The filtered extracts may be added to the actual metal salt solutions with or without heat to produce the nanoparticles of silver or gold.

2.8.1 Pleurotus Sajor Caju (*Fungi*) Extract for Silver Nanoparticles: Fungi filtrate of Pleurotus sajor caju was used to synthesize Ag NPs. The filtrate of the fungi was obtained, as discussed above, and kept at pH 6. For the synthesis of Ag NPs, the cell-free filtrate of fungi was added to 1 mM $AgNO_3$ solution and kept under dark conditions. A color change from pale yellow to brown was observed upon the addition of $AgNO_3$. Characterization of the synthesized sample confirmed mono-dispersed Ag NPs with peak absorbance at 381 nm (Nithya et al. 2009).

2.8.2 Marinobacter Algicola Extract for Gold Nanoparticles: Gold nanoparticles were synthesized using the supernatant solution of Marinobacter algicola. The supernatant solution was prepared by growing bacteria at 30°C for 24 hours, following which the biomass was centrifuged. The supernatant solution of M. algicola was added to 1 mM $HauCl_4$, and after 24 hours, the solution color changed from faded yellow to pinkish purple, indicating the formation of Au NPs. Au NP was further confirmed by the observed UV-Vis absorption peak at 560 nm (Gupta et al. 2018).

3. Physico-Optic Characterization of Nanomaterials Synthesized

The nanoparticles synthesized by reducing the metal salts using various green reducing agents come in various shapes and sizes and possess distinct characteristic properties. The physico-optic properties of the green reduced nanoparticles are often characterized by various spectroscopic and other experimental tools and techniques. The leaf, bark or seed extracts or, in general, the green reducing agent used in reducing a particular metal salt significantly affected the overall morphology and property of the particle formed. By performing simple spectroscopic studies, one can easily observe the spectral signatures, such as peak absorption in the UV-visible spectrum. One can perform XRD to estimate the lattice parameters and, hence, find out the structure of the nanoparticle. Similarly, imaging techniques such as SEM, FESEM or TEM are performed as applicable to estimate the shape and size of the synthesized nanoparticles. Multimodal experimental data yields insight into the role of the green reducer in shaping the nanoparticles. For example, AgNP can be formed using Bay leaf extract, mango, lemon leaf extract or microorganisms (Gupta et al. 2018), but in each of these methods, the particles produced had observably different properties. While some of these properties are listed in the corresponding

Table 3. Summarizing various types of available plant extracts and their role as green-reducing agents.

Extract type	NPs synthesized	Size (nm)	Properties	Indicator	Application/ Purpose	Reference
Root (Berberis Asiatica)	Ag	14	UV: 427 nm, XRD: FCC Crystalline, TEM: Spherical 7.3–14.6 nm	Light yellow-reddish brown	Anti-bacterial activity	(Dangi et al. 2020)
Root (Morinda Citrifoila L)	Au	15	UV: 413 nm, XRD: FCC Crystalline, FESEM: Spherical and Oval, TEM: Spherical and oval 32–55 nm	Pink-Ruby red		(Suman et al. 2014)
Root (Sphagneticola trilobata Linn)	ZnO	65–80	SEM: 65–80 nm, XRD: FCC Crystalline, FTIR: 3,144 cm^{-1}, 1,665 cm^{-1} and 1,640 cm^{-1}	Dark Yellow	Toxic metal removal	(Shaikh et al. 2020)
Root (Mimosa Pudica)	Fe_2O_3	67	UV: 294 nm, XRD: Orthorhombic, FTIR: 3,417 cm^{-1}, 2,922 cm^{-1}, 2,354 cm^{-1}, 1,628 cm^{-1}, 1,415 cm^{-1}, 1,074 cm^{-1}, SEM: Spherical 67 nm, VSM: super paramagnetic	Light brown-dark brown	Green synthesis	(Niraimathee et al. 2016)
Corn starch	Flower shaped Ag	48	UV: 422 nm, XRD: FCC, FTIR: 3,338 cm^{-1}, 1,637 cm^{-1}, TEM: 47.8 nm (Poly dispersed)	Brown-yellow	Synthesis of Flower shaped particles	(Ponsanti et al. 2020)
Cassava starch	Flower shaped Ag	108.1	UV: 426 nm, XRD: FCC, FTIR: 3,338 cm^{-1}, 3,304 cm^{-1}, 2,928 cm^{-1}, 1,650 cm^{-1}, 1,639 cm^{-1}, TEM: 107.5 nm (Mono-dispersed)	Brown-yellow	Synthesis of Flower shaped particles	(Ponsanti et al. 2020)
Sago starch	Flower shaped Ag	114.5	UV: 435 nm, XRD: FCC, FTIR: 3,338, 1,637 cm^{-1}, TEM: 118.9 nm (Mono-dispersed)	Brown-yellow	Synthesis of Flower shaped particles	(Ponsanti et al. 2020)
Starch powder	Nickel oxide		UV: 319 nm, XRD: FCC Structure, FTIR: 424, 1,639 cm^{-1}, 3,425 cm^{-1}, FESEM: Spherical, VSM: Ferromagnetic	Dark blue- Light green-grayish black	Toxicity study	(Sabouri et al. 2018)
Bark extract (Pine)	Ag	32	UV: 430 nm, TEM: Spherical 10–40 nm	Yellowish brown- Dark Brown	Optimization of process	(Iravani et al. 2013)

Bark extract (Cinnamon)	Au	35	UV: 535 nm, TEM: Spherical, 35 nm, SAED: Crystalline and polydispersed	Light yellow-Purple Red	Fluorescence Quencher	(Mitwalli et al. 2020)
Juice extract (Grapes)	Ag	18	UV: 396–419 nm, XRD: FCC, SEM: 10–30 nm, FTIR: 3,255.1 cm^{-1}, 3,270.3 cm^{-1}, 2,844.7 cm^{-1}	Yellowish brown-Dark brown	Anti-microbial and anti-oxidant activity	(Zia et al. 2017)
Juice extract (Tomato)	Ag	12	UV: 396 nm, XRD: FCC, SEM: 10–30 nm, FTIR: 3255.1 cm^{-1}, 3270.3 cm^{-1}, 2,844.7 cm^{-1}	Colorless-Purple-Ruby red	Anti-microbial and anti-oxidant activity	(Zia et al. 2017)
Juice Extract (Lemon)	Au	17	UV: 556 nm, XRD: FCC, SAED: FCC, crystalline TEM: Spherical (15–80 nm), DLS: 32.2 nm, Zeta potential: –45.9	Colorless-Purple-Ruby red	Biological applications	(Sujitha et al. 2017)
Fruit extract (Apple)	Ag	30	UV: 420–450 nm, XRD: FCC Crystalline, FTIR: 2,364.89 cm^{-1}, 2,342.38 cm^{-1}, DLS: 30.25 ± 5.26 nm, Zeta Potential: 5.68 ± 3.28 mV	Colorless-Dark brown	Anti-bacterial	(Ali et al. 2016)
Fruit extract (Pomegranate)	Au	10–50	UV: 536 nm, SEM: Spherical, FTIR: 1,036 cm^{-1}, 1,227 cm^{-1}, 1,630 cm^{-1}, 2,903 cm^{-1} and 3,335 cm^{-1}	Light yellow-dark brown	Anti-bacterial	(Basavegowda et al. 2013)
Fruit extract (Potato)	ZnO	20	XRD: Hexagonal Wurtzite, SEM: Spherical (100–300 nm), FTIR: 3,282 cm^{-1}, 1,636 cm^{-1}, 1,355 cm^{-1}, 1,000 cm^{-1}, 1,050 cm^{-1}, 1,240 cm^{-1}, TEM: Hexagonal structure	Milkish color	Biomedical applications	(Buazar et al. 2016)

places in the preceding section, some common metallic nanoparticles prepared using different green reducing agents, such as leaf/bark/root extracts or microorganism extract, though both are green, are shown in a tabular form below (Table 3). Even when the same kind of extract from different plant sources is used, e.g., starch extract from various plants yielded the same silver nanoparticle but with significantly different morphology from 48 nm with corn starch to 108 nm with cassava starch (Ponsanti et al. 2020).

4. Summary and Discussion

From the above discussion on the synthesis of nanoparticles of different shapes and sizes, it is clear that following the green methods, one can successfully perform the synthesis of nanomaterials. As we have discussed, various spectroscopic methods performed for the confirmation and measurements of the formed nanoparticles supported the synthesis of nanoparticles with sizes ranging from 8 nm to 114 nm. It was found that the naturally present reducing and stabilizing agents inside the prepared green extract contribute to the synthesis of well-stabilized nanoparticles. As these processes exclude the use of lots of chemicals during the synthesis, the residues are not hazardous to the environment. The major disadvantage of this method is that the yield of nanomaterials is very low compared to the chemical method. That is why it prohibits industrial use. Following chemical processes, we can have more yield than following the green reducing agent method. The green reducing method is also not effective for the tunability of noble nano-metals. A concern in using green reducing agents is the significantly less control over the shape or size of the particle compared to the chemically reduced/seed-mediated synthesis of nanoparticles, which can limit the application of the synthesized nanoparticles. In the coming days, there is a lot of scope for the researchers with respect to exploring and establishing the mechanism of formation of nanoparticles via green reducing techniques, including plant extracts discussed here but not limited to that and establishing a more controlled synthetic approach with the naturally available plant extracts in its various forms.

References

Al-Ghamdi, A. Y. (2019). Antimicrobial and catalytic activities of green synthesized silver nanoparticles using bay laurel leaves extract. *Journal of Biomaterials and Nanobiotechnology*, 10(01): 26–39. https://doi.org/10.4236/jbnb.2019.101003.

Ali, Z. A., Yahya, R., Sekaran, S. D. and Puteh, R. (2016). Green synthesis of silver nanoparticles using apple extract and its antibacterial properties. *Advances in Materials Science and Engineering*, 2016: 1–6. https://doi.org/10.1155/2016/4102196.

Baghizadeh, A., Ranjbar, S., Gupta, V. K., Asif, M., Pourseyedi, S., Karimi, M. J. and Mohammadinejad, R. (2015). Green synthesis of silver nanoparticles using seed extract of Calendula officinalis in liquid phase. *Journal of Molecular Liquids*, 207: 159–163. https://doi.org/10.1016/j.molliq.2015.03.029.

Basavegowda, N., Sobczak-Kupiec, A., Fenn, R. I. and Dinakar, S. (2013). Bioreduction of chloroaurate ions using fruit extract *Punica granatum* (Pomegranate) for synthesis of highly stable gold nanoparticles and assessment of its antibacterial activity. *Micro and Nano Letters*, 8(8): 400–404. https://doi.org/10.1049/mnl.2013.0137.

Buazar, F., Bavi, M., Kroushawi, F., Halvani, M., Khaledi-Nasab, A. and Hossieni, S. A. (2016). Potato extract as reducing agent and stabiliser in a facile green one-step synthesis of ZnO nanoparticles.

Journal of Experimental Nanoscience, 11(3): 175–184. https://doi.org/10.1080/17458080.2015.10
39610.

Dangi, S., Gupta, A., Gupta, D. K., Singh, S. and Parajuli, N. (2020). Green synthesis of silver nanoparticles using aqueous root extract of Berberis asiatica and evaluation of their antibacterial activity. *Chemical Data Collections*, 28: 100411. https://doi.org/10.1016/j.cdc.2020.100411.

El Mitwalli, O. S., Barakat, O. A., Daoud, R. M., Akhtar, S. and Henari, F. Z. (2020). Green synthesis of gold nanoparticles using cinnamon bark extract, characterization, and fluorescence activity in Au/eosin Y assemblies. *Journal of Nanoparticle Research*, 22(10): 309. https://doi.org/10.1007/s11051-020-04983-8.

Ghoshal, G. and Singh, M. (2022). Characterization of silver nano-particles synthesized using fenugreek leave extract and its antibacterial activity. *Materials Science for Energy Technologies*, 5: 22–29. https://doi.org/10.1016/j.mset.2021.10.001.

Gupta, R. and Padmanabhan, P. (2018). Synthesis and characterization of Gold nanoparticles by a novel marine bacteria Marinobacter Algicola: Progression from nanospheres to various shapes. *Journal of Microbiology, Biotechnology and Food Sciences*. doi: 10.15414/jmbfs.2018.8.1.732-737.

He, Y., Wei, F., Ma, Z., Zhang, H., Yang, Q., Yao, B., Huang, Z., Li, J., Zeng, C. and Zhang, Q. (2017). Green synthesis of silver nanoparticles using seed extract of Alpinia katsumadai, and their antioxidant, cytotoxicity, and antibacterial activities. *RSC Advances*, 7(63): 39842–39851. https://doi.org/10.1039/C7RA05286C.

Hussein, Nabila H., Shaarawy, H. H., Hawash, S. I. and Amal E. Abdel-Kader. 2018. Synthesis of silver nano particles using fenugreek seeds extract. *ARPN Journal of Engineering and Applied Sciences*, 13(2): 417–22.

Iravani, S. and Zolfaghari, B. (2013). Green synthesis of silver nanoparticles using *Pinus eldarica* bark extract. *BioMed. Research International*, 2013: 1–5. https://doi.org/10.1155/2013/639725.

Jagtap, U. B. and Bapat, V. A. (2013). Green synthesis of silver nanoparticles using Artocarpus heterophyllus Lam. seed extract and its antibacterial activity. *Industrial Crops and Products*, 46: 132–137. https://doi.org/10.1016/j.indcrop.2013.01.019.

Kamath, V., Chandra, P. and Jeppu, G. P. (2020). Comparative study of using five different leaf extracts in the green synthesis of iron oxide nanoparticles for removal of arsenic from water. *International Journal of Phytoremediation*, 22(12): 1278–1294. https://doi.org/10.1080/15226514.2020.1765139.

Kandregula, G. (2015). Green synthesis of TiO_2 nanoparticles using Aloe Vera extract. *International Journal of Advanced Research in Physical Sciences*, 2(1A): 28–34. https://www.researchgate.net/publication/282502800.

Khalil, M. M. H., Ismail, E. H., El-Baghdady, K. Z. and Mohamed, D. (2014). Green synthesis of silver nanoparticles using olive leaf extract and its antibacterial activity. *Arabian Journal of Chemistry*, 7(6): 1131–1139. https://doi.org/10.1016/j.arabjc.2013.04.007.

Khan, M. I., Akhtar, M. N., Ashraf, N., Najeeb, J., Munir, H., Awan, T. I., Tahir, M. B. and Kabli, M. R. (2020). Green synthesis of magnesium oxide nanoparticles using Dalbergia sissoo extract for photocatalytic activity and antibacterial efficacy. *Applied Nanoscience*, 10(7): 2351–2364. https://doi.org/10.1007/s13204-020-01414-x.

Kumar Sur, U., Ankamwar, B., Karmakar, S., Halder, A. and Das, P. (2018). Green synthesis of silver nanoparticles using the plant extract of Shikakai and Reetha. *Materials Today: Proceedings*, 5(1): 2321–2329. https://doi.org/10.1016/j.matpr.2017.09.236.

Nawabjohn, M. S., Sivaprakasam, P., Anandasadagopan, S. K., Begum, A. A. and Pandurangan, A. K. (2022). Green synthesis and characterisation of silver nanoparticles using fe tora seed extract and investigation of antibacterial potential. *Applied Biochemistry and Biotechnology*, 194(1): 464–478. https://doi.org/10.1007/s12010-021-03651-4.

Nazeruddin, G. M., Prasad, N. R., Prasad, S. R., Shaikh, Y. I., Waghmare, S. R. and Adhyapak, P. (2014). Coriandrum sativum seed extract assisted in situ green synthesis of silver nanoparticle and its anti-microbial activity. *Industrial Crops and Products*, 60: 212–216. https://doi.org/10.1016/j.indcrop.2014.05.040.

Niraimathee, V. A., Subha, V., Ernest Ravindran, R. S. and Renganathan, S. 2016. (2016). Green synthesis of iron oxide nanoparticles from Mimosa pudica root extract. *International Journal of Environment and Sustainable Development*, 15.3(2016): 227–240. https://www.researchgate.net/publication/304661263.

Nithya, R. and Ragunathan, R. (2009). Synthesis of silver nano particle using Pleurotus Sajor caju and its antimicrobial study. *Digest Journal of Nanomaterials and Biostructures*, 4(4): 623–629.

Philip, D. (2010). Green synthesis of gold and silver nanoparticles using Hibiscus rosa sinensis. *Physica E: Low-Dimensional Systems and Nanostructures*, 42(5): 1417–1424. https://doi.org/10.1016/j.physe.2009.11.081.

Philip, D. (2011). Mangifera Indica leaf-assisted biosynthesis of well-dispersed silver nanoparticles. *Spectrochimica Acta Part A: Molecular and Biomolecular Spectroscopy*, 78(1): 327–331. https://doi.org/10.1016/j.saa.2010.10.015.

Ponsanti, K., Tangnorawich, B., Ngernyuang, N. and Pechyen, C. (2020). A flower shape-green synthesis and characterization of silver nanoparticles (AgNPs) with different starch as a reducing agent. *Journal of Materials Research and Technology*, 9(5): 11003–11012. https://doi.org/10.1016/j.jmrt.2020.07.077.

Rahimzadeh, C. Y., Barzinjy, A. A., Mohammed, A. S. and Hamad, S. M. (2022). Green synthesis of SiO2 nanoparticles from Rhus coriaria L. extract: Comparison with chemically synthesized SiO_2 nanoparticles. *PLOS ONE*, 17(8): e0268184. https://doi.org/10.1371/journal.pone.0268184.

Sabouri, Z., Akbari, A., Hosseini, H. A. and Darroudi, M. (2018). Facile green synthesis of NiO nanoparticles and investigation of dye degradation and cytotoxicity effects. *Journal of Molecular Structure*, 1173: 931–936. https://doi.org/10.1016/j.molstruc.2018.07.063.

Shaik, A. M., David Raju, M. and Rama Sekhara Reddy, D. (2020). Green synthesis of zinc oxide nanoparticles using aqueous root extract of *Sphagneticola trilobata* Lin and investigate its role in toxic metal removal, sowing germination and fostering of plant growth. *Inorganic and Nano-Metal Chemistry*, 50(7): 569–579. https://doi.org/10.1080/24701556.2020.1722694.

Shanthi, (Mrs). S. and Tharani, S. S. N. (2016). Green synthesis of zirconium dioxide (ZrO2) nanoparticles using acalypha indica leaf extract. *International Journal of Engineering and Applied Sciences*, 3(4).

Sharma, P., Kherb, J., Prakash, J. and Kaushal, R. (2023). A novel and facile green synthesis of SiO_2 nanoparticles for removal of toxic water pollutants. *Applied Nanoscience*, 13(1): 735–747. https://doi.org/10.1007/s13204-021-01898-1.

Shukla, D. and Vankar, P. S. (2012). Synthesis of plant parts mediated gold nanoparticles. *International Journal of Green Nanotechnology*, 4(3): 277–288. https://doi.org/10.1080/19430892.2012.706175.

Siripireddy, B. and Mandal, B. K. (2017). Facile green synthesis of zinc oxide nanoparticles by Eucalyptus globulus and their photocatalytic and antioxidant activity. *Advanced Powder Technology*, 28(3): 785–797. https://doi.org/10.1016/j.apt.2016.11.026.

Sujitha, M. v. and Kannan, S. (2013). Green synthesis of gold nanoparticles using Citrus fruits (Citrus limon, Citrus reticulata and Citrus sinensis) aqueous extract and its characterization. *Spectrochimica Acta Part A: Molecular and Biomolecular Spectroscopy*, 102: 15–23. https://doi.org/10.1016/j.saa.2012.09.042.

Suman, T. Y., Radhika Rajasree, S. R., Ramkumar, R., Rajthilak, C. and Perumal, P. (2014). The green synthesis of gold nanoparticles using an aqueous root extract of Morinda citrifolia L. *Spectrochimica Acta Part A: Molecular and Biomolecular Spectroscopy*, 118: 11–16. https://doi.org/10.1016/j.saa.2013.08.066.

Uddin, M. J., Chaudhuri, B., Pramanik, K., Middya, T. R. and Chaudhuri, B. (2012). Black tea leaf extract derived Ag nanoparticle-PVA composite film: Structural and dielectric properties. *Materials Science and Engineering: B*, 177(20): 1741–1747. https://doi.org/10.1016/j.mseb.2012.09.001.

Zeiri, Y., Elia, P., Zach, R., Hazan, S., Kolusheva, S. and Porat, Z. (2014). Green synthesis of gold nanoparticles using plant extracts as reducing agents. *International Journal of Nanomedicine*, 4007. https://doi.org/10.2147/IJN.S57343.

Zia, M., Gul, S., Akhtar, J., Haq, I. ul, Abbasi, B. H., Hussain, A., Naz, S. and Chaudhary, M. F. (2017). Green synthesis of silver nanoparticles from grape and tomato juices and evaluation of biological activities. *IET Nanobiotechnology*, 11(2): 193–199. https://doi.org/10.1049/iet-nbt.2015.0099.

CHAPTER 3

Role of Stabilizing Agent Role in Nanomaterials (NM)

*Naushad Edayadulla** and *Chandraraj Shanmuga Sundari*

1. Introduction

The incredible applications of nanotechnology in several fields include medicine, cleaning up environmental contamination, information and communication technologies, catalysis, detection, and the production of stronger and lighter materials. Creating and altering materials at the nanoscale requires scaling from a single atom to a collection of them. The necessary nanostructures have been created using a variety of techniques, including physical, chemical, and biological (eco-friendly approaches) (Bayda et al. 2020). Nanomaterials (NMs) have astonishing properties contrary to bulk metals, which frequently exhibit reduced energy state densities, a high surface-to-volume relationship, and sizes in the nanoscale choice; NMs offer an incredible variety of features. Size, shape, and surface morphology have an important impact on these nanoscopic materials' physical, chemical, optical, and electrical properties. In order to prepare nanoparticles, metal ions are typically reduced in solutions or high-temperature gaseous environments. Because of these particles' high surface energies and reactivity, most systems develop aggregation without any passivation or surface protection. However, the acquisition of NMs with the necessary properties and chemicals binds to particle surfaces in a way that the predictable features are preserved or enhanced when needed. Furthermore, in order to create innovative, advanced nanostructures, it is necessary to fully comprehend the role played by each chemical during the growth process. For the synthesis of nanomaterials, there are two methodologies: one is a bottom-up approach, and the other is a top-down approach. The synthesis of nanoparticles is done using a top-down

Department of Chemistry, Vel Tech Rangarajan Dr. Sagunthala, R&D Institute of Science and Technology, Avadi, Chennai, India. https://orcid.org/0000-0002-3970-5775; https://orcid.org/0000-0001-7814-2074.
* Corresponding author: edayam2004@gmail.com

technique, where size reduction is employed as an appropriate initiating material (Jeevanandam et al. 2018). The surface structure of the nanoparticles is imperfect due to the top-down technique, which is a serious constraint because the surface structure considerably impacts the chemistry and other physical characteristics of nanoparticles (Ramanathan et al. 2021). Smaller entities are brought together to create the nanoparticles in a process known as bottom-up synthesis. In this approach, first, smaller entities are formed; after that, these entities are assembled to produce the final particles, and particle sizes are in the nanometer range (Barhoum et al. 2022). These synthesis methods start with the nucleation and development of the NMs, then the metal precursor is chemically reduced, usually in solution. To initiate the transition, one must choose a reducing agent.

Additionally, bottom-up approaches frequently need stabilizing agents to prevent NM aggregation and overgrowth. While the stabilizing agent can selectively promote the preferred exposure of particular surface facets in the created NMs, the reducing agent modifies the speed of chemical reduction (Polte 2015). These two additions were carefully chosen in order to influence the NM morphologies in the solution. Finding the ideal circumstances for seed production and subsequent NM growth is necessary for controlling NM synthesis via mechanochemical methods. In solutions, it is well understood how reducing agents affect these processes (Attarad Ali et al. 2016). As a way to adjust the surface charge and manage the size of the nanoparticles, stabilizing agents were the only ones employed to integrate a favorable shell coating. Understanding how the stabilizing substance affects the surface characteristics of the nanoparticles is crucial. The capping ligand chains covalent bonds with the surfaces of the nanoparticles provide steric hindrance, which gives the nanocomposite its final stability (Javed et al. 2020). At the nanoscale, there is a rise in the fraction of atoms on the surface, which is further augmented by capping. The supported metal nanomaterials are concerned; the stabilizing agent might make it easier for the metal to be anchored to the support, leading to a high metal dispersion.

2. Chemicals as Stabilizing Agents

The selection of stabilizing and reducing agents is one of the numerous experimental variables that affect the bottom-up production of metal NMs. It has been challenging but intellectually rewarding to develop simple, adaptable techniques for the controlled manufacturing of NMs in the size or form of choice. The shape of NMs frequently results from the competing growth of many crystallographic surfaces. This is often accomplished by selectively localizing surface-modifying or capping agents, adjusting the relative growth rates of various facets, and modulating factors for the nucleation and reaction like time, reagent concentration, temperature, and pH (Harish et al. 2022).

Several stabilizers that result in the creation of stable nanoparticles are utilized to stop agglomeration and particle growth. The stabilizer is preferentially absorbed in the location where nanoparticles are growing, which slows down the merging and pace of the development of particle species. The interactions of functional groups of stabilizers with the NM surface and the surrounding medium, as well as the surface energies of the NM and stabilizer, determine how well a stabilizer will bind to it. The right stabilizer choice will determine the stability of NMs. Regarding function, stabilizers are the antithesis of catalysts; whereas enzymes and catalysts speed up chemical processes, stabilizers slow them down. Stabilizers inhibit or modify chemical and molecular processes, such as corrosion, oxidation, and separation.

An amphiphilic molecule with a polar head group and a non-polar hydrocarbon tail serves as the stabilizing or capping agent. Stabilizing agents confer because of their amphiphilic nature, functioning, and compatibility with another phase are improved (Campisi et al. 2016). Various stabilizing agents, such as surfactants, tiny ligands, dendrimers, polymers, polysaccharides, and cyclodextrins, have been utilized in the creation of nanomaterials (Figure 1). These have all been successfully employed as stabilizing and capping agents, and they all have the power to cause minor modifications in nanomaterials that reveal a powerful medicinal and environmental cleansing effect (Javed et al. 2022).

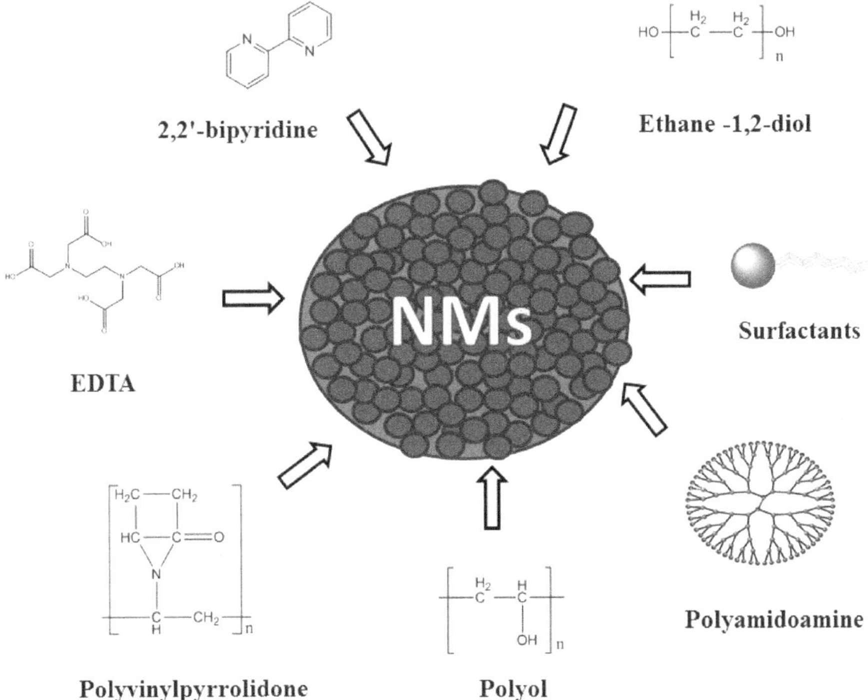

Figure 1. Nanomaterials are stabilized by chemicals.

2.1 Polyvinylpyrrolidone (PVP)

One of the crucial stabilizing agents used in nanotechnology is PVP. It has been utilized to get past issues with the manner in which nanoparticles are generally generated, including their size, toxicity, and aggregation (Kallum M. Koczkur et al. 2015). In a number of investigations, PVP has been used to stabilize metal nanoparticles like zinc (Zn), iron (Fe), gold (Au), silver (Ag), etc.

Depending on the specific synthetic circumstances and chemical system, PVP can function as a surface stabilizer, reducing agent, dispersant, and growth modifier. PVP is a large, non-ionic, non-toxic polymer containing functional groups for CH_2, C-N, and C=O that are frequently utilized in the production of NPs (Jadhav et al. 2013). The pyrrolidone moiety, which is very hydrophilic, and a sizable hydrophobic group are present in the PVP molecule. Water and many non-aqueous liquids act well as PVP solvents because of the pyrrolidone ring's apolar methylene and methine groups as well as the highly polar amide group (Graf et al. 2006). PVP's hydrophobic carbon chains interact with one another in solvents and exert repulsive forces that keep NPs from clumping together, making PVP an excellent stabilizer (steric hindrance effect) (Song et al. 2011).

Through the carbonyl group of the pyrrolidone, polyvinylpyrrolidone served as a structure-directing stabilizing agent in the chemisorption of PVP onto the Pt surface, producing Pt nanocubes. Aliphatic chains give the Pt particles a steric action and create a multilayer shell around them. Finally, PVP can modify the relative growth rate along the direction of 100 relative to 111 to facilitate the construction of Pt nanocubes (Safo et al. 2019). By raising the PVP concentration, more Ag atoms will be covered by PVP to prevent agglomeration, ultimately resulting in smaller silver nanoparticles and pure Ag NPs. Fe_3O_4-PVP nanoparticles have much higher thermal stability and surface charge than uncapped Fe_3O_4 nanoparticles.

2.2 Polyvinyl Alcohol (PVA)

PVA is one of the more advantageous capping substances used in nanotechnology. PVA has been used to stabilize Ag nanoparticles in order to regulate their size, shape, and water resistance. The sol-gel approach was used to produce PVA-coated ZnO nanoparticles with enhanced optical emission, size dispersion, and crystallinity. Due to magnetic forces, iron oxide or magnetite nanoparticles are more prone to clump together, which limits their effectiveness and potential applications. In order to produce iron oxide nanoparticles, glutaraldehyde has been employed as a cross-linker and PVA as a stabilizing agent.

As a capping agent, PVA is functionalized to redisperse cellulose nanofibrils and reduce its wet storage and transport because of its irreversible agglomeration as it increases transportation costs.

2.3 EDTA

By attaching to the divalent metal ions in the extracellular matrix (ECM), the water-soluble polymer EDTA (Ethylenediaminetetraacetic acid) is routinely used as a chelating agent to extract cells from the ECM. Additionally, metal ions have been eliminated using EDTA as a complexing agent. EDTA has grown significantly in nanoscience due to its use as a stabilizer in the production of nanoparticles. Nanoparticles' size and shape can be precisely adjusted. It has been employed as a stabilizing agent in the production of a number of different metal nanoparticles, including those made of gold (Au), chromium (Cr), zinc (Zn), cadmium (Cd), and copper (Cu). Researchers utilizing the co-precipitation method produced EDTA-covered nickel oxide (NiO) nanoparticles, and the results showed that they had better surface magnetization characteristics.

2.4 Surfactant

Surfactants naturally have the capacity to regulate the crystallization of nanomaterials to produce desirable morphologies. Nanoparticle dispersion quality was increased by coating it with various surfactants categorized as anionic (SDS), cationic (CTAB, PEI), and non-ionic (Triton X 100, PEG-6000) surfactants.

It has been suggested that a variety of surfactants, including SDS, polyethylene glycol, Pluronic, Triton, polyvinylpyrrolidone (PVP), and fullerene, can enhance the stability and dispersion of carbon nanotubes. Strong surfactant-to-surfactant and surfactant-to-nanotube interactions, along with the high charge capacity of surfactants, made the MWCNTs have great adsorption capacity and consequent stabilizing behavior. Surfactant concentrations required for maximal dispersibility, the number of surfactant molecules per unit of CNT area, and hydrophobicity (such as micellization) all play a role in CNT stability (Fernandes et al. 2015).

2.5 Polymers and Copolymers

A homopolymer is a polymer created by joining identical monomeric units together. Conversely, copolymers are polymers that are created by combining two different kinds of monomeric components. A list of stabilizing copolymers includes poly(acrylamide), poly(vinyl pyrrolidone), poly(vinyl alcohol), and poly(ethylene glycol) (PEG). Well-known stabilizers include vinyl alcohol-vinyl acetate copolymers (PVAs), which are produced via fractionally hydrolyzing poly(vinyl acetate). The polymer chain's hydrophilic poly(vinyl alcohol) (PVOH) and hydrophobic poly(vinyl alcohol) (PVAc) fragments encourage each other's interaction on an oil-water boundary, which has the effect of sterically stabilizing the polymer.

2.6 Low Molecular Weight Polyols

Ethylene glycol (EG) is a straightforward example of the polyol, a precursor that possesses the hydroxyl functionality. Due to its complexation property, it plays a special role in regulating a particle's nucleation, development, and potential agglomeration during synthesis. The effects of various polyols, including glycerin, 1,3-propanediol, erythritol, and 1,4-butanediol, as stabilizing agents at high pH with multipolar salt concentrations were investigated by researchers. It was discovered that the stabilizing ability was directly correlated with its hydrophilicity. Erythritol is, by far, the best stabilizer because it has the greatest hydroxyl groups, which seem to interrelate with the siloxane units on the cluster area and change how the colloidal silica particles interact with one another. PEG (polyethylene glycol) is indicated as a potential stabilizer for metal nanoparticles like gold (Au), zinc (Zn), and silver (Ag) in order to lessen cytotoxicity, increase stability, and enhance biocompatibility.

Additionally, it was discovered that charge-stabilized particles were less stable than steric-stabilized particles. Researchers found that PEG, as a stabilizing agent, improves the efficiency of dye-sensitized solar cells and current conversion efficiency. PEG is used with Au NP to attain *in vivo* stability and prevent uptake by the reticular endothelial system. In addition to retaining a significant amount of AgNP-PEG's bactericidal power and reducing protein conformational changes, stable PEG-capped AgNP also revealed extremely high hemocompatibility.

2.7 *Ligands*

A ligand attached to the surface of nanoparticles influences their surroundings and maintains colloidal stability by creating an appropriate steric exclusion, electrostatic repulsion, or hydration layer on the surface, which prevents nanoparticle aggregation. The ligand molecule has to be anchored to the particle's surface, employing certain interaction forces. According to research (Okada et al. 2018), the chain length, the hydrophobicity/hydrophilicity ratio, and the expected quantities of ligands all affect the ligand-nanoparticle complexes colloidal stability. In 2000, Warner et al., created gold nanoparticles using a combination of ionic and phosphine ligands, 2-(Dimethylamino)ethanethiol hydrochloride, a cationic ligand, (2-mercaptoethanesulfonate), an anionic ligand or both. The protective nature of the ligand shell was thought to be why the researchers demonstrated aggregation resistance through keeping dispersion of the initial particle (1.4; 0.4 nm) by their small core size and narrow size. The stability and functional characteristics of novel amphiphilic tridentate thiolates for Au NP are outstanding. After converting sodium borohydride with silver nitrate, spherical and triangular silver nanoparticles became stable by utilizing a dithiocarbamate tin complex through a transmetallation process.

2.8 *Dendrimers*

Dendrimers, a class of synthetic macromolecules, each feature a distinctive branching chain topology. They are also three-dimensional, nanometer-sized, and have a variety of capping properties. The interior of a dendrimer features a cavity that can contain pharmaceutical compounds, improving the stability of the nanoparticles in the solution. In addition, drug molecules can be chemically or physically adsorbed to surface dendrimer molecules for uses like targeting and drug delivery. The chemical properties of the dendrimer can be optimized to produce the suitable qualities required for a specific application, such as therapy, targeting, or detection. The amphiphilic Janus dendrimers can bind to drug particles and act as steric stabilizers to prevent tiny molecule aggregation and nucleation (Selin et al. 2018).

By carefully regulating the concentrations of polyacrylic acid (PAA) and dimethylamine borane (DMAB), AgNPs may be produced in a range of hues, from violet to red, forms (spherical, hexagonal, rod-shaped, or cube-shaped), and sizes (from nanometer to micrometer). Rivero et al., showed that a literature publication existed for an experimental matrix presenting colorful silver nanoparticle solutions utilizing both a protective agent (PAA) and a reducing agent (DMAB) (Rivero et al. 2013).

Table 1. Chemicals as stabilizing agents.

S. No.	Nanomaterials	Stablizier	Benefits	References
1.	Au NPs	Polyethylene glycol	Get vivo stability	(Sau and Rogach 2010)
2.	HAP	Polyethylene glycol	Induces the high release of Ca and P	(Sohn et al. 2009)
3.	AgNPs	Polyethylene glycol	H_2O_2 sensor	(Gulati et al. 2018)
4.	Pt NCs	Polyvinylpyrrolidone	Structure-directing capping agent	(Pinzaru et al. 2018)
5.	Fe_3O_4 NPs	Polyvinylpyrrolidone	Crystallinity	(Gharibshahi et al. 2017)
6.	ZnO NPs	Polyvinyl alcohol	Optical and photometric features	(Asadpour et al. 2022)
7.	Magnetite (Fe_3O_4) NPs	Polyvinyl alcohol	Prevent air oxidation	(Pastoriza-Santos and Liz-Marzán 2020)
8.	Au NPs	EDTA	Cancer therapeutic agent	(Junaidi et al. 2017)
9.	NiO NPs	EDTA	Restores the surface magnetization	(Rahal et al. 2017)
10.	Fe_3O_4 NPs	EDTA	Super paramagnetic behavior	(Kyrychenko et al. 2017)
11.	Amylose nanoparticles	Tween80/Span80	Shape control	(Sultana et al. 2020)
12.	Carbon nanotubes	NaDDBs, polyethylene glycol, SDS, Pluronic and Triton	Enhance the stability and dispersion	(Keinänen et al. 2018)
13.	Au NPs	Thiol-terminated poly(EG) monomethyl ethers	Speeding up particle nucleation	(Shimmin et al. 2004)
14.	Ag NPs	Dithiocarbamate tin complex	To obtain spherical and triangular shape	(Okada et al. 2018)
15.	AuNPs	PAMAM dendrimers	Both stabilizers and reducing agents	(Selin et al. 2018)

Table 1 summarizes some chemical stabilizers utilized in the fabrication of nanomaterials.

3. Biogenic Materials as Stabilizing Agents

There are a variety of physicochemical techniques for stabilizing nanomaterials that work based on one of the four guiding concepts: electrostatic, electrosteric stabilizations, steric, and stabilization by a solvent/ligand. Even though the physicochemical techniques for stabilizing nanoparticles are widely known, the chemical stabilizers have the significant drawback of being costly and environmentally hazardous due to the employment of risky chemicals and practices. An additional benefit of using biocomponents to synthesize nanomaterials is their remarkable stability.

3.1 *In Vitro Stabilization*

In vitro stabilization refers to the process of nanomaterial stabilization that occurs when they are created in close vicinity to some biosystems, such as bacteria, fungi, plants, or their extracts (Figure 2). A single biomolecule acts as the 'cap' for the created nanomaterial, stabilizing it. For the *in vitro* stabilization of produced nanomaterials, biomolecules like peptides, proteins, and a particular group of metal-binding compounds known as phytochelatins are utilized. The molecule is more stable due to metal-protein interactions. The units with free amino groups give the metal species higher stability on the nanoscale; the critical stability constants for proteinaceous structures with a free amino group, sulfhydryl group, and disulfide group were compared. Phytochelatins, a unique family of peptides, are responsible for stabilization in the cases of yeast and algae. Three amino acids, particularly glycine, cysteine, and glutamine, make up phytochelatins, which are functional counterparts of metallothioneins. By thiolate coordination, metal ions are bound to create metal complexes, which serve as stabilizing agents. According to previous reports, siderophore, a unique family of molecules produced by algal cells in response to the existence of extracellular media with iron, was used to stabilize nanoparticles in photosynthetic algae (Li et al. 2011).

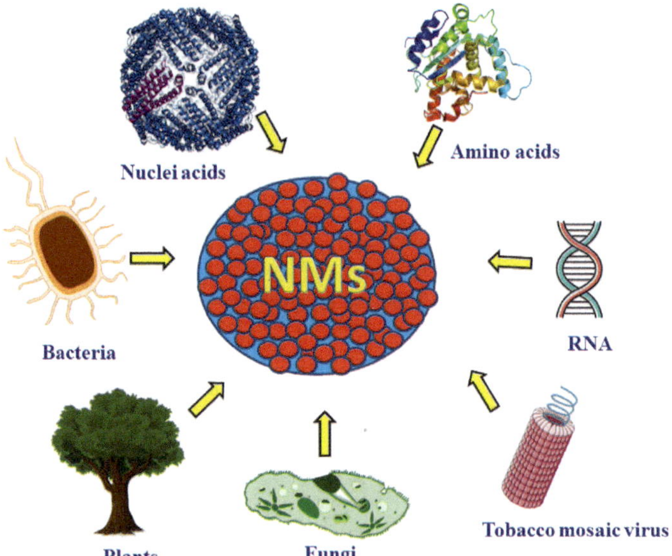

Figure 2. Nanomaterials stabilized by biogenic materials.

3.2 *In Vivo Stabilization*

Viral scaffolds, to be more precise, serve as a prototype for inorganic material formation and nucleation. The components trapped within the highly structured nano-structure of viral scaffolds are likewise on the nanoscale. The cowpea chlorotic mottle virus (CCMV) has an RNA, core, and protein shell. It is possible to extract the

RNA from the core, and the resulting cavity may be utilized to generate nanocrystals. The substance is stabilized due to the neutralization of the charge that developed inside this cavity after the RNA was removed. The synthesis and stability of PbS and CdS have both been accomplished using the tobacco mosaic virus. The mineralization of inorganic minerals has been aided by using cowpea mosaic virus as nucleation cages. Charge replacement was followed in this case by charge neutralization over an cavitated metal ion (Zhang et al. 2018).

3.3 Protein and Amino Acids

As potential mechanisms by which specific bacterial proteins were involved in the NMs' production and stability, the creation of covalent connections with atoms at the exterior of the AgNMs brought on by residues of cysteine (HS-), N-Ag bonds from H_2N-groups, and also other interactions between proteins, were all investigated. In order to create a capping layer that stabilizes the AgNMs, a free protein on the surface of the AgNMs may create hydrogen bonds with water or other proteins. To stabilize various kinds of NMs, albumin and gelatin have been frequently employed. It was discovered that amino acids such as glycine, leucine, tryptophan, and histidine effectively act as a stabilizer for the development of stable and water-soluble gold nanoparticle colloids. According to some studies, the body is unable to generate some amino acids, like arginine. This amino acid is biocompatible and non-harmful, and it can be used to functionalize metal NMs and stabilize them against aggregation and oxidation during or after a synthesis procedure (Sharma et al. 2019).

3.4 Ferritin

Ferritin, a mammalian-derived protein, has also been utilized as a template for the creation of nanomaterials. It is believed that the ferritin cage's ability to stabilize nanoparticles solely through charge distribution. Ferritin has a net negative charge on its inner surface and a net positive potential on its outer surface. Channels that link the inner and outer surfaces offer a route for cations to enter the cavity. The inside surface will then be able to bind cations in the cavity. When the cavity is completely filled with crystals, crystallization stops at the binding sites, which act as nuclei (Sharma et al. 2019).

3.5 Nucleic Acids

Functionalized nanoparticles may be bound to single-stranded DNA, DNA duplexes, and RNAs with chemically altered bases. Because the bases of double-stranded DNA are spaced apart by around 0.34 nm, particles may be placed with sub-nanometer accuracy. As a result, only the DNA template's design and the particle's size affect how the particles are arranged. Because the intermolecular spacing of the nanoparticles may be altered at a whim, engineered DNA strands provide a good template. Additionally, the two produced DNA strands can be connected to one another to create a lengthier DNA strand. This technique has been used successfully by numerous researchers to stabilize the nanoparticles on DNA subsurfaces. It is, therefore, important to comprehend the composition of the microorganisms that contribute to the capping of metals as well as their spatial distribution. A lot may be learned about the stabilization process by looking into the links that have been built between the metal and the stabilizing entity.

3.6 Bacteria-Mediated Synthesis

The production and management of microorganisms are pricier than using plant extracts, even if the microbial production of nanoparticles is simple to scale up, ecologically friendly, and consistent with the use of the nanoparticles for medicinal purposes. Shiying He and colleagues created extracellular gold nanoparticles with a size sort of 10 nm to 20 nm using the bacterium *Rhodopseudomonas capsulata*, which were stable for at least three months with only minor particle aggregation in the solution (He et al. 2007). By interacting with organic molecules in the bacterial cell, silver ions were used to produce silver nanoparticles in a non-enzymatic manner using both Gram-positive and Gram-negative bacteria. Only lactic acid bacteria, such as *Pediococcus pentosaceus*, *Lactobacillus* spp., *Lactococcus garvieae*, and *Enterococcus faecium*, were capable of reducing silver (Sintubin et al. 2009). Thus, a high pH offered reduction speeds that were far quicker than other biological processes and more akin to chemical ones. How the cell membrane regulates the development of the nanoparticles and avoids their coagulation is shown by a technique for producing silver nanoparticles that uses the biomatrix, which is formed by bacteria and serves as a stabilizing and reducing substance.

3.7 Plant-Assisted Synthesis

The metallic ions are thought to be reduced and further stabilized by extracts from plants that are bioactive phenolic acids, proteins, alkaloids, carbohydrates, polyphenols, and terpenoids. In a process known as "plant-assisted synthesis", a mixture of metal salts and plant extract is mixed so as to stabilize the product. The terpenoids and flavonoids in the leaf extract of *Azadirachta indica* function as molecules with surface activity that keep the metal nanoparticles stable (Patil et al. 2022). Aldoses are present in the extract of *Cymbopogan flexuous*, acting as both a reducing and stabilizing agent to cause the creation of gold nanotriangles as a result of the reduction from $AuCl^{4-}$ ions (Khandel et al. 2018). Aldehydes and ketones bond with newly created nanoparticles after reduction, giving them a liquid-like appearance and providing stability at room temperature. The tamarind leaf broth contains a carboxylic acid group, maybe tartaric acid, which serves as a stabilizing agent. Additionally, it was found that this bioreduced gold nanoparticle stability was aided by tartaric acid.

3.8 Secondary Metabolites Involved in Biosynthesis of Nanoparticles

Kasthuri et al. (2009) produced anisotropic gold and silver nanoparticles of quasi-spherical shape using apiin, a flavonoid glycoside obtained from henna leaves. According to the spectral analysis, the apiin compound greatly helped stabilize

Ag and Au nanoparticles through an electrostatic contact between the C=O group and the metal ion. The compound apiin's surface association with the reduced components is responsible for nanoparticles remaining stable in water for over three months (Kasthuri et al. 2009). In order to create saponin-capped triangular silver nanocrystals, Babli Debnath and his coworkers used fenugreek seed extract, which acts as a reducing and stabilizing compound. A few seeds, including *Pisum sativum*, *Cicer arietinum*, and *Vigna radiata*, were used to test the saponin-capped nanocrystals to determine how they influenced root and shoot growth and germination (Debnath et al. 2020). The main constituent of *Szyygium aromaticum* is eugenol, which was found to be associated with the bio-reduction of $HAuCl_4$ as well as $AgNO_3$ to nanomaterial. It was suggested that proton dissociation in the eugenol OH group may produce hybrid structures that could undergo further oxidation. The effective reduction of metal ions plays a role in the nanoparticle's creation. This mechanism involves the interaction of metal ions with plant reducing agents (O-H), resulting in the production of a chemical intermediate (M=O), which then disintegrates to generate metal atoms (M) and then grows and further stabilizes the nanomaterial (Ashwani Kumar et al. 2010). According to a previous report, the hybrid conversion of enol to keto will produce reactive hydrogen that will cause metal ions to become metallic nanoparticles. It has been described that for *Ocimum basilicum* extracts, the transformation of the flavonoids luteolin and rosmarinic acid from enol to keto form causes the Ag ions to produce silver nanoparticles (Pirtarighat et al. 2019). A naturally occurring plant polyphenol called Bayberry tannin (BT) is utilized to make Au@Pd core-shell nanoparticles (Au@Pd NPs) in a single step in aqueous solution at room temperature. Density functional theory (DFT) studies revealed that BT molecules increased the Au NPs surface negative charges, which promoted the interaction of Pd2+ onto Au NPs, causing a Pd shell to form. This was the key component in the generation of Au@Pd NPs (Huang et al. 2011). Some biogenic stabilizers benefits in the synthesis of nanomaterials are listed in Table 2.

4. Effects of other Parameters on the Creation of Nanoparticles Mediated by Plants

Temperature control is essential for controlling the aspect proportion, size, comparative numbers, and shapes of nanoparticles. The shape, size, and optical characteristics of nanoparticles are finely tuned due to temperature variations in reaction circumstances. When compared to chemically synthesized nanoparticles, the reaction ratio is greater or on par with the latter, as shown by the fact that the two plant leaves, M. kobus and D. kaki, have the ability to change gold nanoparticles by greater than 90% during a very short period at a reaction temperature of 95°C (Song et al. 2009). At higher reaction temperatures, gold nanoparticles have been seen to enlarge. This is brought on by an improvement in the metal ions' fusion efficiency when they dematerialize at supersaturation. The pH of the solution has a significant impact on nanoparticle size. Researchers discovered that the number of gold nanoparticles generated by A. sativa plant extract may be greatly affected by altering the pH of the medium (Makarov et al. 2014). Additionally, it has been discovered

Table 2. Natural materials as stabilizing agents.

S. No.	Nanomaterials	Natural materials as stabilizing agents	Benefits	References
1.	Cu NPs	Natural Exudates - almond gum	Colorimetric sensor to detect trace pollutants	(Zeebaree et al. 2022)
2.	Ag NPs	*Kalanchoe daigremontiana* leaf extract	Stable for a period of up to 27 months	(Morales et al. 2019)
3.	Ag NPs	Gelatin	Inhibit the growth of metallic agglomerates	(Javed et al. 2020)
4.	Fe_3O_4 NPs	Citric acid	Biomedical applications due to their small size, stability, and superparamagnetic behavior	(Nigam et al. 2011)
5.	ZnO nanofluids	Gum Arabic		(Pauzi et al. 2020)
6.	Ag NPs	*Lysimachia foenum-graecum* Hance extract		(Chartarrayawadee et al. 2020)
7.	ZnO Nanopowders	Honey		(Hoseini et al. 2015)
8.	Ag NPs Au NPs	Quercetin		(Ozdal et al. 2019)
9.	Ag NPs	The sulfated polysaccharide known as ulvan was taken from green algae of the Ulva sp.	Compared to conventional stabilizers, ulvan considerably increased the stability of the synthesized AgNPs in physiological conditions (Biomedical applications)	(Massironi et al. 2019)
10.	Ag NPs	*Ocimum basilicum-*luteolin and rosmarinic acids	Antibacterial nanoparticles	(Ahmad et al. 2010)
11.	Ag and Au NPs	Henna leaves-Apinin	Nanoparticles are found to be stable in water for more than 3 months	(Kasthuri et al. 2009)

that pH has an effect on the co-precipitation method's ability to produce magnetite nanoparticles. However, in addition to pH and temperature, other variables are also significant in the formation of nanoparticles. In ZnS samples, it was discovered that the bond group energy rose as the dopant concentration increased. This finding was confirmed by using optical absorption spectroscopic methods. Diameters between 11 nm and 32 nm are found in those produced with no inclusion of NaCl solution; however, gold nanoparticles produced with high concentrations of NaCl have sizes between 5 nm and 16 nm. As the concentration of NaCl increases, the size of these nanoparticles decreases. It must be known how chloride, bromide, and iodide impact plant nanoparticle creation. Triangle nanoparticles are encouraged to proliferate by chloride, whereas iodide changes the form of nanoparticles and eventually leads to aggregated spherical nanoparticles. Chloride ions lead to the production of diamond-shaped nanoparticles in the case of copper nanoparticles. When aqueous solutions of Ag or Au metal ions are applied to sun-dried plant biomass, the result is the formation of gold and silver nanoparticles with diameters ranging from 55 nm to 80 nm that are spherical or triangular in shape. *Cinnamonum camphora* leaf might be associated with the relative protectability of secondary metabolites from leaf extracts (Khandel et al. 2018). Most often, it is known that the water-soluble heterocyclic polyol compounds reduce silver ions or chloroaurate ions.

5. Conclusions

This unique review essentially highlights the advantages of various stabilizing agents in nanotechnology. Though it is challenging to create capped nanoparticles that are much more capable of medicinal and environmental cleanup than uncapped nanomaterials, there is a growing literature that emphasizes the unexpectedly advantageous part of stabilizing mediators in various synthetic methodologies. The critical use of nanomaterials depends on the regulated size, shape, and surface composition produced by capping nanomaterials. However, such methods must be improved to confirm the stabilizing phenomena effectively; more repeatable studies must be carried out to eliminate any inconsistencies in obtaining the appropriate stabilizing agents with pure and controlled stabilizing action. Furthermore, it was necessary to use sophisticated characterization techniques to accurately assess the function of stabilizing substances at the boundary of the nanoparticle stabilizer.

References

Ahmad Naheed, Seema Sharma, Alam, Md. K., Singh, V. N., Shamsi, S. F., Mehta, B. R. and Anjum Fatma. (2010). Rapid synthesis of silver nanoparticles using dried medicinal plant of Basil. *Colloids and Surfaces B: Biointerfaces*, 81(1): 81–86. doi:10.1016/j.colsurfb.2010.06.029.

Asadpour Saeid, Ahmad Raeisi vanani, Masoumeh Kooravand and Arash Asfaram. (2022). A review on zinc oxide/poly(vinyl alcohol) nanocomposites: Synthesis, characterization and applications. *Journal of Cleaner Production*, 362(August): 132297. doi:10.1016/j.jclepro.2022.132297.

Ashwani Kumar Singh, Talat Mahe, Singh, D. P. and Srivastava, O. N. (2010). Biosynthesis of gold and silver nanoparticles by natural precursor clove and their functionalization with amine group | SpringerLink. *Journal of Nanoparticle Research*, 12: 1667–1675. doi:https://doi.org/10.1007/s11051-009-9835-3.

Attarad Ali, Hira Zafar, Muhammad Zia, Ihsan ul Haq, Abdul Rehman Phull, Joham Sarfraz Ali and Altaf Hussain. (2016). Synthesis, characterization, applications, and challenges of iron oxide nanoparticles - PMC. https://www.ncbi.nlm.nih.gov/pmc/articles/PMC4998023/.

Barhoum Ahmed, María Luisa García-Betancourt, Jaison Jeevanandam, Eman A. Hussien, Sara A. Mekkawy, Menna Mostafa, Mohamed M. Omran, Mohga S. Abdalla and Mikhael Bechelany. (2022). Review on natural, incidental, bioinspired, and engineered nanomaterials: History, definitions, classifications, synthesis, properties, market, toxicities, risks, and regulations. *Nanomaterials*, 12(2). Multidisciplinary Digital Publishing Institute: 177. doi:10.3390/nano12020177.

Bayda Samer, Muhammad Adeel, Tiziano Tuccinardi, Marco Cordani and Flavio Rizzolio. (2020). The history of nanoscience and nanotechnology: from chemical–physical applications to nanomedicine.

Molecules, 25(1). Multidisciplinary Digital Publishing Institute (MDPI). doi:10.3390/molecules25010112.

Campisi Sebastiano, Marco Schiavoni, Carine Edith Chan-Thaw and Alberto Villa. (2016). Untangling the role of the capping agent in nanocatalysis: recent advances and perspectives. *Catalysts*, 6(12). Multidisciplinary Digital Publishing Institute: 185. doi:10.3390/catal6120185.

Chartarrayawadee Widsanusan, Phattaraporn Charoensin, Juthaporn Saenma, Thearum Rin, Phichaya Khamai, Pitak Nasomjai and Chee On Too. (2020). Green synthesis and stabilization of silver nanoparticles using lysimachia foenum-graecum hance extract and their antibacterial activity. *Green Processing and Synthesis*, 9(1). De Gruyter: 107–118. doi:10.1515/gps-2020-0012.

Debnath Babli, Sumit Sarkar and Ratan Das. (2020). Effects of saponin capped triangular silver nanocrystals on the germination of pisum sativum, cicer arietinum, vigna radiata seeds & their subsequent growth study. *IET Nanobiotechnology*, 14(1): 25–32. doi:10.1049/iet-nbt.2019.0161.

Fernandes Ricardo M. F., Bárbara Abreu, Bárbara Claro, Matat Buzaglo, Oren Regev, István Furó and Eduardo F. Marques. (2015). Dispersing carbon nanotubes with ionic surfactants under controlled conditions: comparisons and insight. *Langmuir: The ACS Journal of Surfaces and Colloids*, 31(40): 10955–10965. doi:10.1021/acs.langmuir.5b02050.

Gharibshahi Leila, Elias Saion, Elham Gharibshahi, Abdul Halim Shaari and Khamirul Amin Matori. (2017). Influence of Poly(Vinylpyrrolidone) concentration on properties of silver nanoparticles manufactured by modified thermal treatment method. *PLOS ONE*, 12(10). Public Library of Science: e0186094. doi:10.1371/journal.pone.0186094.

Graf Christina, Sofia Dembski, Andreas Hofmann and Eckart Rühl. (2006). A general method for the controlled embedding of nanoparticles in silica colloids. *Langmuir*, 22(13). American Chemical Society: 5604–5610. doi:10.1021/la060136w.

Gulati Shivani, Sachdeva, M. and Bhasin, K. K. (2018). Capping agents in nanoparticle synthesis: surfactant and solvent system. *AIP Conference Proceedings*, 1953(1). American Institute of Physics: 030214. doi:10.1063/1.5032549.

Harish Vancha, Md Mustafiz Ansari, Devesh Tewari, Manish Gaur, Awadh Bihari Yadav, María-Luisa García-Betancourt, Fatehy M. Abdel-Haleem, Mikhael Bechelany and Ahmed Barhoum. (2022). Nanoparticle and nanostructure synthesis and controlled growth methods. *Nanomaterials*, 12(18). Multidisciplinary Digital Publishing Institute: 3226. doi:10.3390/nano12183226.

He Shiying, Zhirui Guo, Yu Zhang, Song Zhang, Jing Wang and Ning Gu. (2007). Biosynthesis of gold nanoparticles using the bacteria Rhodopseudomonas Capsulata. *Materials Letters*, 61(18): 3984–3987. doi:10.1016/j.matlet.2007.01.018.

Hoseini Seyed Javad, Majid Darroudi, Reza Kazemi Oskuee, Leila Gholami and Ali Khorsand Zak. (2015). Honey-based synthesis of ZnO Nanopowders and their cytotoxicity effects. *Advanced Powder Technology, Special issue of the 7th World Congress on Particle Technology*, 26(3): 991–996. doi:10.1016/j.apt.2015.04.003.

Huang Xin, Hao Wu, Shangzhi Pu, Wenhua Zhang, Xuepin Liao and Bi Shi. (2011). One-step room-temperature synthesis of Au@Pd core-shell nanoparticles with tunable structure using plant tannin as reductant and stabilizer. *Green Chemistry*, 13(4). The Royal Society of Chemistry: 950–957. doi:10.1039/C0GC00724B.

Jadhav, S. V., Nikam, D. S., Khot, V. M., Thorat, N. D., Phadatare, M. R., Ningthoujam, R. S., Salunkhe, A. B. and Pawar, S. H. (2013). Studies on colloidal stability of PVP-Coated LSMO nanoparticles for magnetic fluid hyperthermia. *New Journal of Chemistry*, 37(10). The Royal Society of Chemistry: 3121–3130. doi:10.1039/C3NJ00554B.

Javed Rabia, Anila Sajjad, Sania Naz, Humna Sajjad and Qiang Ao. (2022). Significance of capping agents of colloidal nanoparticles from the perspective of drug and gene delivery, bioimaging, and biosensing: An insight. *International Journal of Molecular Sciences*, 23(18). Multidisciplinary Digital Publishing Institute: 10521. doi:10.3390/ijms231810521.

Javed Rabia, Muhammad Zia, Sania Naz, Samson O. Aisida, Noor ul Ain and Qiang Ao. (2020). Role of capping agents in the application of nanoparticles in biomedicine and environmental remediation: Recent trends and future prospects. *Journal of Nanobiotechnology*, 18(1): 172. doi:10.1186/s12951-020-00704-4.

Jeevanandam Jaison, Ahmed Barhoum, Yen S. Chan, Alain Dufresne and Michael K. Danquah. (2018). Review on nanoparticles and nanostructured materials: History, sources, toxicity and regulations. *Beilstein Journal of Nanotechnology*, 9(April): 1050–1074. doi:10.3762/bjnano.9.98.

Junaidi Kuwat Triyana, Edi Suharyadi, Harsojo and Linda Y. L. Wu. (2017). The roles of Polyvinyl Alcohol (PVA) as the capping agent on the polyol method for synthesizing silver nanowires. *Journal of Nano Research*, 49(September): 174–180. doi:10.4028/www.scientific.net/JNanoR.49.174.

Kallum M. Koczkur, Stefanos Mourdikoudis, Lakshminarayana Polavarapu and Sara E. Skrabalak. (2015). Polyvinylpyrrolidone (PVP) in nanoparticle synthesis - Dalton Transactions (RSC Publishing). https://pubs.rsc.org/en/content/articlelanding/2015/dt/c5dt02964c.

Kasthuri, J., Veerapandian, S. and Rajendiran, N. (2009). Biological synthesis of silver and gold nanoparticles using apiin as reducing agent. *Colloids and Surfaces B: Biointerfaces*, 68(1): 55–60. doi:10.1016/j.colsurfb.2008.09.021.

Keinänen Pasi, Sanna Siljander, Mikko Koivula, Jatin Sethi, Essi Sarlin, Jyrki Vuorinen and Mikko Kanerva. (2018). Optimized dispersion quality of aqueous carbon nanotube colloids as a function of sonochemical yield and surfactant/CNT ratio. *Heliyon*, 4(9): e00787. doi:10.1016/j.heliyon.2018.e00787.

Khandel Pramila, Ravi Kumar Yadav, Deepak Kumar Soni, Leeladhar Kanwar and Sushil Kumar Shahi. (2018). Biogenesis of metal nanoparticles and their pharmacological applications: Present status and application prospects. *Journal of Nanostructure in Chemistry*, 8(3): 217–254. doi:10.1007/s40097-018-0267-4.

Kyrychenko Alexander, Dmitry A. Pasko and Oleg N. Kalugin. (2017). Poly(Vinyl Alcohol) as a water protecting agent for silver nanoparticles: The role of polymer size and structure. *Physical Chemistry Chemical Physics*, 19(13). The Royal Society of Chemistry: 8742–8756. doi:10.1039/C6CP05562A.

Li Xiangqian, Huizhong Xu, Zhe-Sheng Chen and Guofang Chen. (2011). Biosynthesis of nanoparticles by microorganisms and their applications. *Journal of Nanomaterials*, 2011(September). Hindawi: e270974. doi:10.1155/2011/270974.

Makarov, V. V., Love, A. J., Sinitsyna, O. V., Makarova, S. S., Yaminsky, I. V., Taliansky, M. E. and Kalinina, N. O. (2014). 'Green' nanotechnologies: synthesis of metal nanoparticles using plants. *Acta Naturae*, 6(1). National Research University Higher School of Economics: 35.

Massironi Alessio, Andrea Morelli, Lucia Grassi, Dario Puppi, Simona Braccini, Giuseppantonio Maisetta, Semih Esin, Giovanna Batoni, Cristina Della Pina and Federica Chiellini. (2019). Ulvan as novel reducing and stabilizing agent from renewable algal biomass: application to green synthesis of silver nanoparticles. *Carbohydrate Polymers*, 203(January): 310–321. doi:10.1016/j.carbpol.2018.09.066.

Morales Gladis, Gloria Campillo, Ederley Vélez, Jeaneth Urquijo, César Hincapié and Jaime Osorio. (2019). Kalanchoe daigremontiana leaf extract: A green stabilizing agent in synthesis of silver nanoparticles. *Journal of Physics: Conference Series*, 1247(1). IOP Publishing: 012019. doi:10.1088/1742-6596/1247/1/012019.

Nigam Saumya, Barick, K. C. and Bahadur, D. (2011). Development of citrate-stabilized Fe_3O_4 nanoparticles: Conjugation and release of doxorubicin for therapeutic applications. *Journal of Magnetism and Magnetic Materials*, 323(2): 237–243. doi:10.1016/j.jmmm.2010.09.009.

Okada Yohei, Kodai Ishikawa, Naoya Maeta and Hidehiro Kamiya. (2018). Understanding the colloidal stability of nanoparticle–ligand complexes: Design, synthesis, and structure–function relationship studies of amphiphilic small-molecule ligands. *Chemistry – A European Journal*, 24(8): 1853–1858. doi:10.1002/chem.201704306.

Ozdal Zeliha Duygu, Ertugrul Sahmetlioglu, Ibrahim Narin and Ahmet Cumaoglu. (2019). Synthesis of gold and silver nanoparticles using flavonoid quercetin and their effects on lipopolysaccharide induced inflammatory response in microglial cells. *3 Biotech*, 9(6): 212. doi:10.1007/s13205-019-1739-z.

Pastoriza-Santos Isabel and Luis M. Liz-Marzán. (2020). N,N-Dimethylformamide as a reaction medium for metal nanoparticle synthesis. In *Colloidal Synthesis of Plasmonic Nanometals*. Jenny Stanford Publishing.

Patil Shriniwas P., Rajesh Y. Chaudhari and Mahesh S. Nemade. (2022). Azadirachta indica leaves mediated green synthesis of metal oxide nanoparticles: A review. *Talanta Open*, 5(August): 100083. doi:10.1016/j.talo.2022.100083.

Pauzi Norlin, Norashikin Mat Zain and Nurul Amira Ahmad Yusof. (2020). Gum arabic as natural stabilizing agent in green synthesis of ZnO nanofluids for antibacterial application. *Journal of Environmental Chemical Engineering*, 8(3): 103331. doi:10.1016/j.jece.2019.103331.

PPinzaru Iulia, Dorina Coricovac, Cristina Dehelean, Elena-Alina Moacă, Marius Mioc, Flavia Baderca, Ioana Sizemore, Seth Brittle, Daniela Marti, Cornelia Daniela Calina, Aristidis M. Tsatsakis and Codruţa Şoica. (2018). Stable PEG-coated silver nanoparticles—A comprehensive toxicological profile. *Food and Chemical Toxicology: An International Journal Published for the British Industrial Biological Research Association*, 111(January): 546–556. doi:10.1016/j.fct.2017.11.051.

Pirtarighat Saba, Maryam Ghannadnia and Saeid Baghshahi. (2019). Biosynthesis of silver nanoparticles using ocimum basilicum cultured under controlled conditions for bactericidal application. *Materials Science and Engineering: C*, 98(May): 250–255. doi:10.1016/j.msec.2018.12.090.

Polte, Jörg. (2015). Fundamental growth principles of colloidal metal nanoparticles—A new perspective - CrystEngComm (RSC Publishing) DOI:10.1039/C5CE01014D. https://pubs.rsc.org/en/content/articlehtml/2015/ce/c5ce01014d.

Rahal, H. T., Awad, R., Abdel-Gaber, A. M. and El-Said Bakeer, D. (2017). Synthesis, characterization, and magnetic properties of pure and EDTA-capped NiO nanosized particles. *Journal of Nanomaterials*, 2017(January). Hindawi: e7460323. doi:10.1155/2017/7460323.

Ramanathan Santheraleka, Subash C. B. Gopinath, Md Arshad, M. K., Prabakaran Poopalan and Veeradasan Perumal. (2021). 2—Nanoparticle synthetic methods: Strength and limitations. pp. 31–43. *In*: Subash C. B. Gopinath and Fang Gang (eds.). *Nanoparticles in Analytical and Medical Devices*. Elsevier. doi:10.1016/B978-0-12-821163-2.00002-9.

Rivero Pedro Jose, Javier Goicoechea, Aitor Urrutia and Francisco Javier Arregui. (2013). Effect of both protective and reducing agents in the synthesis of multicolor silver nanoparticles. *Nanoscale Research Letters*, 8(1): 101. doi:10.1186/1556-276X-8-101.

Safo, I. A., Werheid, M., Dosche, C. and Oezaslan, M. (2019). The role of Polyvinylpyrrolidone (PVP) as a capping and structure-directing agent in the formation of Pt nanocubes. *Nanoscale Advances*, 1(8). RSC: 3095–3106. doi:10.1039/C9NA00186G.

Sau Tapan K. and Andrey L. Rogach. (2010). Nonspherical noble metal nanoparticles: Colloid-chemical synthesis and morphology control. *Advanced Materials*, 22(16): 1781–1804. doi:10.1002/adma.200901271.

Selin Markus, Sami Nummelin, Jill Deleu, Jarmo Ropponen, Tapani Viitala, Manu Lahtinen, Jari Koivisto, Jouni Hirvonen, Leena Peltonen, Mauri A. Kostiainen and Luis M. Bimbo. (2018). High-generation amphiphilic janus-dendrimers as stabilizing agents for drug suspensions. *Biomacromolecules*, 19(10). American Chemical Society: 3983–3993. doi:10.1021/acs.biomac.8b00931.

Sharma Deepali, Suvardhan Kanchi and Krishna Bisetty. (2019). Biogenic synthesis of nanoparticles: A review. *Arabian Journal of Chemistry*, 12(8): 3576–3600. doi:10.1016/j.arabjc.2015.11.002.

Shimmin Robert G., Andrew B. Schoch and Paul V. Braun. (2004). Polymer size and concentration effects on the size of gold nanoparticles capped by polymeric thiols. *Langmuir*, 20(13). American Chemical Society: 5613–5620. doi:10.1021/la036365p.

Sintubin Liesje, Wim De Windt, Jan Dick, Jan Mast, David van der Ha, Willy Verstraete and Nico Boon. (2009). Lactic acid bacteria as reducing and capping agent for the fast and efficient production of silver nanoparticles. *Applied Microbiology and Biotechnology*, 84(4): 741–749. doi:10.1007/s00253-009-2032-6.

Sohn Kwonnam, Franklin Kim, Ken C. Pradel, Jinsong Wu, Yong Peng, Feimeng Zhou and Jiaxing Huang. (2009). Construction of evolutionary tree for morphological engineering of nanoparticles. *ACS Nano*, 3(8). American Chemical Society: 2191–2198. doi:10.1021/nn900521u.

Song Jae Yong, Hyeon-Kyeong Jang and Beom Soo Kim. (2009). Biological synthesis of gold nanoparticles using magnolia kobus and diopyros kaki leaf extracts. *Process Biochemistry*, 44(10): 1133–1138. doi:10.1016/j.procbio.2009.06.005.

Song Jee Eun, Tanapon Phenrat, Stella Marinakos, Yao Xiao, Jie Liu, Mark R. Wiesner, Robert D. Tilton and Gregory V. Lowry. (2011). Hydrophobic interactions increase attachment of gum arabic- and PVP-coated Ag nanoparticles to hydrophobic surfaces. *Environmental Science & Technology*, 45(14). American Chemical Society: 5988–5995. doi:10.1021/es200547c.

Sultana Shaheen, Nada Alzahrani, Reem Alzahrani, Wejdan Alshamrani, Waad Aloufi, Amena Ali, Shehla Najib and Nasir Ali Siddiqui. (2020). Stability issues and approaches to stabilised nanoparticles based drug delivery system. *Journal of Drug Targeting*, 28(5). Taylor & Francis: 468–486. doi:10.1 080/1061186X.2020.1722137.

Zeebaree Samie Yaseen Sharaf, Osama Ismail Haji, Rzgar Farooq Rashid, Suhad Abdulrahman Yasin, Aymn Yaseen Sharaf Zeebaree, Amal Jamil Sadiq Albarwary, Ali Yaseen Sharaf Zebari and Husaen Abdalelah Gerjees. (2022). Novel natural exudate as a stabilizing agent for fabrication of copper nanoparticles as a colourimetric sensor to detect trace pollutant. *Surfaces and Interfaces*, 32(August): 102131. doi:10.1016/j.surfin.2022.102131.

Zhang Yu, Yixin Dong, Jinhua Zhou, Xun Li and Fei Wang. (2018). Application of plant viruses as a biotemplate for nanomaterial fabrication. *Molecules*, 23(9). Multidisciplinary Digital Publishing Institute: 2311. doi:10.3390/molecules23092311.

Chapter 4

Stability and Toxicity of Green-Synthesized NM

Raja Murugesan,[1,5,]* *Vadivel Siva,*[2] *Raman Sureshkumar*[3]
and *Dharani Vetriselvan*[4]

1. Introduction

At present, science and technology play a vital role in the emerging ideas in the past few years. In particular, nanomaterials (NMs) pay more attention to the various developments because their physio-chemical properties can make various new technologies and systems for wide fields (Mirzaei and Darroudi 2017). In the last decade, novel synthesis of NMs such as carbon nanotubes (CNTs), quantum dots, graphene and graphene oxide plays a major role in the biomedical field due to their properties of stability. To acquire the NMs of desired sizes, shapes and functionalization properties, two diverse principles of the synthesis process have been reported (Huang et al. 2006, Kim et al. 2007, Laurent et al. 2008). Previously, the nanoparticles (NPs) and NMs were synthesized by various techniques such as ball milling, sputtering, etching, etc. In addition, bottom-up approaches include CVD (chemical vapor deposition), spray pyrolysis and laser pyrolysis. However, suppose these prepared NPs or NMs are involved in huge specific applications; the above-mentioned methods then face many challenges and limitations, such as

[1] Faculty of Pharmacy, Karpagam Academy of Higher Education, Coimbatore 641 021, Tamil Nadu, India.

[2] Department of Physics, Karpagam Academy of Higher Education, Coimbatore 64021, Tamil Nadu, India.

[3] Department of Pharmaceutics, JSS College of Pharmacy, Ooty 643001, Tamil Nadu, India.

[4] Department of Pharmacognosy, RVS College of Pharmaceutical Sciences, Coimbatore 641402, Tamil Nadu, India.

[5] Cancer Research Centre, Karpagam Academy of Higher Education, Coimbatore 641 021, Tamil Nadu, India.

* Corresponding author: rajasai68@gmail.com/raja.murugesan@kahedu.edu.in

Table 1. Synthesis of NPs/NMs by using various techniques.

Synthesis of Nanoparticles (NPs)/Nanomaterials (NMs)		
S. No.	**Bottom-Up methods**	**Top-Down methods**
1.	CVD (Chemical vapor deposition)	Laser ablation
2.	Laser pyrolysis	Chemical etching
3.	Spray pyrolysis	Ball milling
4.	Sol-gel process	Electro-explosion
5.	Aerosol processes	Sputtering
6.	Atomic/molecular condensation	–

stability, modeling factors, bio-compatibility, bioaccumulation and toxicity (Cao 2004). The different syntheses of NMs and NPs are illustrated in Table 1.

Besides, these above-mentioned confines present both foundational challenges and new and great opportunities in this emerging field of research. To overcome those limitations, the new stage of 'green synthesis' methods is getting excessive attention in present research and development on materials science and technology. Fundamentally, green synthesis of NMs, formed through a parameter, clean up and control and remedy process, will directly help upthrow for eco-friendly toward the environment.

Generally, some of the principles of 'green synthesis' play a major role in the various components such as the prevention process of waste products, reduction of pollution and use of nontoxic.

Green synthesis is essential to avoid the synthesis of annoying or destructive by-products through the build-up of consistent, supportable, and eco-friendly production of the procedures. Using supreme solvent systems and natural resources like organic-based systems is vital to accomplish this milestone. Green synthesis of metallic NMs has been assumed to accommodate several biological materials (e.g., fungi, algae, bacteria and various plant extracts). Among the available green methods of synthesis for metal oxide NPs/NMs, consumption of plant extracts is a moderately humble and calm process to synthesizing NMs/NPs at a huge scale virtual to fungi and bacteria-facilitated synthesis. The illustration of green synthesized NMs from microorganisms and plants is shown below, and the merits of green-synthesized NMs are shown in Figure 1.

1. Microorganisms
   - Bacteria
   - Fungus
   - Algae

2. Plants
   - Leaves
   - Root
   - Flower
   - Fruits

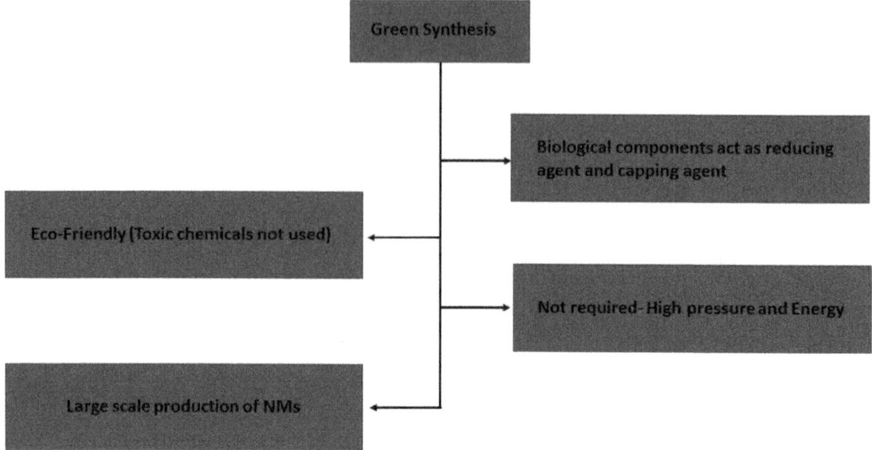

Figure 1. Illustration of merits of green-synthesized NMs.

Green synthesis procedures mainly play a huge role in the biological pioneers depending on the numerous parameters like temperature, solvent and PH conditions. Based on these reasons, the prepared NMs/NPs are considered an effective study due to their chemical constituents, such as aldehydes, ketones, amines, carboxylic acid, phenols and terpenoids (Doble and Krutiventi 2007).

Here, we discussed the up-to-date state of research based on green-synthesized NMs with their advantages and various methods. In addition, we also summarized the role of green synthesized-based drug delivery and diagnosis along with their mechanism, stability and analysis of toxicity. Inclusive, our goal is to systematically describe green synthesis techniques and their associated components that will help researchers elaborate in this incipient field while assisting as a beneficial attendant for the readers with universal attention in this topic.

2. Biological Synthesis of NMs for Drug Delivery and Diagnosis

The biological synthesis comprises various techniques, e.g., bacteria-mediated, yeast-mediated, fungi, etc. These microorganisms, such as fungi and yeast, are involved in a major role in enzyme production and NP synthesis for biomedical applications. Plant-based NMs especially pay more attention to the drug delivery in a particular site.

2.1 Anticancer Activity by Using Green-Synthesized NMs

The green-synthesized NMs have good capability for anticancer activity against various cancer cell lines proven and reported. The summarized NMs-based drug delivery for cancer disease is given in Table 2 below.

Table 2. Summarized microorganism-based drug delivery for the cancer.

colspan
Microorganism-Based Drug Delivery (Cancer)

S. No.	Microbes species	Microbes type	NPs type	Anticancer activity (*in vitro* studies)	Highlights	References
1.	Endophytic Fungus	Fungus	Au	HEP_2 and Vero cells	It has shown good efficiency IC 50:23 µg/mL	(Nachiyar et al. 2015)
2.	Beauvaria bassiana	Fungus	Ag	Hela cell lines	Cell viability (83.5±1.66%)	(Prabakaran et al. 2016)
3.	Fusarium oxysporum	Fungus	Au	MG-63 and primary osteoblast	Toxicity: Normal IC_{50}:0.102 µg/mL	(Iram et al. 2016)
4.	Enterococcus sp	Bacterium	Au	$HepG_2$ and A549	IC_{50}:0.50 µg/mL	(Rajeshkumar et al. 2016)
5.	Haalococcous salifodinae BK18	Bacterium	Se	Hela	No toxicity and huge efficiency	(Srivastava et al. 2014)
6.	Moraxella ostonsis	Bacterium	TiO_2	HaCaT and Hep_2	IC_{50}:55 µg/mL HaCaT and 172 µg/mL of HeP_2	(Valli Naachiyar et al. 2016)
7.	Nocardiopsis sp	Marine antinobacterium	Au	HeLa	IC_{50}:350 µg/mL	(Manivasagan et al. 2015)
8.	Nocardiopsis sp	Marine antinobacterium	Au	HeLa	IC_{50}:250 µg/mL after incubation 24 h	(Manivasagan et al. 2015)
9.	Streptomyces naganishii (MA7)	Actinobacteria	Ag	HeLa	IC_{50}:1.53 µg/Ml LD$_{50}$: 24.39 µg/Ml	(Shanmugasundaram et al. 2013)
10.	Streptomyces rochei	Actinomycetes	Ag	MCF-7; PC-3	IC_{50}:40 µg/ML IC_{50}:48.5 µg/ML	(Abd-Elnaby et al. 2013)
11.	Nocardiopsis valliformis OT1	Alkaliphilllic actinobacterium	Au	HeLa	IC_{50}:100 µg/ML	(Rathod et al. 2016)

Table 2 contd. ...

...Table 2 contd.

Plant-Based NMs to Drug Delivery						
S. No.	**Plant**	**Part used**	**NPs type**	**Anticancer activity**	**Highlights**	**References**
12.	Musa paradisiaca	Stem of the banana	Au	MCF-7 and HEK-293	IC_{50}: More than 80 nm in MCF-7 with no toxicity	(Arunkumar et al. 2013)
13.	Origanum vulgare (oregano)	Leaf	Ag	A549	IC_{50}:100 µg/ML	(Sankar et al. 2013)
14.	Padina gymnospora	Leaf	Au	HepG2	IC_{50}:51.9 nM	(Singh et al. 2015)
15.	Premna serratifolia L	Leaf	Ag	Hepato cancer cell (Swiss albino mice)	500 mg/kg (*in vivo*) significant activity and improved the biochemical factor	(Patra et al. 2015)
16.	Oak fruit hull	Fruit	Ag	MCF-7 and Mononuclear cell (Human)	IC_{50}:50 µg/ML IC_{50}:100 µg/ML in blood cells and no toxic effect	(Heydari et al. 2015)
17.	Piper pedicellatum	Fruit	Au-rGo	PC-3	IC_{50}:12.02 µg/ML	(Saikia et al. 2016)
18.	Potentilla fuigens	Root	Ag	MCF-7 and U-87	IC_{50}:4.91 µg/ML and IC_{50}:8.23 µg/ML	(Mittal et al. 2015)
19.	Rosa indica	Petal	Ag	HCT-15	IC_{50}:~30 µg/ML	(Manikandan et al. 2015)
20.	Sargassum muticum	Seaweed	Au	CEMss	IC_{50}:55.2 µg/ML	(Devi et al. 2013)

2.2 Cancer Diagnosis by Using Bio-Compatible NMs

At present, nano-medical developments facilitated the improvement of NPs/NMs for diagnostic and therapeutic approaches. Mainly, NPs can play a huge role in cancer diagnosis at an early stage by allowing imaging (visualization) of cancer cells. Cancer diagnostic devices used regularly in preclinical research and clinical practice comprise ultrasound, optical imaging, positron emission tomography, photo-acoustic imaging, MRI, CT and single-photon emission CT (SPECT). In this regard, due to their vital properties such as principles (physical method), specificity and sensitivity to the various agents. Moreover, it plays a major role in tissue penetration and tissue contrast. Strangely, AFP (Alpha fetal protein) acts as a biomarker in the diagnosis purpose of liver cancers. ELISA techniques involve the diagnosis in the presence of serum by colorimetric detection process (Baetke et al. 2015, Xuan et al. 2016).

2.3 NMs-Based Nanomedicine in the Future

A huge range of future approaches to metal-based NP/NM therapeutics is probable. However, there are still numerous challenges to be faced and, eventually, repetitive clinical practice. Moreover, the functionalization-based NMs can easily reach a particular site for drug delivery without side effects.

3. Mechanism of Stability and Toxicity

The green-synthesized NMs have huge capabilities for drug carriers to target particular sites due to their stability such as diffusion, penetration, toxicological, biodegradability and clearance.

3.1 Diffusion and Penetration of NMs

Generally, the diffusive resistance of different tissues such as the intestine, nose, mucosal membranes, lungs and vagina make it capable of drug delivery, and also researchers proved it scientifically by the diffusion and sedimentation method. Specifically, it is also shown that the cellular uptake and penetration are independent of size, density, surface coating, initial dose concentration and surface morphology of NMs (Cho et al. 2011).

3.2 Toxicological Aspects of NMs

Mukherjee and co-authors proved the nontoxic nature of these biogenic NPs by using female mice C57BL6/J (*In vivo* method) for the evaluation of various parameters. The results show no toxic effect, and the life span is increased in the cancer-bearing mice; moreover, a huge efficiency level is shown against cancer activity. At present, the researchers are focused on the therapeutic and pharmacokinetic responses (Mukherjee et al. 2013). Furthermore, biodegradability and clearance are determined using green-synthesized NMs for drug delivery systems due to polymeric complexation/coating along with the pharmaceutical ingredients. The lysosome is the most frequent in the intracellular site of NP sequestration and degradation due to the majority of endocytic routes of NM cell uptake converting upon it. Recent research suggests that some NMs can also trigger autophagy in addition to the endo-lysosomal pathway. The lysosome performs a variety of physiological tasks, including the degradation of damaged organelles, proteins, and intracellular pathogens via the autophagy (macro-autophagy) pathway. Hence, the induction of autophagy by nanoparticles may be an effort to destroy everything the cell identifies as alien or abnormal. While the autophagy and endo-lysosomal pathways may affect how NMs are disposed of, a growing body of research indicates that bio-persistent

NPs may also adversely affect these pathways. There is plenty of proof of the bio-persistence of the toxicological base system.

4. Analysis of NM Toxicity

At present, green-synthesized NMs act as both diagnosis and drug delivery, but at the same time, nano-toxicology or toxicity studies are required for clinical studies. So, before the formulation, we have to check the toxicity level by using *in silico* method. After that, the results correlated with the *in vitro* and *in vivo* methods.

4.1 **In Silico** *Method*

Hence, this account presents a coherent process for the amalgamation of in silico NM toxicity and destiny and analyses of transport for the impacts of environmental. These approaches required knowledge of NM toxicity and various exposures to environmental-based concentrations. Since the huge number of various types of NMs, and those numbers are likely to increase, there is a crucial essential to the accelerated determination of their toxicity studies and the assessment of their probable delivery in the atmosphere. The progress in HTS (high throughput screening) is now enabling the quick generation of large data sets for NM toxicity valuation. However, this type of analysis needs to create reliable toxicity impacts, specifically when HTS comprises data from a greater number of assays and organisms/cell lines. Beginning toxicity of metrics with HTS data required the latest data processing methods in order to clearly recognize the significant level of biological effects related to acquaintance with NMs. HTS data can make validated studies by using *in silico* toxicity models, e.g., "quantitative structure activity" (Cohen et al. 2013).

4.2 **In Vitro** *Method*

As a result of various factors that influence NP and NM toxicity, as formerly revealed, there is no particular method for identifying or determining toxicity levels. First, to evaluate *in vitro* toxicity studies, one ordinary characterization method to determine the NMs and NPs toxicity *in vitro* studies is MTT (3-[4,5-dimethylthiazole-2-yl]-2,5-diphenyltetrazolium bromide) assays, which calculate the cells' mitochondrial function by detecting mitochondrial dehydrogenase through an enzymatic reduction process. Alternative standard characterization methods for toxicity determination are examining ROS formation within the cells, which shows oxidative stress and interferes with cell function. In order to portion intracellular ROS, which acts as an indicator for the fluorescent signal process. This signal can be determined by spectroscopy or confocal microscopy (Kuppusamy et al. 2016).

4.3 **In Vivo** *Method*

Additionally, there are apprehensions about unintentional particle accretion in organs and induced toxicity for *in vivo* studies. However, the NMs/NPs easily make for the risk of environmental factors directly or indirectly in the presence of living organisms along with the various capacities of toxicological parameters. Numerous NPs, such as gold (AU) NPs and other metal-based NPs, have shown the toxic effects and accumulation of organs. Nevertheless, the toxicity pathways such as the kidney, spleen, lungs, intestine and heart are huge accumulation sites.

In this regard, the accumulation mainly depends on their size and surface, e.g., carbon-based NMs mostly accumulate in the liver and QD (quantum dots), which can easily cross the BBB (Blood-brain barrier), so it produces few problems during the *in vivo* experimental (Philip et al. 2010, Baghizadeh et al. 2015). Moreover, at present, the latest technology is involved in the assessment of toxicity and determination of the efficient and homogeneous distribution of aerosols into the human cells cultured on the membrane surface such as perfusion chamber, e.g., MINUCELL and Artificial intelligence (AI); also the Karlsruhe exposure system and CULTEX was implicated for analyzing of toxicity level in the NMs.

5. Pharmacokinetic (Pk) Parameters of NMs

Pk studies mainly involve the determination of drug concentration and elimination rate via ADME parameters.

5.1 Absorption

5.2 Distribution

5.3 Metabolism

5.4 Excretion

Almost all the Pk studies are shown from the perception of NMs structure in the systemic circulation evaluated by analysis of blood and plasma concentrations. It is commonly presumed that a compound must influence the systemic circulation in directive toward the tissues by distribution process and excretion by the kidney.

After this method is applied to drugs, a deterioration in blood concentrations over time generally interprets into excretion by the urine and feces. In these conditions, T1/2 relates to the time a drug is present in the body. The dependence of T1/2 on volume of distribution (Vd) indicates that degradation from blood may be very slow if broad tissue dispersion occurs, as is the case for many lipophilic pollutants like chlorinated hydrocarbons. T1/2 may be further increased by ineffective elimination from the body (low Cl). The only way to eliminate a molecule is through metabolism into a more hydrophilic and excretable chemical state. However, the transfer of the substance into tissues where further excretion does not take place (e.g., stuck in the reticuloendothelial (RE) system), bound to tissue proteins, post-distributional aggregation) maybe the cause of the decline in blood concentrations for nanomaterials. Blood half-life in these circumstances may paradoxically be rather brief despite the lengthy body persistence. In addition to blood samples, nanomaterials may also be carried into the lymphatic system, which complicates pharmacokinetic analyses of them and exposes lymphoid tissue to larger quantities than would otherwise be the case. Novel approaches mainly depend on the optimization of biodistribution to target organs and controlling the elimination/ excretion process involved in the release kinetics for the pharmacokinetic system based on bio-availability and biodistribution. The use of traditional interpretations of the significance of fundamental ADME measures must be done with caution due to the peculiarities of nanomaterials compared to most medications or small-molecule xenobiotics (Shyh-Dar et al. 2012).

6. Effect of Physicochemical Properties of NMs on Toxicity

6.1 Size and Shape

Physicochemical factors and biological barriers affect the NPs' capacity to penetrate the cell. They can penetrate through the cell membrane to pass biological barriers due to their small size and high surface area to volume ratio, which allows them to deliver the medication inside the cell. It has been discovered that an optimal size range with high tissue permeability and minimum accumulation is between 10 nm and 12 nm. In making an NP, picking the proper size is crucial since it determines which excretion pathway the drug will be inside. For instance, particles with a smaller HD of 5.5 nm go through the kidney to be excreted, but larger NPs go through the liver. Additionally, choose a reliable carrier of small-molecule theragnostic agents. In addition to protecting the medication from inactivation and/or degradation, the drug-loaded carrier can regulate the effectiveness of drug delivery, which can lessen any negative consequences. The design of the NPs has an impact on how well they are absorbed by cells; for instance, elongated NPs absorb better than spherical ones (Wang and Malik 2013, Wei et al. 2012, Moghimi et al. 2012, Salatin et al. 2015).

6.2 Surface Morphology

The physical, chemical, and biological character of all the molecules in biological systems is greatly impacted by changes on NPs' surfaces. NPs can produce positive or negative charges on their surface, which alters how the cell membrane interacts with them and influences how well they are absorbed and distributed. Electrostatic interactions cause positively charged NPs to be more absorbent than negatively charged NPs. As demonstrated by NPs coated with polyethylene glycol (PEG) on their surface, altering the NP surface with a neutral non-ionic polymer confers stability to the NP by reducing opsonization and lengthening blood circulation time. When the NPs are dissolved in biological fluids (such as blood), surface property also plays a significant effect. When in contact with the biological fluid, NPs develop a protein layer (protein corona) on their surface. The attraction of the NP to the cell membrane is greatly influenced by this layer. Each type of NP has a varied affinity for a certain protein in a biological fluid and alters the physicochemical properties, which would then affect the rate and extent of biodistribution. Diverse NPs generate different protein corona as a result (Salatin et al. 2015).

6.3 The Effect of Coating Materials and Compositions of Materials

Numerous therapeutic NPs are made up of a variety of parts with distinct geometries or conformations, including core-shell, core-satellite, linear, and hyper-branched structures. Their *in vivo* performance, such as absorption, biodistribution, elimination, and targeting ability, can be affected by minute variations in geometry or conformation. Additionally, toxicity can be considerably impacted by the geometry/conformation changes of NPs and their disintegration in the *in vivo* environments.

The chemical makeup of theragnostic NPs determines how easily they degrade. Biodegradable polymeric NPs have hydrolyzable linkages like ester, ortho-ester, and anhydride in their backbones. Biodegradable polymers can help the body eliminate NPs more quickly and lessen their long-term toxicity. The majority of inorganic NPs, on the other hand, are not biodegradable. Given that these inorganic NPs are larger and more hydrophobic than tiny molecules and stay in the body for a longer period of time, questions have been raised about their potential long-term toxicity (Li and Huang et al. 2008, Choi and Frangioni et al. 2010, Wang et al. 2012).

7. Future Perspective

The targeting and treatment of numerous diseases have been successfully associated with a wide variety of theranostic NPs. Anticancer drugs can be implanted on or encapsulated within various NP structures for specialized imaging and treatment. In reality, the majority of therapeutic NPs/NMs feature multifunctionality, a potent EPR effect, long-term blood circulation, and special size-dependent optical or magnetic properties. However, NPs/NMs fall short of meeting key crucial requirements for clinical translation, including high physiological stability, effective clearance, little accumulation in unintended tissues and organs, and quick transport to diverse organs and tissues. The majority of NPs/NMs' clinical uses have also been hindered by an incomplete understanding of their unique toxicities. To answer this topic, more research in the area of preclinical and predictive toxicity is necessary.

The bulk of therapeutic methods have been used to treat cancer by killing malignant cells with chemicals, heat, and gene transfections. A promising theranostic use of multifunctional NPs has just recently been suggested for the diagnosis and treatment of long-term neurodegenerative diseases like Alzheimer's, Parkinson's, and strokes. Multifunctional NPs, for instance, can be used to treat brain diseases that are brought on by iron overload. Numerous neurodegenerative illnesses, including Alzheimer's disease, are thought to have oxidative stress as one of its primary causes, which is brought on by metals like iron that build up excessively in the brain. Iron chelators have been widely utilized to treat certain illness states by removing the extra iron from the brain.

8. Conclusion

One common kind of nanomaterials that has contributed to the development of nanotechnology is nanoparticles with various properties. Due to recent advancements in technology, scientists interested in such methods have recently built their own nanocomposites and researched the applications and properties of new nanomaterials. This article defined nanotechnology and detailed how metals, metal oxides, graphene oxides, and polymers can be used to create nanomaterials. Intriguing greenways that produce nanoparticles with less or no harm than conventional techniques include plant extracts and microorganism biomolecules. The production and use of diverse nanomaterials are now open to new possibilities thanks to this book chapter.

References

Abd-Elnaby, H. M., Abo-Elala, G. M. and Abdel-Raouf Hamed, U. M. (2016). Antibacterial and anticancer activity of extracellular synthesized silver nanoparticles from marine Streptomyces Rochei MHM13. *Egypt. J. Aquatic Res.*, 42(3): 301–312.

Arunkumar, P., Vedagiri, H. and Premkumar, K. (2013). Rapid bioreduction of trivalent aurum using banana stem powder and its cytotoxicity against MCF-7 and HEK293 cell lines. *J. Nanopart. Res.*, 15(3): 1–8.

Baetke, S. C., Lammers, T. and Kiessling, F. (2015). Applications of nanoparticles for diagnosis and therapy of cancer. *Br. J. Radiol.*, 88(1054): 2015–2023.

Baghizadeh, A., Ranjbar, S., Gupta, V. K., Asif, M., Pourseyedi, S., Karimi, M. J. and Mohammadinejad, R. (2015). Green synthesis of silver nanoparticles using seed extract of Calendula officinalis in liquid phase. *J. Mol. Liq.*, 207: 159–163.

Cao, G. (2004). Nanostructures and Nanomaterials—Synthesis, Properties and Applications. Singapore: World Scientific.

Cho, E. C., Zhang, Q. and Xia, Y. (2011). The effect of sedimentation and diffusion on cellular uptake of gold nanoparticles. *Nat. Nanotechnol.*, 6(6): 385–391.

Choi, H. S. and Frangioni, J. V. (2010). Nanoparticles for biomedical imaging: Fundamentals of clinical translation. *Molecular Imaging*, 9: 291–310.

Cohen, Y., Rallo, R., Liu, R. and Liu, H. H. (2013). *In silico* analysis of nanomaterials hazard and risk. *ACC Chem. Res.*, 19; 46(3): 802–12.

Devi, J. S., Bhimba, B. V. and Peter, D. M. (2013). Production of biogenic silver nanoparticles using Sargassum Longifolium and its applications. *Indian J. Geo-Mar. Sci.*, 42(1): 125–130.

Doble, M. and Kruthiventi, A. K. (2007). Green Chemistry and Engineering. Cambridge: Academic Press.

Heydari, R. and Rashidipour, M. (2015). Green synthesis of silver nanoparticles using extract of oak fruit hull (Jaft): Synthesis and *in vitro* cytotoxic effect on MCF-7 cells. *Int. J. Breast Cancer*, 33(16): 45–54.

Huang, X., El-Sayed, I. H., Qian, W. and El-Sayed, M. A. (2006). Cancer cell imaging and photothermal therapy in the near-infrared region by using gold nanorods. *J. Am. Chem. Soc.*, 128: 2115–20.

Iram, S., Khan, S., Ansary, A. A., Arshad, M., Siddiqui, S., Ahmad, E., Khan, R. H. and Khan, M. S. (2016). Biogenic terbium oxide nanoparticles as the vanguard against Osteosarcoma. *Spectrochim. Acta*, Part A 168(6): 123–131.

Kim, J. S., Kuk, E. and Yu, K. N. (2007). Antimicrobial effects of silver nanoparticles. *Nanomed. Nanotechnol. Biol. Med.*, 3: 95–101.

Kuppusamy, P., Yusoff, M. M., Maniam, G. P. and Govindan, N. (2016). Biosynthesis of metallic nanoparticles using plant derivatives and their new avenues in pharmacological applications—An updated report. *Saudi Pharm. J.*, 24: 473–484.

Laurent, S., Forge, D. and Port, M. (2008). Magnetic iron oxide nanoparticles: Synthesis, stabilization, vectorization, physicochemical characterizations, and biological applications. *Chem. Rev.*, 108: 2064–110.

Li, S. D. and Huang, L. (2008). Pharmacokinetics and biodistribution of nanoparticles. *Mol. Pharm.*, 5: 496–504.

Manikandan, R., Manikandan, B., Raman, T., Arunagirinathan, K., Prabhu, N. M., Basu, M. J., Perumal, M., Palanisamy, S. and Munusamy, A. (2015). Biosynthesis of silver nanoparticles using ethanolic petals extract of rosa indica and characterization of its antibacterial, anticancer and anti-inflammatory activities. *Spectrochim. Acta*, Part A, 138(7): 120–129.

Manivasagan, P. and Oh, J. (2015). Production of a novel fucoidanase for the green synthesis of gold nanoparticles by Streptomyces sp. and its cytotoxic effect on HeLa cells. *Mar. Drugs*, 13(11): 6818–6837.

Manivasagan, P., Alam, M. S., Kang, K. H., Kwak, M. and Kim, S. K. (2015). Extracellular synthesis of gold bionanoparticles by Nocardiopsis sp. and evaluation of its antimicrobial, antioxidant and cytotoxic activities. *Bioprocess Biosyst. Eng.*, 38(6): 1167–1177.

Mirzaei, H. and Darroudi, M. (2017). Zinc oxide nanoparticles: Biological synthesis and biomedical applications. *Ceramics Int.*, 43(1): 907–914.

Mittal, A. K., Tripathy, D., Choudhary, A., Aili, P. K., Chatterjee, A., Singh, J. P. and Banerjee, U. C. (2015). Bio-synthesis of silver nanoparticles using potentilla fulgens wall. ex hook and its therapeutic evaluation as anticancer and antimicrobial agent. *Mater. Sci. Eng: C*, 53(15): 120–127.

Moghimi, S. M., Hunter, A. C. and Andresen, T. L. (2012). Factors controlling nanoparticle pharmacokinetics. *Annu. Rev. Pharmacol. Toxicol.*, 52: 481–503.

Mukherjee, S., Vinothkumar, B., Prashanthi, S., Bangal, P. R., Sreedhar, B. and Patra, C. R. (2013). Potential therapeutic and diagnostic applications of one-step *in situ* biosynthesized gold nanoconjugates (2-in-1 System) in cancer treatment. *RSC Adv.*, 3(7): 2318–2329.

Nachiyar, V., Sunkar, S. and Prakash, P. (2015). Biological synthesis of gold nanoparticles using endophytic fungi. *Der. Pharma Chem.*, 7(55): 31–38.

Patra, S., Mukherjee, S., Barui, A. K., Ganguly, A., Sreedhar, B. and Patra, C. R. (2015). Green synthesis, characterization of gold and silver nanoparticles and their potential application for cancer therapeutics. *Mater. Sci. Eng: C*, 53: 298–309.

Philip, D. (2010). Green synthesis of gold and silver nanoparticles using Hibiscus Rosa Sinensis. *Phys. E*, 42: 1417–1424.

Prabakaran, K., Ragavendran, C. and Natarajan, D. (2016). Mycosynthesis of silver nanoparticles from beauveria bassiana and its larvicidal, antibacterial, and cytotoxic effect on human cervical cancer (HeLa) cells. *RSC Adv.*, 6(51): 44972–44986.

Rajeshkumar, S. (2016). Anticancer activity of eco-friendly gold nanoparticles against lung and liver cancer cells. *J. Genet. Eng. Biotechnol.*, 14(1): 195–202.

Rathod, D., Golinska, P., Wypij, M., Dahm, H. and Rai, M. (2016). A new report of nocardiopsis valliformis strain OT1 from alkaline lonar crater of india and its use in synthesis of silver nanoparticles with special reference to evaluation of antibacterial activity and cytotoxicity. *Med. Microbiol. Immunol.*, 205(5): 435–447.

Saikia, I., Sonowal, S., Pal, M., Boruah, P. K., Das, M. R. and Tamuly, C. (2016). Biosynthesis of gold decorated reduced graphene oxide and its biological activities. *Mater. Lett.*, 178(14): 239–242.

Salatin, S., Dizaj, S. M. and Khosroushahi, A. Y. (2015). *Cell Biol. Int.* 2015.10.1002/cbin.10459.

Sankar, R., Karthik, A., Prabu, A., Karthik, S., Shivashangari, K. S. and Ravikumar, V. (2013). Origanum vulgare mediated biosynthesis of silver nanoparticles for its antibacterial and anticancer activity. *Colloids Surf. B*, 108(11): 80–84.

Shanmugasundaram, T., Radhakrishnan, M., Gopikrishnan, V., Pazhanimurugan, R. and Balagurunathan, R. (2013). A study of the bactericidal, anti-biofouling, cytotoxic and antioxidant properties of actinobacterially synthesised silver nanoparticles. *Colloids Surf. B*, 111(21): 680–687.

Shyh-Dar, L. and Leaf, H. (2008). Pharmacokinetics and biodistribution of nanoparticles. *Mol. Pharmaceutics*, 5: 496–504.

Singh, M., Saurav, K., Majouga, A., Kumari, M., Kumar, M., Manikandan, S. and Kumaraguru, A. (2015). The cytotoxicity and cellular stress by temperature-fabricated polyshaped gold nanoparticles using marine macroalgae, padina gymnospora. *Biotechnol. Appl. Biochem.*, 62(3): 424–432.

Srivastava, P., Braganca, J. M. and Kowshik, M. (2014). *In vivo* synthesis of selenium nanoparticles by halococcus salifodinae BK18 and their anti-proliferative properties against HeLa cell line. *Biotechnol. Prog.*, 30(6): 1480–1487.

Valli Nachiyar, C., Vijayalakshmi, R., Nivedha, K., Bavanilatha, M. and Swetha, S. (2016). Moraxella osloensis mediated synthesis of TiO$_2$ nanoparticles. *Int. J. Pharmacy Pharmac. Sci.*, 8(5): 397–400.

Wang, Z. and Malik, A. B. (2013). Mechanisms regulating endothelial permeability. *Ther. Deliv.*, 4: 131–133.

Wang, S., Kim, G., Lee, Y. E., Hah, H. J., Ethirajan, M., Pandey, R. K. and Kopelman, R. (2012). Supramolecular systems in biomedical field. *ACS Nano.*, 6: 6843–6851

Wei, A., Mehtala, J. G. and Patri, A. K. (2012). *J. Controlled Release*, 164: 236–246.

Xuan, Z. H., Li, M. M., Rong, P. F., Wang, W., Li, Y. J. and Liu, D. B. (2016). Plasmonic ELISA based on the controlled growth of silver nanoparticles. *Nanoscale*, 8(39): 17271–17277.

CHAPTER 5

Biodegradable Waste Products as Reducing Agents for Green Nanomaterial Synthesis and their Applications

Niloy Chatterjee[1,2] *and Pubali Dhar*[1,2,*]

1. Introduction

The increasing demand for nanoparticles in various applications has led to the development of several methods for their synthesis. However, conventional methods have several drawbacks, such as high cost, toxicity, and environmental impacts. In recent years, the use of waste materials as precursors or templates for nanoparticle synthesis has emerged as a promising approach. Waste materials, such as agricultural, industrial, and municipal waste, are abundant, cost-effective, and readily available sources for nanoparticle synthesis (Jamkhande et al. 2019, Khan et al. 2019).

The commercial viability of recycling and reusing different kinds of waste items is closely related to the economic value of the final goods (such as nanoparticles). Therefore, creating "value-added" nanomaterials from waste materials using such "clean manufacturing" technologies appears as a desirable combination for accomplishing both circularity and sustainability goals. Such cutting-edge technologies might encourage the adoption of circular economy models, which are

[1] Laboratory of Food Science and Technology, Food and Nutrition division, University of Calcutta, 20B Judges Court Road. Kolkata 700027, West Bengal, India.
Email: chatterjeenil99@gmail.com
[2] Centre for Research in Nanoscience and Nanotechnology, University of Calcutta, JD 2, Sector III, Salt Lake City, Kolkata 700 098, West Bengal, India.
* Corresponding author: pubalighoshdhar23@gmail.com, pubalighoshdhar@yahoo.co.in

urgently needed to satisfy the world's environmental and resource-oriented goals. With this "waste-to-wealth" strategy, less material is used, materials, goods, and services are redesigned to utilize fewer resources, and "waste" is recovered as a valuable resource to create new goods and materials. Industrially and economically, these processes and activities are restorative or regenerative by design, allowing resources used in them to retain their highest value for as long as possible, and they aim to eliminate waste through the superior design of materials, products, and systems. In order to handle the climate catastrophe, rebuild the economy, safeguard the environment, enhance social justice, and improve economics and sustainability, such contemporary techniques must be adopted more and more (Puntillo et al. 2021, Kaur et al. 2018).

The development of nanoparticles has led to significant advancements in various fields, such as electronics, energy, biomedical, and catalysis. Conventionally, nanoparticles are synthesized using costly and toxic precursors, which result in negative environmental impacts (Ealia and Saravanakumar 2017). Therefore, the development of sustainable and eco-friendly methods for nanoparticle synthesis is highly desirable. The use of waste materials as precursors or templates for nanoparticle synthesis has emerged as a promising approach for sustainable nanoparticle synthesis. This approach not only reduces the cost of nanoparticle synthesis but also provides a sustainable solution for waste management (Ali et al. 2021, Nabi et al. 2018). Despite the potential benefits of waste-derived nanoparticles, several challenges associated with their synthesis need to be addressed, such as the presence of impurities, reproducibility, and scalability (Lizundia et al. 2022). The use of advanced techniques, such as green chemistry principles and process intensification, can overcome these challenges and make the synthesis of waste-derived nanoparticles a sustainable and efficient process. Therefore, developing reliable and reproducible methods for synthesizing nanoparticles from waste materials is crucial for sustainable utilization (Arif et al. 2023, Dhanya et al. 2023). This review article provides a comprehensive overview of recent advancements in the synthesis of nanoparticles from waste materials. The article discusses the various techniques used for nanoparticle synthesis from waste materials, their potential applications, and the challenges associated with their synthesis. The article aims to provide insights into the potential of waste materials as a sustainable source for nanoparticle synthesis and to highlight the importance of sustainable nanoparticle synthesis for environmental and economic benefits.

2. What is Nanotechnology?

Nanotechnology is a branch of science and technology that deals with the study and manipulation of materials on an atomic and molecular scale. It involves the design, production, and application of structures, devices, and systems with dimensions ranging from 1 to 100 nanometers (nm). At this scale, the properties of materials differ from those of the bulk material due to quantum mechanical effects, making them unique and highly valuable for various applications (Nasrollahzadeh et al. 2019, Madkour and Madkour 2019).

Nanotechnology finds applications in a wide range of fields, including medicine, electronics, energy, materials science, and environmental science. For example, in medicine, nanoparticles can be designed to target specific cells or tissues, enabling more efficient drug delivery and imaging. In electronics, nanotechnology enables the production of smaller and more efficient electronic devices, such as transistors and memory chips. In materials science, nanotechnology enables the development of more robust and more durable materials, such as composites and coatings (Oluwasanu et al. 2019, Park et al. 2020).

The development of nanotechnology has led to significant advancements in various fields and has the potential to revolutionize how we live and work. However, it also raises concerns regarding the potential health and environmental risks associated with the production and use of nanomaterials. Therefore, careful consideration and regulation of the development and application of nanotechnology are necessary to ensure its safe and sustainable use (Pokrajac et al. 2021, Pandey and Jain 2020).

3. What are Nanoparticles?

Nanoparticles are particles with dimensions ranging from 1 to 100 nanometers (nm), which is approximately 1/1,000th of the diameter of a human hair. Due to their small size, nanoparticles have unique physicochemical properties that differ from those of bulk materials. These properties include a high surface area to volume ratio, enhanced reactivity, and quantum size effects, which make them highly valuable for various applications in fields such as biomedicine, electronics, energy, and materials science (Modena et al. 2019, Mansoori 2017).

Nanoparticles can be synthesized using various methods, including physical, chemical, and biological methods. Physical methods involve the formation of nanoparticles through mechanical, thermal, or electrical means, such as milling, laser ablation, and sputtering. Chemical methods involve the reduction of metal salts or the oxidation of metal precursors in the presence of reducing agents, stabilizers, or surfactants, such as sol-gel, precipitation, and hydrothermal methods. Biological methods involve the use of microorganisms or biological molecules, such as proteins, enzymes, and DNA, to synthesize nanoparticles, such as bacterial synthesis, green synthesis, and biomineralization (Jamkhande et al. 2019, Chakraborty and Pradeep 2017).

The properties of nanoparticles can be tailored by controlling their size, shape, composition, and surface properties. For example, changing the size and shape of nanoparticles can affect their optical and magnetic properties, while changing their composition can affect their chemical reactivity and stability. Moreover, modifying the surface of nanoparticles can improve their biocompatibility, stability, and targeting efficiency (Shnoudeh et al. 2019, Saleh 2020).

Nanoparticles have various applications in biomedicine, such as drug delivery, imaging, and biosensing. In electronics, nanoparticles are used to produce of smaller and more efficient electronic devices, such as transistors and memory chips. In energy, nanoparticles are used in the development of more efficient solar cells and catalysts for fuel cells. In materials science, nanoparticles are used to enhance the properties of various materials, such as composites, coatings, and sensors (Sriram and Suttee 2020, Pandey 2022).

Despite their potential benefits, the production and use of nanoparticles raise concerns regarding their potential health and environmental impacts. Therefore, careful consideration and regulation of the development and application of nanoparticles are necessary to ensure their safe and sustainable use (Patel et al. 2021).

4. What are Waste Materials?

Waste materials refer to any unwanted, discarded, or unused materials generated from human activities that are no longer considered useful or valuable. Waste materials can be classified into various categories, including municipal solid waste, industrial waste, hazardous waste, electronic waste, and agricultural waste (Samant et al. 2018).

Municipal solid waste refers to household waste, such as food waste, paper, plastics, and metals, generated from households, commercial establishments, and institutions. Industrial waste refers to waste generated from industrial processes, such as manufacturing, mining, and construction, which includes various types of solid, liquid, and gaseous waste. Hazardous waste refers to waste that poses a significant risk to human health or the environment due to its chemical, physical, or biological properties. Electronic waste refers to discarded electronic devices, such as computers, smartphones, and televisions, which contain hazardous materials like lead, cadmium, and mercury. Agricultural waste refers to waste generated from agricultural activities, such as crop residues, animal manure, and agricultural chemicals (Tulebayeva et al. 2020).

The management of waste materials is crucial to protect human health and the environment. Improper waste management can lead to pollution, the spread of diseases, and environmental degradation. Therefore, waste materials must be properly collected, transported, treated, and disposed of to minimize their negative impacts (Ferronato and Toretta 2019).

Moreover, waste materials can also be a potential source of valuable resources, such as energy, materials, and nutrients. Various technologies and processes, such as waste-to-energy, recycling, and composting, can be used to recover valuable resources from waste materials. The utilization of waste materials as a resource can reduce the reliance on virgin materials, conserve natural resources, and reduce the environmental impacts of waste disposal (Ferronato and Toretta 2019, Zeller et al. 2019).

In conclusion, waste materials refer to any unwanted or unused materials generated from human activities, which can pose significant risks to human health and the environment. Proper waste management is crucial to minimize the negative impacts of waste, and the utilization of waste materials as a resource can provide significant environmental and economic benefits.

5. Waste Materials as Resource for Synthesis of Nanoparticles

Nanoparticles can be synthesized from a variety of materials, including both natural and synthetic sources. One reason why waste materials are increasingly being explored as a source for nanoparticle synthesis is that it presents an opportunity to repurpose and recycle materials that might otherwise go to waste (Figure 1) (Abd Elkodous et al. 2019, Ortega et al. 2022).

Waste materials can be rich in metals and other elements that are useful for nanoparticle synthesis. For example, agricultural waste such as rice husks, sugarcane bagasse, and fruit peels contain high levels of silica, which can be used to produce silica nanoparticles. Other types of waste, such as industrial byproducts or electronic waste, can contain metals like copper, silver, and gold that are commonly used in nanoparticle synthesis (Ahmed et al. 2022).

By using waste materials for nanoparticle synthesis, researchers can reduce the environmental impact of waste disposal while also producing valuable materials for various applications. Additionally, waste-derived nanoparticles may have unique properties that differ from those synthesized using traditional methods, which can open up new opportunities for research and development (Abdelbasir et al. 2020).

From a scientific point of view, the use of waste materials for nanoparticle synthesis offers several advantages. Firstly, waste materials are often readily available and can be obtained at a low cost, which makes them an attractive alternative to expensive raw materials. Additionally, the use of waste materials for nanoparticle synthesis can help reduce the environmental impact of waste disposal by repurposing materials that might otherwise be discarded (Samaddar et al. 2018).

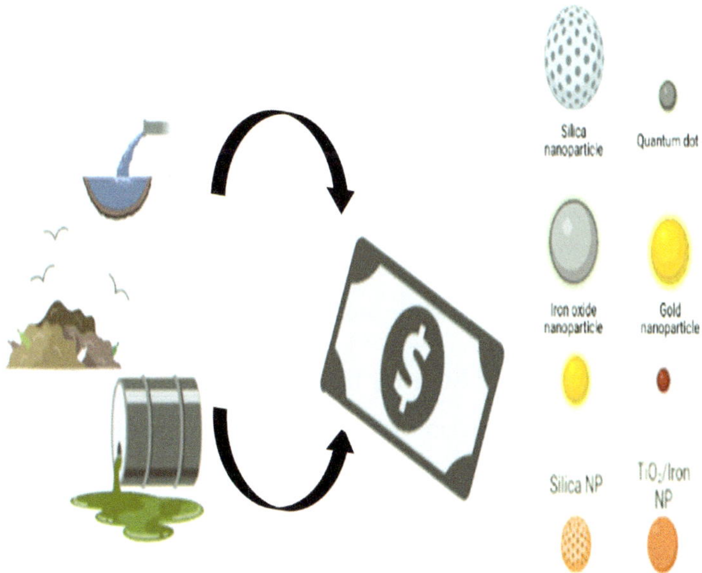

Figure 1. Waste to wealth-synthesis of nanoparticles from waste materials.

Moreover, some waste materials contain specific elements or compounds that are well-suited for nanoparticle synthesis. For instance, agricultural waste such as plant matter, sawdust, or wood chips is rich in cellulose, a natural polymer that can be used to produce cellulose-based nanoparticles. Similarly, industrial waste such as fly ash or slag contains significant amounts of metal ions that can be used to synthesize metal nanoparticles (Liu et al. 2021, Lochab et al. 2014).

Using waste materials for nanoparticle synthesis can also provide a way to tailor the properties of the nanoparticles. For example, waste-derived nanoparticles may have different sizes, shapes, and surface chemistries than those synthesized using traditional methods. This can be due to the presence of impurities or other chemical components in the waste material, which can influence the growth and formation of nanoparticles during the synthesis process (Omran and Baek 2022). Overall, the use of waste materials for nanoparticle synthesis is an innovative and sustainable approach that has the potential to yield valuable materials with unique properties.

From an environmental perspective, synthesizing nanoparticles from waste materials offers several advantages. Firstly, it helps to reduce the amount of waste that goes to landfills or is incinerated, which can have negative environmental impacts. By repurposing waste materials for nanoparticle synthesis, we can reduce the demand for new raw materials and conserve natural resources (Lange 2021).

Additionally, nanoparticle synthesis using waste materials can help to reduce the carbon footprint associated with the production of nanoparticles. This is because waste-derived nanoparticles can be synthesized using less energy compared to traditional synthesis methods, which typically require high temperatures, pressure, and toxic chemicals. By reducing the amount of energy and resources required for nanoparticle synthesis, we can minimize greenhouse gas emissions and other environmental impacts associated with nanoparticle production (Samaddar et al. 2018, Panes et al. 2022).

Moreover, the use of waste-derived nanoparticles can also lead to the development of more sustainable and eco-friendly products. For example, waste-derived nanoparticles can be used to produce biodegradable materials that break down more easily in the environment compared to traditional plastics. This can help to reduce plastic pollution and other environmental problems associated with non-biodegradable materials (Abdur Rahman et al. 2023).

Overall, synthesizing nanoparticles from waste materials offers a promising approach to reducing waste, conserving natural resources, and minimizing the environmental impacts of nanoparticle production (Singh et al. 2018).

6. Waste Separation or Sorting for Nanoparticle Synthesis

Waste material or garbage includes a variety of heterogeneous products, and all of them are generally unsuitable for nanoparticle synthesis. Therefore, they are to be very carefully sorted or separated into different elements as per requirement. Separating waste is the innovation of traditional waste collection and disposal, but it is essential for nanomaterials synthesis (Lubongo and Alexandridis 2022, Neo et al. 2022, Ji et al. 2022). The common methods of sorting and preparing waste are as follows:

a. Manual Separation: Before beginning mechanical processing, bulky materials are removed by hand or with a sorting belt. Although laborious and time-consuming, it does allow for the sorting of waste in a way that helps to improve waste separation and the recovery of recyclable elements.

b. Air Separation: Fans produce an upward-moving air column separating materials with various densities.

c. Size Reduction: Two different sorts of tools are employed to reduce or crush the trash. For materials that are challenging to shred, hammer mills (rotating sets of swinging steel hammers) and shear shredders are used. Both require a lot of energy and maintenance.

d. Drying: The drying procedure lowers the waste's moisture content and stops leachate from forming. Any partially decomposed trash should be dried, ideally using hot air, the sun, or a mix of the two.

e. Magnetic Separation: In this process, electro-magnets are utilized, which may be turned on or off to remove the metals that have accumulated.

f. Non-Magnetic Separation: Eddy current separators, also known as non-ferrous separators, separate non-magnetic metals using the current created by tiny swirls on a big conductor.

g. Modern Automated Techniques: In this approach, various automated machines controlled by computers, without human involvement, are employed. Nowadays, even the machine learning approach is being applied in this field. However, it requires highly sophisticated instrumentation and is yet to be established in this domain.

7. Advantages of Nanoparticle Synthesis From Waste Materials

There are several advantages of synthesizing nanoparticles from waste materials, including (Abdelbasir et al. 2020, Abid et al. 2021, Sasidharan et al. 2019):

a. Sustainable and Eco-Friendly: The use of waste materials for nanoparticle synthesis is an environmentally friendly approach as it repurposes materials that would otherwise go to waste.

b. Cost-Effective: Waste materials are often readily available and can be obtained at a low cost, making them a cost-effective alternative to expensive raw materials.

c. Energy Efficient: Synthesizing nanoparticles from waste materials can be less energy-intensive compared to traditional methods, reducing the carbon footprint associated with nanoparticle production.

d. Unique Properties: Waste-derived nanoparticles may have unique properties compared to those synthesized using traditional methods, providing opportunities for novel applications.

e. Tailored Properties: The use of waste materials for nanoparticle synthesis can lead to the formation of nanoparticles with tailored properties, such as different sizes, shapes, and surface chemistries.

f. Reduction of Waste: Repurposing waste materials for nanoparticle synthesis reduces the amount of waste that goes to landfills or is incinerated, helping to mitigate negative environmental impacts.

g. Conservation of Natural Resources: Using waste materials for nanoparticle synthesis, we can reduce the demand for new raw materials and conserve natural resources.

Overall, synthesizing nanoparticles from waste materials is a promising approach that offers numerous advantages in terms of sustainability, cost-effectiveness, and tailored properties.

8. Disadvantages of Nanoparticle Synthesis From Waste Materials

While there are several advantages to synthesizing nanoparticles from waste materials, there are also some potential disadvantages, including (Biswas et al. 2022, Ojo et al. 2021, Kharissova et al. 2019):

a. Variable Quality: The quality of waste materials can be variable, which can affect the quality and consistency of the resulting nanoparticles.

b. Contaminants: Waste materials may contain contaminants that can affect the properties and safety of the resulting nanoparticles, which can require additional purification and testing.

c. Limited Availability: Depending on the type of waste material used, the availability may be limited, which could limit the scalability and commercial viability of the nanoparticle synthesis process.

d. Additional Processing Steps: Synthesizing nanoparticles from waste materials may require additional processing steps, such as extraction or purification, which can increase the cost and complexity of the process.

e. Environmental Impact of Waste Collection: Collecting waste materials for nanoparticle synthesis can have its own environmental impacts, such as increased transportation emissions or disturbance of natural habitats.

f. Health and Safety Risks: The use of waste materials for nanoparticle synthesis can pose health and safety risks to workers and researchers, particularly if the waste material contains hazardous materials or chemicals.

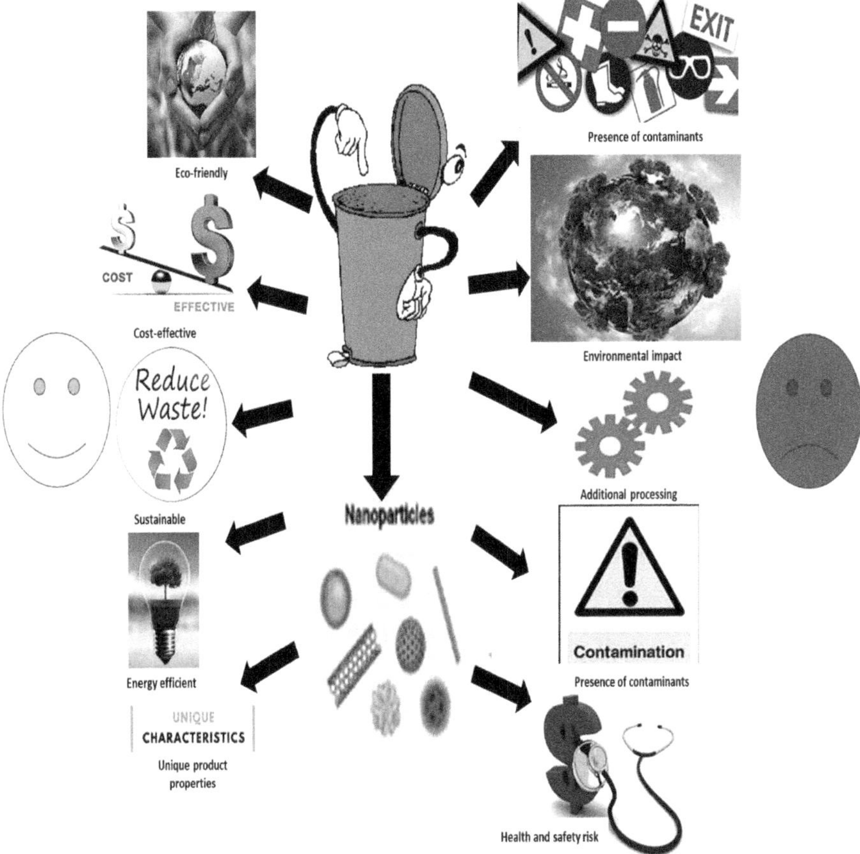

Figure 2. Advantage and disadvantage of synthesis of nanoparticles from waste materials.

While synthesizing nanoparticles from waste materials offers many advantages, there are also potential disadvantages that must be carefully considered and addressed to ensure the safety, quality, and scalability of the process.

9. Research Gaps

There are still several research gaps and challenges that need to be addressed in the field of nanoparticle synthesis from waste materials (Matussin et al. 2020, Gilbertson et al. 2015, Brar et al. 2022, Nkele and Fabian 2022):

a. Understanding the Mechanisms of Nanoparticle Formation: More research is needed to better understand the mechanisms of nanoparticle formation using waste materials, including the role of impurities and other chemical components in the waste material.

b. Optimization of Synthesis Conditions: More work is needed to optimize the synthesis conditions for different types of waste materials to maximize the yield, quality, and consistency of the resulting nanoparticles.

c. Characterization of Nanoparticles: Comprehensive characterization of the properties of waste-derived nanoparticles, including their size, shape, surface chemistry, and toxicity, is critical for ensuring their safety and efficacy.

d. Toxicity and Environmental Impact: Additional research is needed to fully understand the toxicity and environmental impact of waste-derived nanoparticles, particularly in terms of their potential accumulation in the environment and effects on ecosystems.

e. Scalability and Commercial Viability: The scalability and commercial viability of nanoparticle synthesis from waste materials must be further investigated to determine their economic feasibility and potential for large-scale production.

f. Comparative Studies: Comparative studies between waste-derived nanoparticles and those synthesized using traditional methods can help to determine the potential advantages and limitations of using waste materials for nanoparticle synthesis.

Overall, addressing these research gaps and challenges can help advance the field of nanoparticle synthesis from waste materials and unlock the full potential of this innovative and sustainable approach.

10. Challenges Associated With Nanoparticle Synthesis From Waste Materials

There are several challenges and potential loopholes associated with nanoparticle synthesis from waste materials, including (Sahajwalla and Hossain 2020, Ali et al. 2021, Rani et al. 2021):

a. Quality Control: Ensuring the consistent quality and purity of the nanoparticles synthesized from waste materials can be challenging due to the variability of waste materials.

b. Contaminants: Waste materials can contain contaminants, such as heavy metals or organic pollutants, that can affect the properties and safety of the resulting nanoparticles.

c. Scale Up: Scaling up nanoparticle synthesis from waste materials to industrial levels can be difficult due to the limited availability and variability of waste materials.

d. Energy Efficiency: Although synthesizing nanoparticles from waste materials can be energy efficient, it may not always be the most energy-efficient approach, particularly if waste materials need to be transported over long distances.

e. Toxicity: Waste-derived nanoparticles may still pose toxicity risks, particularly if the waste material is contaminated or if the synthesis process involves toxic chemicals.

f. Regulatory Issues: The use of waste materials for nanoparticle synthesis may face regulatory challenges, particularly if the waste material is classified as hazardous or if there are concerns about the potential environmental or health impacts of the nanoparticles.

g. Lack of Standardization: There is currently a lack of standardization in the synthesis and characterization of waste-derived nanoparticles, making it difficult to compare and evaluate different approaches.

Addressing these challenges and potential loopholes is critical for ensuring the safety, efficacy, and commercial viability of nanoparticle synthesis from waste materials.

11. Green Synthesis of Nanoparticles From Waste

Green synthesis of nanoparticles from garbage and waste materials involves using biodegradable waste materials as starting materials for synthesizing nanoparticles. The process involves converting waste materials that are benign and generally non-toxic into usable and valuable products, reducing the amount of waste generated and minimizing environmental impact. This approach provides a sustainable and cost-effective way to produce nanoparticles with several benefits (Mahantesh et al. 2022). The approach is known as green chemistry, and the biodegradable waste or garbage used as the precursor raw material for nanoparticle synthesis is considered green waste. This strategy is a philosophy of chemical research and engineering that promotes the creation of goods and procedures that reduce the use and production of dangerous compounds while simultaneously having a positive environmental impact. Waste materials are a highly heterogeneous renewable resource that can be chemically coupled with other non-hazardous technologies to produce desired products at the nanoscale. It is affordable, readily available, renewable, sustainable, reasonable, and possesses typical wealth. This method requires little instrumentation and is easy, dependable, cost-effective, and environmentally friendly. Here, the natural source's precursor can be decreased, repurposed, and reused. The green synthesis method reduces costs and streamlines the synthetic process by doing away with the requirement for additional capping or stabilizing agents. This green chemistry approach is one of the promising areas of research in contemporary times and the near future (Adelere and Lateef 2016, Soto et al. 2019).

The use of waste materials for nanoparticle synthesis offers several advantages over traditional synthesis methods. Waste materials are readily available and often cost-effective, reducing the need for expensive raw materials and minimizing the carbon footprint associated with their production. Additionally, the use of waste materials can reduce waste generation, promoting sustainable waste management and reducing environmental impact (Baroi et al. 2023).

Garbage and waste materials can be used as a source of raw materials for nanoparticle synthesis. For example, agricultural waste, such as rice husks or wheat straw, can be used as a silica source for synthesizing silica nanoparticles. Municipal waste, such as food waste or plastic waste, can also be used to produce nanoparticles. For instance, food waste can be used as a source of carbon for the synthesis of carbon-based nanoparticles, while plastic waste can be converted into carbon nanotubes (Devatha and Thalla 2018).

The creation of nanoparticles from trash is frequently accomplished by utilizing natural and eco-friendly processes, such as using microbes or plant extracts. Flavonoids and polyphenols, two biomolecules found in plant extracts, can function as reducing and stabilizing agents in the creation of nanoparticles. Microbial synthesis is a technique that uses microorganisms like bacteria and fungi to create nanoparticles (Bahrulolum et al. 2021).

Green synthesis of nanoparticles from waste materials has several potential applications. For instance, these nanoparticles can be used in water treatment and purification to remove contaminants from wastewater, such as heavy metals and organic pollutants. They can also be used in agriculture to enhance crop growth and

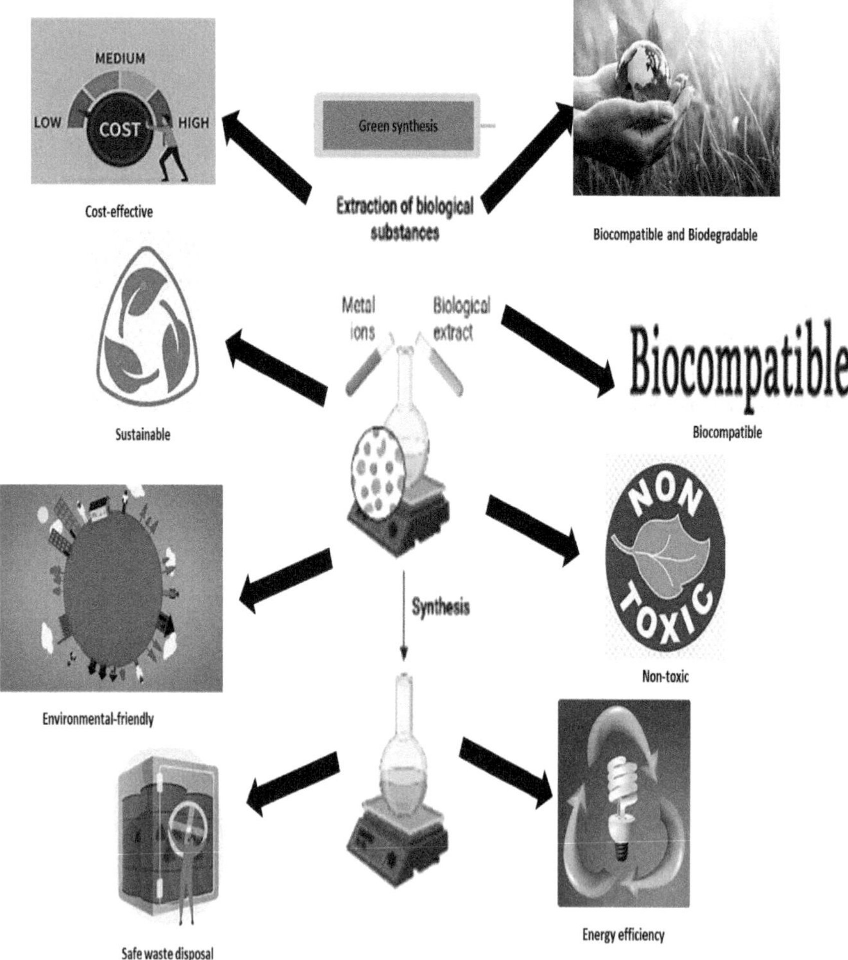

Figure 3. Advantages of green synthesis of nanoparticles from waste materials.

protect against pests and diseases. Additionally, waste-derived nanoparticles can be used in biomedical applications, such as drug delivery and cancer therapy, where they offer several advantages, including biocompatibility and targeted delivery (Taran et al. 2021, Ying et al. 2022).

Green synthesis of nanoparticles from waste materials is a promising approach that offers several advantages from a scientific perspective. The use of waste materials as starting materials for nanoparticle synthesis can offer several benefits, including (Aswathi et al. 2022, Gour and Jain 2019):

i. Sustainable and Environmentally Friendly: Using waste materials can reduce waste, minimize environmental impact, and provide a more sustainable source of raw materials for nanoparticle synthesis.

ii. Cost-Effective: Using waste materials as starting materials for nanoparticle synthesis can be more cost-effective than traditional raw materials, such as metal salts or organic compounds.

iii. Tunable Properties: Waste materials can contain a range of organic and inorganic compounds that can influence the properties of the resulting nanoparticles, allowing for fine-tuning of their physical and chemical properties.

iv. Biocompatible, Biodegradable, and Non-Toxic: Waste-derived nanoparticles are often biocompatible and non-toxic, making them suitable for biomedical applications. They are also biodegradable, reducing the environmental impact of their disposal and potential accumulation in ecosystems.

v. Versatile: Waste-derived nanoparticles can be synthesized using a wide range of waste materials, including agricultural waste, industrial waste, and municipal waste, making this approach highly versatile.

vi. Easy to Scale Up: Green synthesis of nanoparticles from waste materials can be easily scaled up to industrial levels, making it a commercially viable approach.

vii. Potential for Novel Applications: The use of waste materials for nanoparticle synthesis can lead to the development of novel applications, such as waste water treatment or environmental remediation, as well as the removal of heavy metals or organic pollutants from contaminated soils or water.

viii. Reduced Environmental Impact: The use of waste materials can reduce the environmental impact of nanoparticle synthesis by minimizing the use of traditional raw materials, such as metal salts, that can be environmentally harmful.

ix. Energy Efficiency: Green synthesis of nanoparticles from waste materials can be more energy efficient than traditional synthesis methods since waste materials are readily available and may not require energy-intensive purification processes.

x. Safe Disposal of Waste: The use of waste materials for nanoparticle synthesis can provide a safe disposal method for waste materials that may otherwise pose environmental and health risks.

Therefore, green synthesis of nanoparticles from waste materials offers several scientific advantages and can potentially revolutionize the field of nanoparticle synthesis by providing a sustainable, cost-effective, and versatile approach. It offers several environmental advantages and has the potential to contribute to sustainable development by promoting waste reduction and minimizing environmental impact. However, it is important to carefully evaluate the environmental impact of waste-derived nanoparticles and ensure their safety and efficacy before widespread adoption.

12. Common Methods for Nanoparticle Synthesis From Waste Materials

There are several methods for synthesizing nanoparticles from waste materials. Some of the commonly used methods are (Sharma et al. 2019, Nadaroglu et al. 2017, Seroka et al. 2022):

a. Chemical reduction: In this method, waste materials are treated with a reducing agent to convert metal ions into nanoparticles. The reducing agents used can be a variety of organic or inorganic compounds, such as sodium borohydride or hydrazine hydrate. This method is widely used for synthesizing metal-based nanoparticles from waste materials, such as metal ions from electronic waste. Again, silica nanoparticles can be synthesized from rice husk ash by reacting with an alkaline solution such as sodium hydroxide. This chemical reaction produces a colloidal suspension of silica nanoparticles. Silver nanoparticles can be synthesized from silver nitrate using banana peel extract as the reducing agent.

b. Green Synthesis: This method uses natural and eco-friendly agents, such as plant extracts or microorganisms, for nanoparticle synthesis. Plant extracts contain various biomolecules, such as flavonoids and polyphenols, which can act as reducing and stabilizing agents in nanoparticle synthesis. Microorganisms, such as bacteria and fungi, can also synthesize nanoparticles through a process called microbial synthesis. This method is gaining popularity as it is sustainable, cost-effective, and does not require harsh chemicals. For example, silver nanoparticles can be synthesized from banana peels using a green synthesis approach that involves boiling banana peels in water and then adding silver nitrate to the resulting solution. The plant compounds in the banana peels act as reducing agents, converting the silver ions into silver nanoparticles. Again, gold nanoparticles can be synthesized from e-waste using a biological synthesis approach that involves using the bacterium *Acidithiobacillus ferrooxidans* (Order: *Acidithiobacillales*, Family: *Acidithiobacillaceae*) to reduce the gold ions in the waste material into gold nanoparticles. Copper nanoparticle synthesis using coffee waste extract is an example of green synthesis.

c. Electrochemical Synthesis: This method involves using electricity to reduce metal ions into nanoparticles. Waste materials are used as the source of metal ions, and the synthesis is carried out in an electrochemical cell with appropriate electrodes and electrolytes. This method is widely used for synthesizing metal-based nanoparticles from electronic waste, such as copper and gold nanoparticles. The zinc

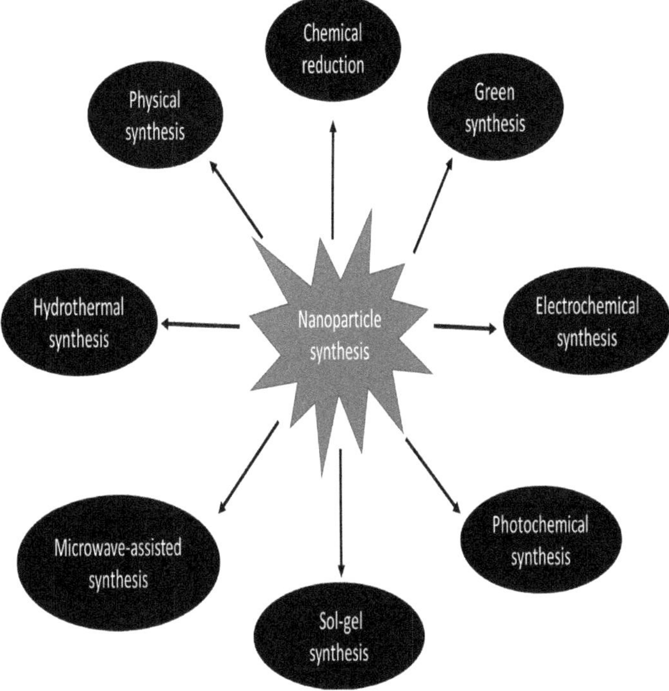

Figure 4. Different methods for nanoparticle synthesis from waste materials.

oxide nanoparticles can be synthesized from spent alkaline batteries. Spent alkaline batteries contain a high concentration of zinc oxide, which can be recovered and used as a precursor for nanoparticle synthesis.

d. Photochemical Synthesis Method: In this method, waste materials are treated with light to induce a chemical reaction that results in nanoparticle synthesis. Light is absorbed by the waste materials, which then release electrons, leading to the formation of nanoparticles. This method is used to synthesize various types of nanoparticles, such as metal oxides and carbon-based nanoparticles, from waste materials—the synthesis of gold nanoparticles from wastewater generated from a jewelry manufacturing process.

e. Sol-Gel Synthesis Method: This method involves converting waste materials into a gel-like substance, which is then finally used to synthesize nanoparticles. Waste materials. The sol-gel method involves the conversion of a liquid precursor into a solid material by a series of chemical reactions. This method is widely used to synthesize metal oxide nanoparticles from waste materials such as fly ash, silica fumes, and rice husk ash. In this method, the waste material is dissolved in a solvent, and a precursor solution is added to initiate the gelation process. The resulting gel is dried and heated to form nanoparticles. Silica nanoparticles can be prepared by using this method.

f. Microwave-Assisted Synthesis: Microwave-assisted synthesis is a rapid and efficient method for synthesizing nanoparticles from waste materials. This method involves using microwave radiation to heat the reaction mixture, resulting in the formation of nanoparticles. This method has the advantages of shorter reaction times, lower energy requirements, and the ability to control particle size and shape – the synthesis of iron oxide nanoparticles from industrial wastewater.

g. Hydrothermal Synthesis: Hydrothermal synthesis is a high-temperature and high-pressure method for synthesizing nanoparticles from waste materials. In this method, the waste materials are mixed with a solvent and heated under high-pressure conditions to induce the formation of nanoparticles. This method is particularly useful for synthesizing metal oxide nanoparticles from waste materials. The zinc oxide nanoparticles can be synthesized from waste zinc oxide (ZnO) sludge generated from the semiconductor industry.

h. Physical Synthesis: Physical synthesis involves using physical methods, such as milling or laser ablation, to produce nanoparticles. This method can produce nanoparticles with unique sizes and shapes but can be time-consuming and expensive. For example, carbon nanoparticles can be synthesized from sugarcane bagasse using a physical synthesis approach that involves milling the bagasse into small particles and then pyrolyzing the particles in a furnace at high temperatures. This process produces carbon nanoparticles that can be used in various applications. The silver nanoparticles are synthesized from electronic waste (e-waste).

In summary, several methods exist for synthesizing nanoparticles from waste materials, including chemical synthesis, green synthesis, physical synthesis, and biological synthesis. The choice of method depends on the type of waste material and the desired properties of the nanoparticles.

13. Different Types and Examples of Waste Materials Used for Nanoparticle Synthesis

Nanoparticles are tiny particles with dimensions ranging from 1 to 100 nanometers (nm). They have unique properties that make them useful in various fields, such as medicine, electronics, and energy. Nanoparticles can be synthesized from a wide range of materials, including waste materials. The different types of waste materials that can be used for nanoparticle synthesis, along with examples and detailed explanations (Samaddar et al. 2018, Saratale et al. 2018, Joudeh and Linke 2022).

a. Agricultural Waste: Agricultural waste includes various byproducts generated during crop cultivation and food processing. Some examples of agricultural waste that can be used for nanoparticle synthesis are rice husk, sugarcane bagasse, and wheat straw. These materials contain cellulose, hemicellulose, and lignin, which can be extracted and used as precursors for nanoparticle synthesis. For instance, rice husk ash can be used to synthesize silica nanoparticles, while sugarcane bagasse can be used to synthesize carbon nanoparticles.

b. Industrial waste: Industrial waste refers to the byproducts generated during manufacturing processes. Examples of industrial waste that can be used for nanoparticle synthesis include fly ash, red mud, and sludge. Fly ash is a waste material generated during coal combustion, which contains various metal oxides that can be used to synthesize metal oxide nanoparticles. Red mud is a waste material generated during alumina production, which contains iron, titanium, and other metal oxides that can be used to synthesize metal oxide nanoparticles. Sludge is a waste material generated during wastewater treatment, which contains organic and inorganic compounds that can be used to synthesize nanoparticles.

c. Electronic waste: Electronic waste or e-waste refers to discarded electronic devices, such as computers, phones, and televisions. E-waste contains various metals, such as gold, silver, copper, and palladium, which can be recovered and used to synthesize nanoparticles. For instance, gold nanoparticles can be synthesized from discarded computer components using a green synthesis approach that involves using plant extracts as reducing agents.

d. Biological waste: Biological waste includes organic waste generated from living organisms, such as food waste, animal waste, and plant waste. These materials contain various compounds, such as proteins, lipids, and carbohydrates that can be used to synthesize nanoparticles. For instance, shrimp shells can be used to synthesize chitosan nanoparticles, while banana peels can be used to synthesize silver nanoparticles.

In summary, waste materials from various sources can be used for nanoparticle synthesis, including agricultural waste, industrial waste, electronic waste, and biological waste. These waste materials contain various compounds that can be extracted and used as precursors for nanoparticle synthesis. The use of waste materials for nanoparticle synthesis has several advantages, including reducing waste generation and providing a cost-effective approach for nanoparticle synthesis.

14. Different Nanoparticles Synthesized from Waste Materials

Nanoparticles have a wide range of applications in various fields such as medicine, electronics, energy, and environmental remediation. Using waste materials as precursors for nanoparticle synthesis is an attractive approach because it can help reduce waste and provide a cost-effective and environmentally friendly method of producing nanoparticles. Some nanoparticles and their waste precursors along with application and usages is provided in Table 1. Here are some examples of nanoparticles synthesized from waste materials (Abdelbasir et al. 2020, Kumari et al. 2021, Zamare et al. 2021, Dutta and Das 2021):

i. Carbon Nanoparticles: Carbon nanoparticles can be synthesized from agricultural waste such as rice straw, wheat straw, and corn stover. The process involves burning the waste material to produce carbon nanoparticles under controlled conditions.

ii. Silver Nanoparticles: E-waste, such as discarded computer components and cell phones, contain metals such as silver that can be extracted and used to synthesize silver nanoparticles. The process involves dissolving the e-waste in acid and then reducing the silver ions to form nanoparticles. They can also be synthesized from waste tea leaves, which are rich in polyphenols. The process involves boiling the tea leaves in water and then adding a silver salt solution, which is reduced by the polyphenols to form nanoparticles.

iii. Iron Oxide Nanoparticles: Iron oxide nanoparticles can be synthesized from industrial waste, such as steel slag, which is a byproduct of the steel-making process. The process involves reacting the steel slag with an acid to release the iron oxide, which is then purified and used to form nanoparticles. They can also be synthesized from waste iron, such as iron filings or rust. The process involves reacting the iron with an acid to release the iron oxide, which is then purified and used to form nanoparticles.

iv. ZnO Nanoparticles: ZnO nanoparticles can be synthesized from waste zinc, such as zinc ash, which is a byproduct of the galvanizing industry. The process involves reacting the zinc ash with an acid to release the ZnO, which is then purified and used to form nanoparticles.

v. Copper Nanoparticles: Copper nanoparticles can be synthesized from waste copper, such as copper wire scrap. The process involves dissolving the copper in acid and then reducing the copper ions to form nanoparticles. They can also be synthesized from waste-printed circuit boards, which contain copper as a major component. The process involves grinding the boards to a fine powder and then using a reducing agent to form copper nanoparticles.

vi. Titanium Dioxide Nanoparticles: Titanium dioxide nanoparticles can be synthesized from waste slag, which is a byproduct of the titanium dioxide industry. The process involves treating the slag with acid to extract the titanium dioxide, which is then purified and used to form nanoparticles.

vii. Gold Nanoparticles: Gold nanoparticles can be synthesized from waste banana peels, which are rich in polyphenols. The process involves boiling the banana peels in water and then adding a gold salt solution, which is reduced by the polyphenols to form nanoparticles.

viii. Iron Nanoparticles: Iron nanoparticles can be synthesized from waste oil, which is rich in fatty acids. The process involves heating the oil to produce a biochar, which is then activated with a base and used to form nanoparticles.

ix. Magnesium Oxide Nanoparticles: Magnesium oxide nanoparticles can be synthesized from waste eggshells, which are rich in calcium carbonate. The process involves calcining the eggshells to produce calcium oxide, which reacts with magnesium sulfate to form magnesium oxide nanoparticles.

x. Zinc and ZnO Nanoparticles: Zinc can be synthesized from waste zinc-carbon batteries, containing zinc as an active material. The process involves extracting the zinc from the batteries and then using a reducing agent to form zinc nanoparticles.

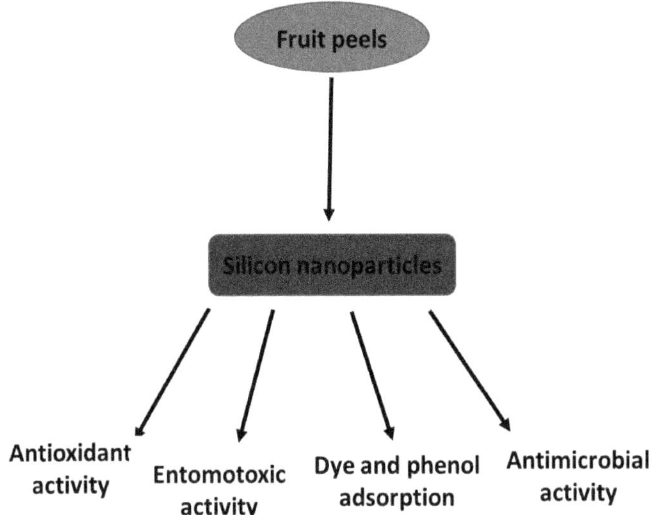

Figure 5. Silicon nanoparticle synthesis method and application (Abomughaid 2022, Mohamad et al. 2019, Osman and Sapawe 2019).

Again, oxide nanoparticles can be synthesized from waste tires, which contain ZnO as a component of their rubber. The process involves pyrolyzing the tires to release the ZnO, which is then purified and used to form nanoparticles.

xi. Silicon Nanoparticles: Silicon nanoparticles can be synthesized from waste glass, which is rich in silicon dioxide. The process involves heating the glass to a high temperature and then reacting it with a reducing agent to form silicon nanoparticles.

These are just a few examples of nanoparticles that can be synthesized from waste materials. There are many other waste materials that can be used to produce nanoparticles, and researchers are continuing to explore new methods and applications for these materials.

15. Downstream Processing Techniques for Nanoparticles Synthesized From Waste Materials

Downstream processing of purification techniques depends on the production process of nanoparticles, the type of nanoparticle being produced, and most importantly the types of waste utilized, contaminants present, and architecture available for purification (Patwardhan et al. 2018, Gopakumar et al. 2019, Salem et al. 2022).

a. Precipitation-Based Methods: Nanoparticles are usually precipitated from impurities and separated based on size and density. By exploiting the precipitation conditions, particularly in relation to precipitating agent concentration, temperature, and reaction time, the purifying factors can be maximized. Precipitation purification, however, is not a selective process.

b. Centrifugation-Based Methods: The size and density of the nanoparticles were used to refine them. These treatments often rely on density gradient and ultracentrifugation

Table 1. Different nanoparticles synthesized from various waste materials and their usage.

Type of nanoparticle synthesized	Waste material utilized	Application and usage	Reference
Silver (Ag)	Outer peels of onion	Acetylation reaction	Yap et al. 2020
Silver (Ag)	Outer peels of sapota	Antimicrobial activities	Vishwasrao et al. 2019
Silver (Ag)	Outer peels of lemon	Antibacterial efficacy	Kaviya et al. 2011
Silica (Si)	Peel the ash of the banana	Dye remediation contamination removal of water	Mohamad et al. 2019
Copper oxide (CuO)	Outer peels of papaya	Photodegradation for oil mill effluent	Phang et al. 2021
Silver (Ag)	Shell of coconut	Antibacterial efficacy	Sinsinwar et al. 2018
Carbon nanonotubes (CNTs)	Rice husk	Electrochemical electrodes and reduce waste biomass	Asnawi et al. 2018
Silica (Si)	Rice husk	Acts as composite materials, fillers, and carriers	Van Hai Le et al. 2013
Cellulose nanocrystals (CNCs)	Areca nut husk	Nanocomposites and packaging films	Bhandari et al. 2021
Palladium (Pd)	Areca nut husk	α-keto imides and Stilbenes	Hegde et al. 2021
Cellulose nanocrystals (CNCs)	Waste paper	Polymer nanocomposite fillers	Mohamed et al. 2016
Calcium oxide (CaO)	Outer shell of eggs	Wastewater treatment	Habte et al. 2019

techniques. Although commonly used, ultracentrifugation procedures take a long time, are challenging to scale up, and have very low recoveries. This is mostly because particle disintegration occurs when pressure forces or osmotic shock are applied.

c. Membrane Separation Methods: These are widely used to divide nanoparticles in a fluid stream based on the difference in hydrodynamic radii. Commercially accessible membranes come in a wide range of pore sizes, making membrane separations relatively flexible processes.

d. Chromatographic Methods: Since it enables high recovery rates and high-purity products, chromatography has grown to be a very common methodology in downstream processing. In addition, it is simple to scale up and provides a solid foundation for massive manufacturing.

There is no single, ideal purifying technique that can handle the wide variety of nanoparticulate products made from waste materials. The mechanism for selecting the optimum method to purify the desired nanoparticles is now a problem for researchers, laboratories, and businesses. The kind, size, and place of manufacture of the product, as well as the final recovery yield, must all be considered when choosing the best downstream process.

16. Applications of Nanoparticles in Diverse Fields

Due to their small size, nanoparticles exhibit unique physical, chemical, and biological properties distinct from those of larger particles. This makes them highly useful in a wide range of fields, including medicine, electronics, energy, and environmental remediation. Here are some examples of how nanoparticles are used in various applications (Ijaz et al. 2020, Han et al. 2019, Salem and Fouda 2021, El- Nasr et al. 2020, Fayomi et al. 2020):

a. Medicine: Nanoparticles are used in medicine for a variety of purposes, including drug delivery, medical imaging, and cancer treatment. Because they are so small, nanoparticles can easily penetrate cell membranes and reach target tissues, making them ideal for delivering drugs directly to specific sites within the body. Additionally, certain types of nanoparticles can be designed to selectively bind to cancer cells, allowing for more targeted and effective cancer therapies.

b. Electronics: Nanoparticles are used in the electronics industry to enhance the performance of electronic devices such as computers and smartphones. For example, nanoparticles can improve the conductivity and stability of materials used in electronic circuits, resulting in faster and more efficient devices. Nanoparticles are also used to produce high-quality display screens, solar cells, and sensors.

c. Energy: Nanoparticles are used in various energy applications, including the production of fuel cells and batteries. Nanoparticles can be used as catalysts to increase the efficiency of chemical reactions in these devices, leading to more efficient energy conversion. Additionally, nanoparticles can be used to enhance the performance of solar cells, allowing them to convert more sunlight into electricity.

d. Environmental Remediation: Nanoparticles are being used in environmental remediation to remove pollutants from soil and water. For example, certain types of nanoparticles can be used to absorb heavy metals and other contaminants, making them easier to remove from polluted sites. Additionally, nanoparticles can be used to break down organic pollutants, such as oil spills, into harmless byproducts.

e. Food and Agriculture: Nanoparticles are used in food and agriculture to improve food quality and safety, as well as to increase crop yields. For example, nanoparticles can be used to create food packaging materials that are more effective at preserving food and preventing spoilage. Additionally, nanoparticles can be used to deliver nutrients and other beneficial substances directly to plants, leading to healthier and more productive crops.

f. Cosmetics: Nanoparticles are used in cosmetics to improve product performance and enhance their appearance. For example, nanoparticles can be used to create sunscreens that are more effective at blocking harmful UV radiation, or to create moisturizers that penetrate the skin more effectively. Nanoparticles can also be used to create cosmetic pigments with enhanced color and texture.

g. Textiles: Nanoparticles are used in textiles to improve their properties, such as water repellency, stain resistance, and UV protection. For example, nanoparticles can be used to create fabrics that are more durable, breathable, and comfortable to wear. Nanoparticles can also be used to create smart textiles that change color or texture in response to changes in temperature or humidity.

h. Water Treatment: Nanoparticles are used in water treatment to remove impurities and contaminants from drinking water. For example, nanoparticles can be used to remove heavy metals, bacteria, and viruses from water, making it safe to drink. Nanoparticles can also be used to improve the efficiency of water treatment processes, such as desalination and wastewater treatment.

i. Construction: Nanoparticles are used in construction to improve the properties of building materials, such as strength, durability, and fire resistance. For example, nanoparticles can be used to enhance the strength of concrete, making it more resistant to cracking and erosion. Nanoparticles can also be used to create insulation materials that are more effective at regulating temperature and sound.

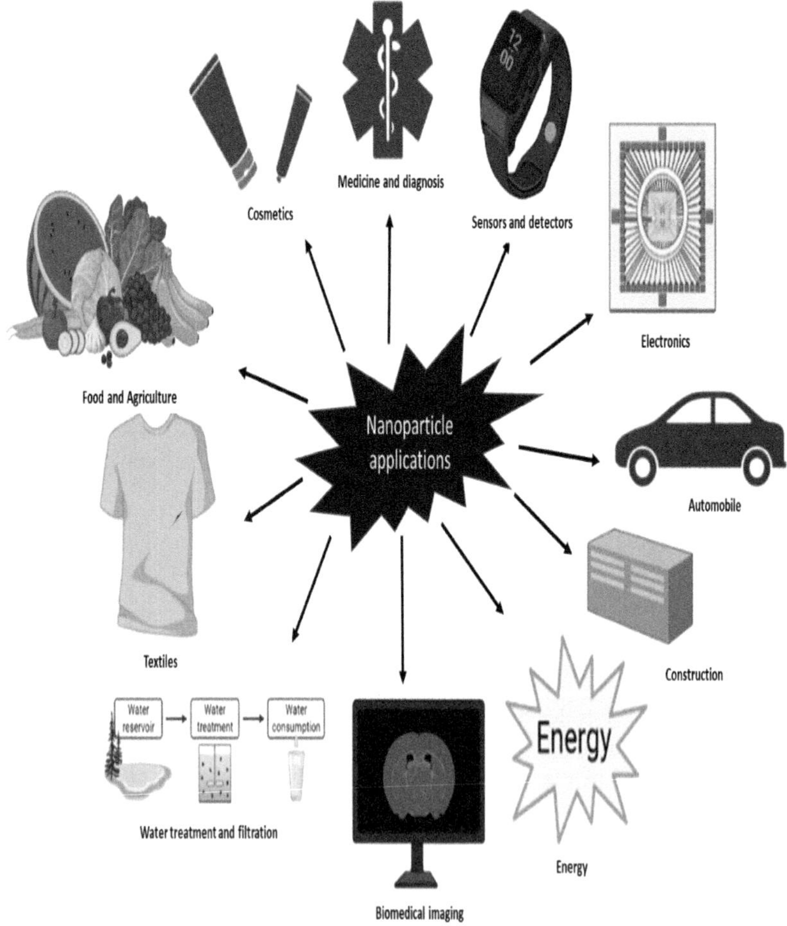

Figure 6. Multifaceted applications of nanoparticles obtained from waste materials.

j. Automotive: Nanoparticles are used in the automotive industry to improve fuel efficiency, reduce emissions, and enhance safety. For example, nanoparticles can be used to create lightweight materials that improve fuel efficiency without sacrificing strength or durability. Nanoparticles can also be used to create sensors that detect hazardous gases or other potential safety hazards in vehicles.

k. Biomedical Imaging: Nanoparticles have unique optical properties that are ideal for biomedical imaging. For example, nanoparticles can be used as contrast agents in magnetic resonance imaging (MRI) to improve image resolution and sensitivity. Nanoparticles can also be used in fluorescence imaging to detect specific biomolecules in cells and tissues.

l. Sensors: Nanoparticles can be used in sensors for various applications, such as environmental monitoring and medical diagnostics. For example, gold nanoparticles can be used in biosensors to detect specific biomolecules in blood or other bodily fluids, while carbon nanotubes can be used in gas sensors to detect hazardous gases.

m. Aerospace and Defense: Nanoparticles can be used in aerospace applications to improve the performance and durability of materials used in aircraft and spacecraft. For example, carbon nanotubes can be used to make lightweight and strong composite materials for use in aircraft and spacecraft structures. They also have numerous applications in defense technologies, including developing high-performance materials, sensors, and energy storage devices. For example, nanoparticles can be used to make lightweight and strong body armor, while carbon nanotubes can be used to improve the performance of electronics and sensors used in military applications.

17. Future Prospects

The future prospects of nanoparticle synthesis from waste materials are promising, as it offers an eco-friendly, sustainable, and cost-effective approach to producing nanoparticles for various applications. Here are some details on the potential prospects of this technology (Duarah et al. 2020, Xu et al. 2019, Pensupa et al. 2018):

a. Environmentally Sustainable Approach: Nanoparticle synthesis from waste materials has the potential to reduce the environmental impact of waste materials significantly. Using waste materials as a source for nanoparticle synthesis reduces the amount of waste generated and promotes the concept of circular economy. This approach also reduces the energy consumption and carbon footprint associated with traditional nanoparticle synthesis methods.

b. Cost-Effective Approach: Using waste materials for nanoparticle synthesis can significantly reduce the production cost. Since waste materials are abundant and easily accessible, they can be used as a low-cost raw material source. This approach also offers a potential economic benefit to waste management industries by creating value-added products from waste.

c. Diverse Applications: Nanoparticles synthesized from waste materials have diverse applications in various fields, such as medicine, catalysis, energy storage, and environmental remediation. For example, metal nanoparticles synthesized from waste materials have shown promising applications in cancer therapy, drug delivery, and water treatment. Additionally, metal oxide nanoparticles synthesized from waste materials can be used as catalysts for various chemical reactions, leading to the development of greener and sustainable chemical processes.

d. Development of New Materials: Nanoparticles synthesized from waste materials have unique properties due to the presence of impurities, which can lead to the development of new materials. For example, silica nanoparticles synthesized from rice husk ash have shown improved thermal stability and mechanical properties, making them a potential candidate for various industrial applications.

e. Technological Advancements: Advancements in technology and research will further enhance the synthesis of nanoparticles from waste materials. New approaches, such as microwave-assisted synthesis and biosynthesis, have shown promising results in synthesizing nanoparticles with unique properties. Additionally, developing new analytical techniques will enable the characterization of nanoparticles synthesized from waste materials at a molecular level, leading to a better understanding of their properties and potential applications.

18. Summary

The future prospects of nanoparticle synthesis from waste materials are promising, as it offers an environmentally sustainable, cost-effective, and diverse approach to producing nanoparticles with unique properties for various applications. Further research and development in this area can lead to the development of new materials and the creation of value-added products from waste.

Nanoparticle synthesis from waste materials is an emerging area of research that involves using waste materials as a source to produce nanoparticles. This approach offers several benefits, such as reduced environmental impact, cost-effectiveness, and diverse applications.

Various methods have been developed for nanoparticle synthesis from waste materials, including chemical reduction, biosynthesis, microwave-assisted synthesis, hydrothermal synthesis, and sol-gel synthesis. These methods involve using waste materials such as industrial waste water, electronic waste, mine tailings, and agricultural waste to produce nanoparticles.

The nanoparticles synthesized from waste materials have unique properties due to the presence of impurities, which can lead to the development of new materials. The applications of these nanoparticles are diverse, including medicine, catalysis, energy storage, and environmental remediation. For example, metal nanoparticles synthesized from waste materials have shown promising applications in cancer therapy, drug delivery, and water treatment. Metal oxide nanoparticles synthesized from waste materials can be used as catalysts for various chemical reactions, leading to the development of greener and more sustainable chemical processes.

Advancements in technology and research will further enhance the synthesis of nanoparticles from waste materials. New approaches, such as microwave-assisted synthesis and biosynthesis, have shown promising results in synthesizing nanoparticles with unique properties. Additionally, developing new analytical techniques will enable the characterization of nanoparticles synthesized from waste materials at a molecular level, leading to a better understanding of their properties and potential applications.

In conclusion, nanoparticle synthesis from waste materials offers an environmentally sustainable, cost-effective, and diverse approach to producing nanoparticles with unique properties for various applications. Further research and development in this area can lead to the creation of value-added products from waste and the development of new materials.

19. Conclusion

The synthesis of nanoparticles from waste materials offers a sustainable and eco-friendly approach to nanoparticle production. Waste materials, such as agricultural residues, food waste, and industrial byproducts, can be used as low-cost and abundant sources of raw materials for nanoparticle synthesis. This approach not only reduces the environmental impact of waste disposal but also helps address the growing demand for nanoparticles in various fields, including medicine, electronics, and energy.

Moreover, using waste materials for nanoparticle synthesis can also lead to the developing of new and unique nanoparticle properties that may not be achievable with traditional synthesis methods. For example, the chemical composition of waste materials can impart unique functional properties to nanoparticles, such as antimicrobial, antioxidant, and antifungal properties.

However, there are some challenges associated with the synthesis of nanoparticles from waste materials, such as the need for specialized equipment and the optimization of reaction conditions. In addition, the quality and consistency of the nanoparticles produced from waste materials may vary depending on the composition and quality of the waste materials used.

Despite these challenges, the synthesis of nanoparticles from waste materials holds great promise for developing sustainable and cost-effective nanoparticle production methods. Further research in this area can help to overcome the existing challenges and expand the potential applications of waste-based nanoparticles in various fields.

Acknowledgments

The authors thank the editors of the book as well as the publisher for their support and encouragement.

References

Abd Elkodous, M., El-Husseiny, H. M., El-Sayyad, G. S., Hashem, A. H., Doghish, A. S., Elfadil, D., Radwan, Y., El-Zeiny, H. M., Bedair, H., Ikhdair, O. A. and Hashim, H. (2021). Recent advances in waste-recycled nanomaterials for biomedical applications: Waste-to-wealth. *Nanotechnology Reviews*, 10(1): 1662–1739.

Abdelbasir, S. M., McCourt, K. M., Lee, C. M. and Vanegas, D. C. (2020). Waste-derived nanoparticles: Synthesis approaches, environmental applications, and sustainability considerations. *Frontiers in Chemistry*, 8: 782.

Abdur Rahman, M., Haque, S., Athikesavan, M. M. and Kamaludeen, M. B. (2023). A review of environmental friendly green composites: Production methods, current progresses, and challenges. *Environmental Science and Pollution Research*, 1–25.

Abid, N., Khan, A. M., Shujait, S., Chaudhary, K., Ikram, M., Imran, M., Haider, J., Khan, M., Khan, Q. and Maqbool, M. (2022). Synthesis of nanomaterials using various top-down and bottom-up approaches, influencing factors, advantages, and disadvantages: A review. *Advances in Colloid and Interface Science*, 300: 102597.

Abomughaid, M. M. (2022). Bio-fabrication of bio-inspired silica nanomaterials from orange peels in combating oxidative stress. *Nanomaterials*, 12(18): 3236.

Adelere, I. A. and Lateef, A. (2016). A novel approach to the green synthesis of metallic nanoparticles: The use of agro-wastes, enzymes, and pigments. *Nanotechnology Reviews*, 5(6): 567–587.

Ahmed, M. M., Badawy, M. T., Ahmed, F. K., Kalia, A. and Abd-Elsalam, K. A. (2022). Fruit peel waste-to-wealth: Bionanomaterials production and their applications in agroecosystems. *Agri-Waste and Microbes for Production of Sustainable Nanomaterials*, 231–257.

Ali, S., Chen, X., Shah, M. A., Ali, M., Zareef, M., Arslan, M., Ahmad, S., Jiao, T., Li, H. and Chen, Q. (2021). The avenue of fruit wastes to worth for synthesis of silver and gold nanoparticles and their antimicrobial application against foodborne pathogens: A review. *Food Chemistry*, 359: 129912.

Arif, Z., Sethy, N. K., Mishra, P. K. and Kumar, P. (2023). Green synthesis of nanomaterials from biomass waste for biodiesel production. In *NanoBioenergy: Application and Sustainability Assessment*, 211–234. Singapore: Springer Nature Singapore.

Asnawi, M., Azhari, S., Hamidon, M. N., Ismail, I. and Helina, I. (2018). Synthesis of carbon nanomaterials from rice husk via microwave oven. *Journal of Nanomaterials*, 2018: 1–5.

Aswathi, V. P., Meera, S., Maria, C. A. and Nidhin, M. (2022). Green synthesis of nanoparticles from biodegradable waste extracts and their applications: A critical review. *Nanotechnology for Environmental Engineering*, 1–21.

Bahrulolum, H., Nooraei, S., Javanshir, N., Tarrahimofrad, H., Mirbagheri, V. S., Easton, A. J. and Ahmadian, G. (2021). Green synthesis of metal nanoparticles using microorganisms and their application in the agrifood sector. *Journal of Nanobiotechnology*, 19(1): 1–26.

Baroi, A. M., Sieniawska, E., Świątek, Ł. and Fierascu, I. (2023). Grape waste materials—An attractive source for developing nanomaterials with versatile applications. *Nanomaterials*, 13(5): 836.

Bhandari, G., Bagheri, A. R., Bhatt, P. and Bilal, M. (2021). Occurrence, potential ecological risks, and degradation of endocrine disrupter, nonylphenol, from the aqueous environment. *Chemosphere*, 275: 130013.

Biswas, M. C., Chowdhury, A., Hossain, M. M. and Hossain, M. K. (2022). Applications, drawbacks, and future scope of nanoparticle-based polymer composites. In *Nanoparticle-Based Polymer Composites* (pp. 243–275). Woodhead Publishing.

Brar, K. K., Magdouli, S., Othmani, A., Ghanei, J., Narisetty, V., Sindhu, R., Binod, P., Pugazhendhi, A., Awasthi, M. K. and Pandey, A. (2022). Green route for recycling of low-cost waste resources for the biosynthesis of nanoparticles (NPs) and nanomaterials (NMs)—A review. *Environmental Research*, 207: 112202.

Chakraborty, I. and Pradeep, T. (2017). Atomically precise clusters of noble metals: Emerging link between atoms and nanoparticles. *Chemical Reviews*, 117(12): 8208–8271.

Devatha, C. P. and Thalla, A. K. (2018). Green synthesis of nanomaterials. In Synthesis of Inorganic Nanomaterials (pp. 169–184). Woodhead Publishing.

Dhanya, B. S., Mishra, A., Chandel, A. K. and Verma, M. L. (2020). Development of sustainable approaches for converting the organic waste to bioenergy. *Science of the Total Environment*, 723: 138109.

Duarah, P., Haldar, D. and Purkait, M. K. (2020). Technological advancement in the synthesis and applications of lignin-based nanoparticles derived from agro-industrial waste residues: A review. *International Journal of Biological Macromolecules*, 163: 1828–1843.

Dutta, D. and Das, B. M. (2021). Scope of green nanotechnology towards amalgamation of green chemistry for cleaner environment: A review on synthesis and applications of green nanoparticles. *Environmental Nanotechnology, Monitoring and Management*, 15: 100418.

Ealia, S. A. M. and Saravanakumar, M. P. (2017). A review on the classification, characterisation, synthesis of nanoparticles and their application. In *IOP Conference Series: Materials Science and Engineering*, 263, 3: 032019. IOP Publishing.

El-Nasr, R. S., Abdelbasir, S. M., Kamel, A. H. and Hassan, S. S. (2020). Environmentally friendly synthesis of copper nanoparticles from waste printed circuit boards. *Separation and Purification Technology*, 230: 115860.

Fayomi, O. S. I., Owodolu, T., Agboola, O., Oyebanji, J. and Popoola, A. P. I. (2020). A paradigm shift on the impact of synthetic Agro waste nanoparticles materials for engineering diversity: A mini overview. In *AIP Conference Proceedings* (Vol. 2307, No. 1, p. 020041). AIP Publishing LLC.

Ferronato, N. and Torretta, V. (2019). Waste mismanagement in developing countries: A review of global issues. *International Journal of Environmental Research and Public Health*, 16(6): 1060.

Gilbertson, L. M., Zimmerman, J. B., Plata, D. L., Hutchison, J. E. and Anastas, P. T. (2015). Designing nanomaterials to maximize performance and minimize undesirable implications guided by the Principles of Green Chemistry. *Chemical Society Reviews*, 44(16): 5758–5777.

Gopakumar, D. A., Pai, A. R., Pasquini, D., Ben, L. S. Y., HPS, A. K. and Thomas, S. (2019). Nanomaterials—state of art, new challenges, and opportunities. *Nanoscale Materials in Water Purification*, 1–24.

Gour, A. and Jain, N. K. (2019). Advances in green synthesis of nanoparticles. *Artificial Cells, Nanomedicine, and Biotechnology*, 47(1): 844–851.

Habte, L., Shiferaw, N., Mulatu, D., Thenepalli, T., Chilakala, R. and Ahn, J. W. (2019). Synthesis of nano-calcium oxide from waste eggshell by sol-gel method. *Sustainability*, 11(11): 3196.

Han, X., Xu, K., Taratula, O. and Farsad, K. (2019). Applications of nanoparticles in biomedical imaging. *Nanoscale*, 11(3): 799–819.

Hegde, R. V., Ghosh, A., Jadhav, A. H., Nizam, A., Patil, S. A., Peter, F. and Dateer, R. B. (2021). Biogenic synthesis of Pd-nanoparticles using Areca Nut Husk Extract: A greener approach to access α-keto imides and stilbenes. *New Journal of Chemistry*, 45(35): 16213–16222.

Ijaz, I., Gilani, E., Nazir, A. and Bukhari, A. (2020). Detail review on chemical, physical and green synthesis, classification, characterizations and applications of nanoparticles. *Green Chemistry Letters and Reviews*, 13(3): 223–245.

Jamkhande, P. G., Ghule, N. W., Bamer, A. H. and Kalaskar, M. G. (2019). Metal nanoparticles synthesis: An overview on methods of preparation, advantages and disadvantages, and applications. *Journal of Drug Delivery Science and Technology*, 53: 101174.

Ji, T., Fang, H., Zhang, R., Yang, J., Fan, L. and Li, J. (2022). Automatic sorting of low-value recyclable waste: a comparative experimental study. *Clean Technologies and Environmental Policy*, 1–13.

Joudeh, N. and Linke, D. (2022). Nanoparticle classification, physicochemical properties, characterization, and applications: a comprehensive review for biologists. *Journal of Nanobiotechnology*, 20(1): 262.

Kaur, G., Uisan, K., Ong, K. L. and Lin, C. S. K. (2018). Recent trends in green and sustainable chemistry and waste valorisation: Rethinking plastics in a circular economy. *Current Opinion in Green and Sustainable Chemistry*, 9: 30–39.

Kaviya, S., Santhanalakshmi, J., Viswanathan, B., Muthumary, J. and Srinivasan, K. (2011). Biosynthesis of silver nanoparticles using Citrus sinensis peel extract and its antibacterial activity. *Spectrochimica Acta Part A: Molecular and Biomolecular Spectroscopy*, 79(3): 594–598.

Khan, I., Saeed, K. and Khan, I. (2019). Nanoparticles: Properties, applications and toxicities. *Arabian Journal of Chemistry*, 12(7): 908–931.

Kharissova, O. V., Kharisov, B. I., Oliva González, C. M., Méndez, Y. P. and López, I. (2019). Greener synthesis of chemical compounds and materials. *Royal Society Open Science*, 6(11): 191378.

Kumari, S. C., Dhand, V. and Padma, P. N. (2021). Green synthesis of metallic nanoparticles: A review. *Nanomaterials*, 259–281.

Lange, J. P. (2021). Managing plastic waste-sorting, recycling, disposal, and product redesign. *ACS Sustainable Chemistry and Engineering*, 9(47): 15722–15738.

Liu, C., Luan, P., Li, Q., Cheng, Z., Xiang, P., Liu, D., Hou, Y., Yang, Y. and Zhu, H. (2021). Biopolymers derived from trees as sustainable multifunctional materials: A review. *Advanced Materials*, 33(28): 2001654.

Lizundia, E., Luzi, F. and Puglia, D. (2022). Organic waste valorisation towards circular and sustainable biocomposites. *Green Chemistry*, 24(14): 5429–5459.

Lochab, B., Shukla, S. and Varma, I. K. 2014. Naturally occurring phenolic sources: Monomers and polymers. *RSC Advances*, 4(42): 21712–21752.

Lubongo, C. and Alexandridis, P. (2022). Assessment of performance and challenges in use of commercial automated sorting technology for plastic waste. *Recycling*, 7(2): 11.

Madkour, L. H. and Madkour, L. H. (2019). Introduction to nanotechnology (NT) and nanomaterials (NMs). *Nanoelectronic Materials: Fundamentals and Applications*, 1–47.

Mahanthesh, A. B., Haldar, S. and Banerjee, S. (2022). Biogeneration of valuable nanomaterials from food and other wastes. *Biotechnology for Zero Waste: Emerging Waste Management Techniques*, 361–368.

Mansoori, G. A. (2017). An introduction to nanoscience and nanotechnology. *Nanoscience and Plant–Soil Systems*, 3–20.

Matussin, S., Harunsani, M. H., Tan, A. L. and Khan, M. M. (2020). Plant-extract-mediated SnO_2 nanoparticles: Synthesis and applications. *ACS Sustainable Chemistry and Engineering*, 8(8): 3040–3054.

Modena, M. M., Rühle, B., Burg, T. P. and Wuttke, S. (2019). Nanoparticle characterization: What to measure? *Advanced Materials*, 31(32): 1901556.

Mohamad, D. F., Osman, N. S., Nazri, M. K. H. M., Mazlan, A. A., Hanafi, M. F., Esa, Y. A. M., Rafi, M. I. I. M., Zailani, M. N., Rahman, N. N., Abd Rahman, A. H. and Sapawe, N. (2019). Synthesis of mesoporous silica nanoparticle from banana peel ash for removal of phenol and methyl orange in aqueous solution. *Materials Today: Proceedings*, 19: 1119–1125.

Mohamed, M. A., Salleh, W. N. W., Jaafar, J., Ismail, A. F., Abd Mutalib, M., Sani, N. A. A., Asri, S. E. A. M. and Ong. C. S. (2016). Physicochemical characteristic of regenerated cellulose/N-doped TiO_2 nanocomposite membrane fabricated from recycled newspaper with photocatalytic activity under UV and visible light irradiation. *Chemical Engineering Journal*, 284: 202–215.

Nabi, G., Khalid, N. R., Tahir, M. B., Rafique, M., Rizwan, M., Hussain, S., Iqbal, T. and Majid. A. (2018). A review on novel eco-friendly green approach to synthesis TiO_2 nanoparticles using different extracts. *Journal of Inorganic and Organometallic Polymers and Materials*, 28: 1552–1564.

Nadaroglu, H., GÜNGÖR, A. A. and Selvi, İ. N. C. E. (2017). Synthesis of nanoparticles by green synthesis method. *International Journal of Innovative Research and Reviews*, 1(1): 6–9.

Nasrollahzadeh, M., Sajadi, S. M., Sajjadi, M. and Issaabadi, Z. (2019). An introduction to nanotechnology. In *Interface Science and Technology*, 28: 1–27. Elsevier.

Neo, E. R. K., Yeo, Z., Low, J. S. C., Goodship, V. and Debattista, K. (2022). A review on chemometric techniques with infrared, Raman and laser-induced breakdown spectroscopy for sorting plastic waste in the recycling industry. *Resources, Conservation and Recycling*, 180: 106217.

Nkele, A. C. and Ezema, F. I. (2020). Diverse synthesis and characterization techniques of nanoparticles. *Thin Films*.

Ojo, O. A., Olayide, I. I., Akalabu, M. C., Ajiboye, B. O., Ojo, A. B., Oyinloye, B. E. and Ramalingam, M. (2021). Nanoparticles and their biomedical applications. *Biointerface Res. Appl. Chem.*, 11(1): 8431–8445.

Oluwasanu, A. A., Oluwaseun, F. A. P. O. H. U. N. D. A., Teslim, J. A., Isaiah, T. T., Olalekan, I. A. and Chris, O. A. (2019). Scientific applications and prospects of nanomaterials: A multidisciplinary review. *African Journal of Biotechnology*, 18(30): 946–961.

Omran, B. A. and Baek, K. H. (2022). Valorization of agro-industrial biowaste to green nanomaterials for wastewater treatment: Approaching green chemistry and circular economy principles. *Journal of Environmental Management*, 311: 114806.

Ortega, F., Versino, F., López, O. V. and García, M. A. (2022). Biobased composites from agro-industrial wastes and by-products. *Emergent Materials*, 5(3): 873–921.

Osman, N. S. and Sapawe, N. (2019). Waste material as an alternative source of silica precursor in silica nanoparticle synthesis—A review. *Materials Today: Proceedings*, 19: 1267–1272.

Pandey, G. and Jain, P. (2020). Assessing the nanotechnology on the grounds of costs, benefits, and risks. *Beni-Suef University Journal of Basic and Applied Sciences*, 9: 1–10.

Pandey, P. (2022). Role of nanotechnology in electronics: A review of recent developments and patents. *Recent Patents on Nanotechnology*, 16(1): 45–66.

Panes, P., Macariola, M. A., Niervo, C., Maghanoy, A. G., Garcia, K. P. and Ignacio, J. J. (2022). A bibliometric approach for analyzing the potential role of waste-derived nanoparticles in the upstream oil and gas industry. *Cleaner Engineering and Technology*, 8: 100468.

Park, W., Shin, H., Choi, B., Rhim, W. K., Na, K. and Han, D. K. (2020). Advanced hybrid nanomaterials for biomedical applications. *Progress in Materials Science*, 114: 100686.

Patel, P., Vyas, N. and Raval, M. (2021). Safety and toxicity issues of polymeric nanoparticles: A serious concern. *Nanotechnology in Medicine: Toxicity and Safety*, 156–173.

Patwardhan, S. V., Manning, J. R. and Chiacchia, M. (2018). Bioinspired synthesis as a potential green method for the preparation of nanomaterials: Opportunities and challenges. *Current Opinion in Green and Sustainable Chemistry*, 12: 110–116.

Pensupa, N., Leu, S., Hu, Y., Du, C., Liu, H Jing, H., Wang, H. and Lin, C. S. K. (2018). Recent trends in sustainable textile waste recycling methods: current situation and future prospects. *Chemistry and Chemical Technologies in Waste Valorization*, 189–228.

Phang, Y. K., Aminuzzaman, M., Akhtaruzzaman, M., Muhammad, G., Ogawa, S., Watanabe, A. and Tey, L. H. (2021). Green synthesis and characterization of CuO nanoparticles derived from papaya peel extract for the photocatalytic degradation of palm oil mill effluent (POME). *Sustainability*, 13(2): 796.

Pokrajac, L., Abbas, A., Chrzanowski, W., Dias, G. M., Eggleton, B. J., Maguire, S., Maine, E., Malloy, T., Nathwani, J., Nazar, L. and Sips, A. (2021). Nanotechnology for a sustainable future: Addressing global challenges with the international network4sustainable nanotechnology, 18608–18623.

Puntillo, P., Gulluscio, C., Huisingh, D. and Veltri, S. (2021). Reevaluating waste as a resource under a circular economy approach from a system perspective: Findings from a case study. *Business Strategy and the Environment*, 30(2): 968–984.

Rani, M., Yadav, J., Chaudhary, S. and Shanker, U. (2021). An updated review on synthetic approaches of green nanomaterials and their application for removal of water pollutants: Current challenges, assessment and future perspectives. *Journal of Environmental Chemical Engineering*, 9(6): 106763.

Sahajwalla, V. and Hossain, R. (2020). The science of microrecycling: A review of selective synthesis of materials from electronic waste. *Materials Today Sustainability*, 9: 100040.

Saleh, T. A. (2020). Nanomaterials: Classification, properties, and environmental toxicities. *Environmental Technology and Innovation*, 20: 101067.

Salem, S. S. and Fouda, A. (2021). Green synthesis of metallic nanoparticles and their prospective biotechnological applications: An overview. *Biological Trace Element Research*, 199: 344–370.

Salem, S. S., Hammad, E. N., Mohamed, A. A. and El-Dougdoug, W. (2022). A comprehensive review of nanomaterials: Types, synthesis, characterization, and applications. *Biointerface Res. Appl. Chem.*, 13(1): 41.

Samaddar, P., Ok, Y. S., Kim, K. H., Kwon, E. E. and Tsang, D. C. (2018). Synthesis of nanomaterials from various wastes and their new age applications. *Journal of Cleaner Production*, 197: 1190–1209.

Samant, M., Pandey, S. C. and Pandey, A. (2018). Impact of hazardous waste material on environment and their management strategies. In *Microbial Biotechnology in Environmental Monitoring and Cleanup* (pp. 175–192). IGI Global.

Saratale, R. G., Saratale, G. D., Shin, H. S., Jacob, J. M., Pugazhendhi, A., Bhaisare, M. and Kumar, G. (2018). New insights on the green synthesis of metallic nanoparticles using plant and waste biomaterials: current knowledge, their agricultural and environmental applications. *Environmental Science and Pollution Research*, 25: 10164–10183.

Sasidharan, S., Raj, S., Sonawane, S., Sonawane, S., Pinjari, D., Pandit, A. B. and Saudagar, P. (2019). Nanomaterial synthesis: Chemical and biological route and applications. In *Nanomaterials Synthesis* (pp. 27–51). Elsevier.

Seroka, N. S., Taziwa, R. T. and Khotseng, L. (2022). Extraction and synthesis of silicon nanoparticles (SiNPs) from sugarcane bagasse ash: A mini-review. *Applied Sciences*, 12(5): 2310.

Sharma, D., Kanchi, S. and Bisetty, K. (2019). Biogenic synthesis of nanoparticles: A review. *Arabian Journal of Chemistry*, 12(8): 3576–3600.

Shnoudeh, A. J., Hamad, I., Abdo, R. W., Qadumii, L., Jaber, A. Y., Surchi, H. S. and Alkelany, S. Z. (2019). Synthesis, characterization, and applications of metal nanoparticles. In *Biomaterials and Bionanotechnology* (pp. 527–612). Academic Press.

Singh, J., Dutta, T., Kim, K. H., Rawat, M., Samddar, P. and Kumar, P. (2018). 'Green' synthesis of metals and their oxide nanoparticles: Applications for environmental remediation. *Journal of Nanobiotechnology*, 16(1): 1–24.

Sinsinwar, S., Sarkar, M. K., Suriya, K. R., Nithyanand, P. and Vadivel, V. (2018). Use of agricultural waste (coconut shell) for the synthesis of silver nanoparticles and evaluation of their antibacterial activity against selected human pathogens. *Microbial Pathogenesis*, 124: 30–37.

Soto, K. M., Quezada-Cervantes, C. T., Hernández-Iturriaga, M., Luna-Bárcenas, G., Vazquez-Duhalt, R. and Mendoza, S. (2019). Fruit peels waste for the green synthesis of silver nanoparticles with antimicrobial activity against foodborne pathogens. *LWT*, 103: 293–300.

Sriram, P. and Suttee, A. (2020). Nanotechnology advances, benefits, and applications in daily life. In *Nanotechnology* (pp. 23–44). CRC Press.

Taran, M., Safaei, M., Karimi, N. and Almasi, A. (2021). Benefits and application of nanotechnology in environmental science: an overview. *Biointerface Research in Applied Chemistry*, 11(1): 7860–7870.

Tulebayeva, N., Yergobek, D., Pestunova, G., Mottaeva, A. and Sapakova, Z. 2020. Green economy: Waste management and recycling methods. In *E3S Web of Conferences* (Vol. 159, p. 01012). EDP Sciences.

Van Hai Le, C. N. H. and Thuc, H. H. T. (2013). Synthesis of silica nanoparticles from Vietnamese rice husk by sol–gel method. *Nanoscale Research Letters*, 8(1).

Vishwasrao, C., Momin, B. and Ananthanarayan, L. (2019). Green synthesis of silver nanoparticles using sapota fruit waste and evaluation of their antimicrobial activity. *Waste and Biomass Valorization*, 10: 2353–2363.

Xu, C., Nasrollahzadeh, M., Selva, M., Issaabadi, Z. and Luque, R. (2019). Waste-to-wealth: Biowaste valorization into valuable bio (nano) materials. *Chemical Society Reviews*, 48(18): 4791–4822.

Yap, Y. H., Azmi, A. A., Mohd, N. K., Yong, F. S. J., Kan, S. Y., Thirmizir, M. Z. A. and Chia, P. W. (2020). Green synthesis of silver nanoparticle using water extract of onion peel and application in the acetylation reaction. *Arabian Journal for Science and Engineering*, 45: 4797–4807.

Ying, S., Guan, Z., Ofoegbu, P. C., Clubb, P., Rico, C., He, F. and Hong, J. (2022). Green synthesis of nanoparticles: Current developments and limitations. *Environmental Technology and Innovation*, 26: 102336.

Zamare, D., Vutukuru, S. S. and Babu, R. (2016). Biosynthesis of nanoparticles from agro-waste: A sustainable approach. *Int. J. Eng. Appl. Sci. Technol.*, 1(12): 85–92.

Zeller, V., Towa, E., Degrez, M. and Achten, W. M. (2019). Urban waste flows and their potential for a circular economy model at city-region level. *Waste Management*, 83: 83–94.

CHAPTER 6

Scale-Up Process of Green Nanomaterials

Vennila Selvaraj,[1] *Suresh Sagadevan*[2] and *Gurunathan Karuppasamy*[1,*]

1. Introduction

The production of nanoparticles is a complex and developing area. An effective, environmentally responsible, and biocompatible approach to solving human issues is the manufacturing of green nanoparticles. These health issues pose the greatest threat to people. Due to their medicinal characteristics, herbal medicines and therapies have long been respected. Because of their therapeutic qualities, chemicals originating from plants can be employed to create nanoparticles that are helpful in the biomedical sector. Scaling up desired nanoparticle architectures is a typical issue in nanoparticle synthesis. Nanotechnology is a branch of science that focuses on nanostructured materials, which are generally between one and one hundredth nano-meter ranges in size. It operates at the nanoscale and provides a variety of essential insights into a wide range of scientific fields, including dentistry, medicine, and bioengineering (Rafique et al. 2017). Green chemistry will be crucial for future uses of nanomaterials. This area of nanoscience must develop environmentally friendly, risk-free NPs if it is to be accepted by the nano community (Varma et al. 2012). The dimension, shape, and physiochemical properties of incorporated particles are significantly influenced by the solvent and reducing operators used during the decrease of nanoparticles (NPs), and this influence affects how well NPs are used.

The two distinct methods utilized for NP amalgamation are "top-down" and "bottom-up." In the top-to-bottom process, a suitable bulk material is reduced in size using a variety of methods, including crushing, grinding, sputtering, and thermal

[1] Nanofunctional Material Lab, Department of Nanoscience and Technology, Science Campus, Alagappa University, Karaikudi, Tamil Nadu, India.
[2] Nanotechnology and Catalysis Research Centre, University of Malaya, 50603 Kuala Lumpur, Malaysia.
* Corresponding author: kgnathan27@rediffmail.com

or laser ablation. NPs are produced by the self-assembly of atoms into new nuclei, which then expand into nanosized particulars by the "bottom-to-top" technique, which employs biological and chemical processes (Mathur et al. 2017). Chemical reduction, electrochemical processes, and sono-decomposition are examples of "bottom-up" techniques. Recent developments claim to have produced metallic nanoparticles (NPs) that are really resistant to surfactants and other chemicals (Nadagouda et al. 2008). Traditional methods have been used for a while, but research has demonstrated that greener methods are more successful at manufacturing NPs because they reduce failure risks, are more affordable, and are simpler to describe (Zhang et al. 2016). Because of their toxic consequences, the physical and chemical procedures for making NPs include a number of environmental stresses. A plant-based biosynthesis of NPs that is finished in a few minutes to a few hours at typical room temperature was created by combining metallic salts and plant extracts. In the past ten years, the most intriguing aspect of this technique has been demonstrated by employing silver (Ag) and gold (Au) NPs, which are safer than some other metallic NPs. Green technologies are both practical for increasing NP output and prudent financial decisions. Due to their outstanding features, greenly linked NPs are increasingly chosen over conventionally administered ones. Using dangerous and destructive chemicals on the environment and humans can increase particle reactivity and toxicity, which may have unintended negative impacts on health (Abdelghany et al. 2018, Hussain et al. 2016). The creation of various metal nanoparticles utilizing bioactive elements, including plant matter, microbes, and a range of bio wastes, such as vegetable and fruit peels, shells, and agricultural waste, is called "green synthesis of nanomaterials." Researchers are looking at using microbes derived from plants and other biomaterials to manufacture metal nanoparticles to address the rising need for "green" and affordable manufacturing methods. The advantages of the green synthesis method are shown in Figure 1, and the many sources used to produce green synthesis nanoparticles are shown in Figure 2.

Material scientists have recently found carbon-based materials and mineral elemental mixes with prospective photoelectric and geometrical abilities that are better than most of their counterparts (Pérez-Beltrán et al. 2016, Goudarzi et al. 2019). The arrangement of carbon in liposomes, carbon nanotubes, dendrimers, and polymeric nanoparticles are instances of organic nanoparticles (NPs), in opposition to magnet, silver, and semiconductors NPs, which are instances of nanoparticles (NPs) (Rather et al. 2021, Chakka et al. 2006, Mafuné et al. 2004, Barbillon et al. 2014). Because it is challenging to pinpoint certain properties in isolated molecules, metallic NPs are crucial for study (Vossen et al. 1991). Phytoconstituents (leaf, flower, bark, seed, peel, etc.), microorganisms, bacteria, and enzymes for the creation of nanoparticles are examples of environmentally friendly substances that offer a variety of advantages of environmental and suitability for pharma and other biomedical (Nasrollahi et al. 2011). Additionally, these materials' production procedure did not involve any harmful byproducts. Microbes present in industrial and pharmaceutical processes are inhibited by nanoparticles for an extended period (Gokulakrishnan et al. 2012). The construction of metallic NP is a significant problem in nanotechnology, both theoretically and, more importantly, in "applied research" (Islam et al. 2013).

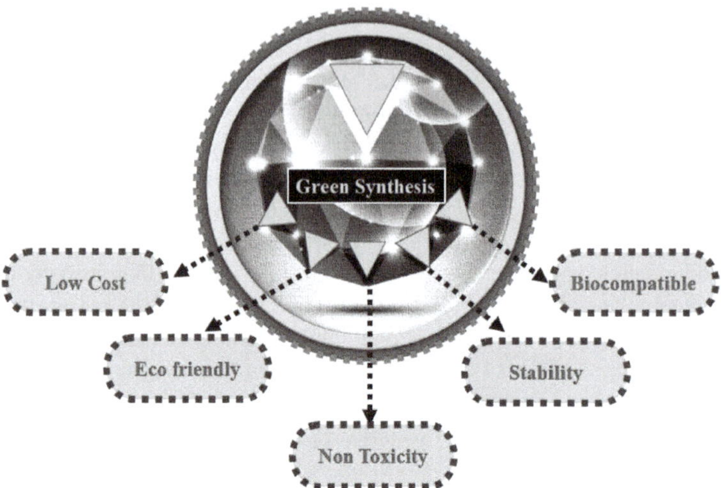

Figure 1. Advantages of green synthesis nanoparticles.

Figure 2. Green synthesis nanoparticles from various sources.

Many green synthesis techniques that employ reducing agents manufactured from eco-friendly materials are covered in this chapter.

2. Why are NPs Biologically Synthesized?

Because of their distinctive qualities, NPs created by biochemical functions are favored above those created through physicochemical mechanisms. Physical and

chemical processes can result in nanoparticles (Singh et al. 2015). These expensive techniques have a number of shortcomings, such as the use of risky solvents, the production of risky byproducts, and poor surface structures (Li et al. 2011). Various chemical species or compounds are routinely used in chemical processes, increasing particle reactivity or toxicity and posing risks to both the environment and human health (Li et al. 2011). Particles formed via green synthesis are different from those created by physicochemical methods. In a bottom-up method known as green synthesis, metal or oxide nanoparticles are created using a natural extraction from a commodity, such as fruit or leaf from a tree or crop, rather than a costly chemical reducing agent. The ability of biological entities to produce NPs is very promising. Economical (Mittal et al. 2013), environmentally friendly (Dhandapani et al. 2012), durable (Gopinath et al. 2014), free of toxic contamination (Chandran et al. 2006), and suitable for mass manufacturing (Huang et al. 2007) are all attributes of the bioactive reduction of the metal precursor to the matching NPs. Additionally, the recycling of pricey metal salts present in waste streams, including gold and silver, is made possible by the biological creation of NPs. These metals have limited deposits, making their prices unstable (Mittal et al. 2013). The needed traits are present in green NPs. Due to the biological components that stabilize nanoparticles, such as proteins, enzymes, carbohydrates, and entire cells, may interact easily with other biomolecules (Iravani 2011). Their interactions with bacteria are enhanced, and their antimicrobial properties are increased. Since they are created biologically, nanoparticles are easily removed from the reaction media or condensed by centrifugation (Wang et al. 2009). When compared to NPs made chemically, the antibacterial activity of biologically generated silver is 20 times higher (Botes 2010). Based on the biological components' intrinsic value, plant extracts are employed to make NPs. Spirulina platensis algal cells were selected because they offer medicinal and nutraceutical benefits in addition to containing a reducing agent (Sintubin et al. 2009, Sintubin et al. 2011, Govindaraju et al. 2008).

3. Methods of Synthesis

The three primary classified techniques for producing nanoparticles are physical, chemical, and biological synthesis. These methods have advantages and disadvantages in terms of yield, scalability, particle density and distribution, shape uniformity, and associated costs. In Figure 3, samples of the synthesizing are displayed. Metal NPs have been made using mechanical, chemical, and biological techniques (Krishnappa et al. 2003, Winterer et al. 2003).

3.1 Green Reducing Scale-Up Synthesis Process

Plants provide promise for heavy metal detoxification. However, further study is still being done on the production of NP from plants. In comparison to microbial synthesis, the use of plant-based NP synthesis, despite ongoing research, holds significant promise compared to the established method of microbial synthesis. These advantages include an increase in the speed of synthesizing, an improvement in dispersibility, more steady NPs, and the lack of a complex treatment process (such

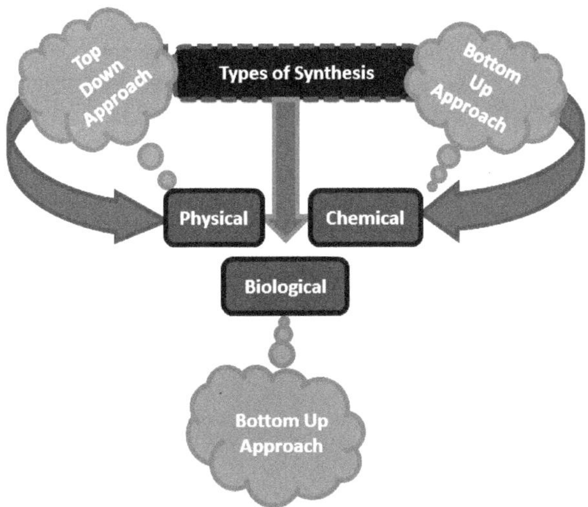

Figure 3. Types of nanoparticles synthesis.

as microorganisms solitude, cell cultures, and maintenance), which is necessary for microbial synthesis. The only reactive substances that lower a plant's capacity are terpenes, carbohydrates, tannins, sugars, protein, and amino acids. Numerous plant components, including seeds, blossoms, leaves, and stems, can also be used in the bio-reduction process (Iravani et al. 2011). Heavy metals may assemble in plants in a variety of ways and to variable degrees. As a result, methods of biosynthesis utilizing plant extracts have come to be recognized as useful, easy-to-use, and affordable alternatives to traditional techniques for creating nanoparticles. A "one-pot" synthesis process may be used to decrease and stabilize metallic nanoparticles utilizing a variety of plants. To further investigate the many uses of metal/metal oxide nanoparticles, several researchers have used green synthesis strategies to create particles utilizing plant leaf extracts. Plants may decrease metal ions on their surface and in numerous organs and tissues far from the ion penetrating point. Plants, especially those with significant metal-ions hyperaccumulating and reduction capacities, are important in this respect. Many different biological molecules with antioxidant properties may be found in plant extracts. The biomolecules crucial for the bio-reduction, modification, and stabilization of nanoparticles include phenolic acids, starches, sugars, terpenoids, phytosterols, alkaloids, polyphenols, enzymes, proteins, and coenzymes. The ability of phytochemicals to function as reduction agents or singlet oxygen scavengers, thus decreasing metal salts to nanoparticles by scavenging electrons from them, is largely responsible for their antioxidant capabilities (Sekatawa et al. 2021).

Different functional groups found in flavonoids have an improved capacity to decrease metal ions. Due to tautomeric changes in flavonoids, where the enol form is transformed into the keto-form, the reactive hydrogen atom is liberated. Metal ions are reduced into metal nanoparticles to carry out this operation. Enol-to-keto transformation in sweet basil (*Ocimum basilicum*) extracts is crucial for producing biogenic silver nanoparticles (Ahmad et al. 2010). Biomolecules have the ability to transform metal salts like coenzymes, proteins, and carbohydrates into nanoparticles. For the creation of AuNPs and Ag-NPs, plant extract-assisted synthesis was initially investigated, similar to previous biosynthesis strategies. Figure 4 includes the decrease of green source extracts and the creation of metal or metal oxide ions. In Table 1, Various nanoparticles derived from different green sources were compared to improve the large-scale synthesis analysis.

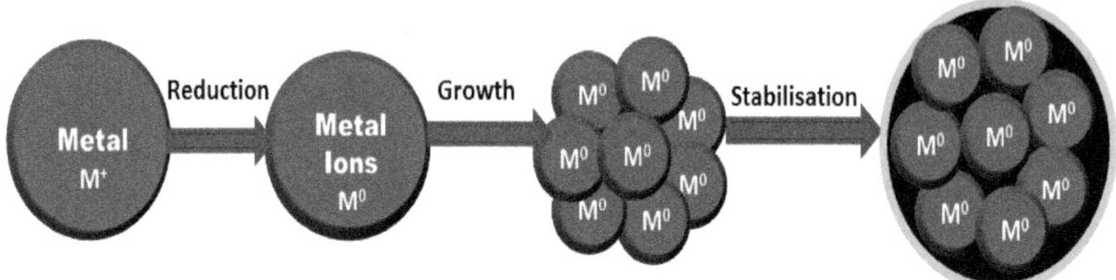

Figure 4. Metal/metal oxide ion formation from green source extracts.

Table 1. Comparative analysis of green synthesis nanoparticles from various plant sources.

S. No.	NP	Plant part	Size	Structure	Application	Reference
1.	Au	*Gelidium pusillum*	12 nm	Spherical	Anticancer activity	(Jeyarani et al. 2020)
2.	Au	*Litsea cubeba*	18 nm	Spherical	Catalytic reduction of 4-nitrophenol	(Doan et al. 2020)
3.	Ag	*Cestrum nocturnum*	20 nm	Spherical	Antioxidant and antibacterial	(Keshari et al. 2020)
4.	Pd	*Cotton boll peels*	9 nm	Spherical	Catalytic activity against toxic azo-dyes	(Narasaiah et al. 2020)
5.	TiO_2	*Lemon peel extract*	80 nm	Spherical	Photocatalytic activity	(Nabi et al. 2020)
6.	Pt	*Tragia involucrata*	10 nm	Spherical	Biomedical and pharmaceutical application	(Selvi et al. 2020)
7.	ZnO	*Calotropis gigantea*	31 nm	Hexagonal and pyramidal	Nitrite sensing, photocatalytic, and antibacterial	(Kumar et al. 2020)
8.	Cu	*Anacardium occidentale*	< 20 nm	Irregular spherical	Efficient removal of uranium	(Chandra et al. 2020)
9.	Fe	*Cupressus sempervirens leaf*	9 nm	Nanoclusters	Dye removal	(Ebrahiminezhad et al. 2018)
10.	Fe	*Eucalyptus leaf*	20 nm	Spheroidal	Eutrophic wastewater	(Wang et al. 2014)
11.	NZVI	*Lemon* (Citrus Limon)	3–300 nm	Spherical cylindrical	Removal of Cr (VI)	(Sekatawa et al. 2021)

3.2 Large-Scale Solar Cell Manufacture

The process of scaling up synthesis and characterization (Smitha et al. 2009) is challenging and involves creating nanostructures for bigger or commercial manufacturing *V. agnus-castus* nanoparticles were used to produce green BiI_3 and AgI particles. The nanomaterials of BiI_3 and AgI caused the response time between BiI_3 and MAI, in addition to AgI and BiI_3 to be accelerated when the area of contact between BiI_3 and MBI and AgI and BiI_3 increased. Using an environmentally friendly method that uses *Vitex agnus-castus* plant extract, BiI_3-NPs, and AgI-NPs were created (VAC). Under illumination circumstances and high relative humidity, a low-cost HTM-free solar cell was later built utilizing charcoal as the opposing electrode and perovskite layers made of $AgBi_2I_7$, $(CH_3NH_3)_3Bi_2I_9$, and BiI_3 (Arjmand et al. 2020). The aloe vera extract, $PbCl_2$, and anhydride were utilized to create lead sulfide (PbS) nanoparticles using the chemical precipitation technique.

According to the C-V analysis, the film formed with a mg mL 1 solution solution guarantees a better interface between the perovskite and nearby metal contacts. With Voc = 0.94 V, Jsc = 27.75 mA cm^2, and FF = 76%, the numerical simulation produced the highest PCE of 27.1%. According to the results, PbS nanoparticles may be used as an HTL to produce stable, highly efficient solar cells (Islam et al. 2022). It has been asserted that a WO3-based clock in DSSC has a high energy conversion (Hamideh et al. 2018). DSSCs produced a substantial PCE of 7.61% using a zinc thiosulfate CE catalyst, which is equivalent to a Pt-sputtered CE's (7.04%) value. Manganese tungstate-carbon electrodes from DSSCs and copper tungstate-carbon electrodes showed good PCEs of 7.33% and 6.52%, respectively. The method used to produce carbon nanoparticles from biomass is clearly shown in Figure 5. During large-scale green synthesis, huge batches may be used to swiftly, frequently, and scalably flow plant extracts in amounts ranging from a few milliliters to several liters. However, only a small number of batches are widely accessible commercially. Future developments in vital machinery are expected to fundamentally alter the environmentally friendly production of metallic nanoparticles (Silva et al. 2015).

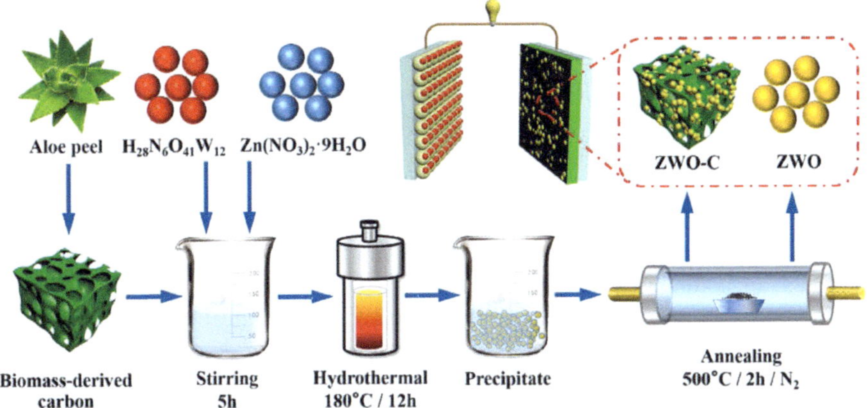

Figure 5. Biomass-derived carbon tungsten oxide for DSSC (Copyright from figure by Hamideh et al. 2018).

3.3 Wastewater Treatments for Green Products

Every day, large numbers of eggshells are created as biowaste across the world. In addition to roughening eggs and enticing flies, its odor causes a variety of excellent benefits to be lost (Hamideh et al. 2018). Bakeries, restaurants, and homes produce most of the eggshell waste. The majority of eggshells, which may be manufactured into expensive things, are formed entirely of calcium carbonate and are not porous. The sol-gel technique may be used to create nano-calcium oxide from used eggshells (Habte et al. 2019). The eggshell membrane is utilized to manufacture magnetic $CuFe_2O_4$ particles with multifunctional capabilities, including catalytic and bacterial abilities, which are exploited in industrial wastewater treatment because of their hierarchical and porosity structure (Zhang et al. 2021). As a result, a less expensive catalyst system that may be used for oxidization, hydrogen evolution processes, hydrolysis, and the degradation of pollutants has been created. This has favorable effects since it is affordable, ecologically responsible, and waste-free (Orooji et al. 2022). It comes from a natural source as well.

4. Green Biocatalytic Synthesis of Nanoscale Materials

4.1 Product Optimization

Modifying one variable at a time, such as air temp, time, pH, mineral composition, and inoculums, while keeping both these variables constant, is a conservative method for optimizing the biogenic synthesis of nanoparticles that is challenging, dull, time-consuming, ineffective, and incapable of control the mutual interactive properties of different variables. The material produced using plant extracts and plant metabolites demonstrated the stability of NPs (in an aqueous medium). After the creation of the metal nanoparticles, zeta potentiometers were used to measure the stability of the Ag-NPs produced through biosynthesis at various pH levels. The stability of the synthesized Ag-NPs across a wide pH range, from 6 to 12, can be recognized. Ag-NPs' zeta potential increases with increasing pH, going from 22.3 mV to 65.0 mV. Ag-NPs that have been synthesized by the use of leaf extract have strong negative zeta potential values, making them stable across a wide pH range (Sadeghi et al. 2015). The most extensively changed mathematical and arithmetic program for optimizing numerous concurrently changing elements is called surface response methodology (RSM). The pairwise cumulative influence of the elements on the answer may be inferred using RSM (Kong et al. 2014). Three multivariate strategies that may be utilized in experimental designs are the Box-Behnken Design (BBD), Central Composite Design (CCD), or Doehlert Matrix (DM). CCD is more efficient in construction than BBD and DM (Jung et al. 2015) – one more example of optimization of surface area to play various applications. Calcium phosphates (CaPs) are common in nature and the bones and teeth of vertebrates. They are highly biocompatible and show promise in a number of biological sectors. Due to their high specific surface area, pH-responsive degradability, high capacity for drug/gene/protein loading, and sustained release performance, nanostructured calcium phosphates (NCaPs) are regarded as potential nanocarriers for the delivery of drugs, genes, and proteins. Numerous biomolecules with high biocompatibility, including nucleic acids, proteins, peptides, liposomes, and biomolecules containing phosphorus, are employed in the manufacture of NCaPs in order to control the structure and surface characteristics. The development of distinct NCaPs with varying sizes and morphologies is also a result of the crucial functions that biomolecules play in the synthesis processes (Qi et al. 2019). The optimization of nanoparticles in various aspects is shown in Figure 6.

The fabrication of nanoparticles has frequently exploited photochemical sources as reducing and capping agents. (NPs). It is necessary to design a method in which specific and optimized photochemicals produce the needed NPs in less time and at a lower cost with a greater repeatability rate. However, morphology-controlled synthesis and shape/size-dependent uses of these NPs need to be researched further. One example, Fe_3O_4 NPs, synthesis chemistry must inevitably move toward environmentally friendly practices, such as those that use plant extracts. Due to the complexity of the components that make up plant extracts, it is difficult to manage size and shape. Compounds such as amino acids, polyphenols, nitrogenous bases, and reducing sugars are abundant in plant extracts. In the creation of magnetite nanoparticles, these compounds act as reducing and stabilizing agents to control size and morphology (López et al. 2020). Another example is in order to create eco-friendly europium titanate ($Eu_2Ti_2O_7$) nanoparticles, natural capping agents such as glucose, fructose, galactose, lactose, and starch were used in an aqueous solution. $Eu_2Ti_2O_7$ controlled the shape and particle size of these nanoparticles (Sobhani-Nasab et al. 2016).

The crystallinity of green nanoparticles one reported results in simple aqueous conditions; crystalline bismuth nanoparticles (BiNPs) were synthesized quickly, cheaply, and environmentally friendly using lemon juice as a reducing and capping

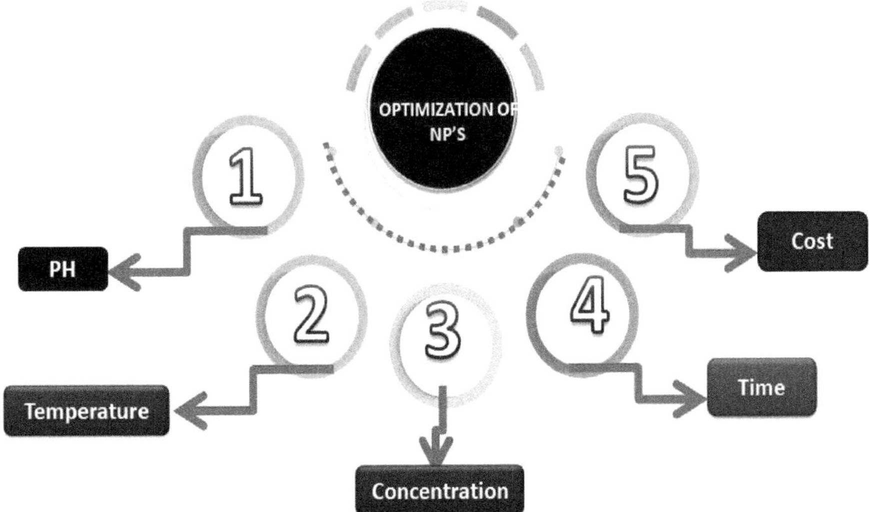

Figure 6. Optimization of nanoparticles in various aspects.

agent. BiNPs are spherical, have a rhombohedral crystalline structure, are stabilized in aqueous media, and are capped with photochemical. The $NaBH_4$ and BiNPs successfully catalyze the conversion of 4-nitrophenol to 4-aminophenol. We were eager to learn more about the catalytic performance of BiNPs in the reduction of nitroaromatic compounds, a problem pollutant that is produced by explosives, analgesic and antipyretic medications, and dyes. This was due to the successful production of BiNPs (Mahiuddin et al. 2021). These are important parameters to improve large-scale synthesis.

4.2 Low Temperature

Using a scalable, green, low-temperature, one-step technique, PdNPs were synthesized as spherical particles with customizable diameters and constant dispersion. The suggested process uses water as a solvent, CS as a stabilizing agent, and vitamin C as a reducing agent. As the temperature rose from 20°C to 95°C, the PdNPs produced experienced a spherical shape change and hydrodynamically shrunk from 64 nm to 29 nm. Regarding size, shape, and aqueous solution stability, the resultant PdNPs were mostly homogenous (Saleh et al. 2018a). Since higher temperatures cause proteins to evaporate, bio-nanoparticles cannot be used for applications like pharmaceuticals, solar cells, or thin film materials; temperature is a crucial component in disclosing nanoparticles, especially for green nanomaterials. Low temperatures are employed for bio-capped materials since biomolecules only have antibacterial and antibiotic properties.

4.3 Preservation of Nanomaterials

Food items' color, flavor, texture, taste, nutritional content, and shelf life are all impacted by these enzymatic and chemical interactions. Oxygen and light are two more variables that have an impact on food quality. Similar hydrolytic lipid mechanisms lead to rancidity. In the presence of air and other oxidizing agents, unsaturated fat is transformed into peroxide molecules. This is avoided by e-donating, reductive, and antioxidant substances. Some pathogenic bacteria that cause spoilage may produce different toxins or toxic cell parts to human health. So, shielding food from illness and microbial activity may be a good food preservation approach. Numerous physical and chemical techniques can be used to get rid of microorganisms. The most effective physical method for this is heating. However, heating at a high temperature may alter the nutritional value of the food. In order to solve this issue, barrier technology is now being used. Microorganisms can be killed by a number of selective nanoparticles (Ghosh et al. 2019).

5. Large-Scale Production of Nanoparticles

5.1 Manufacture Risk of Nanomaterials

The field provides guidelines for the proper use of nanomaterials manufactured using a green chemical approach and ensures that the atmosphere is eco-friendly for scientists, employees, and customers. The risk associated with producing nanoparticles may be evaluated and controlled using the following approach:

- ✓ Specify the use of the nanoparticles.
- ✓ Analyze the chemical composition, solubility, crystallinity, porosity, density, and surface area of nanomaterials, as well as their size distribution and mean size.
- ✓ Determine the interaction between the nanomaterial and the biological system at the cellular, tissue, molecular, systemic, and ecological levels using *in silico, in vivo,* and *in vitro* experiments.
- ✓ It is crucial to inform and educate the public on controlling nanostructured materials and nanoparticles (Saleh et al. 2018b).

5.2 *Engineered Bionanomaterials*

The finest substitutes for conventional particles for biomedical purposes have been found to be bionanomaterials. Nanomaterials created from biomolecules or used to encapsulate or immobilize other kinds of nanomaterials are known as bionanomaterials. Band gap tuning and optical properties so interplay play between the Biomolecule and metal nanoparticle toward the preparation of the green synthesis process. One example of a band gap playing the role of green-silver nanoparticles was created via a process called plant-mediated synthesis. As a reducing and stabilizing agent, *cyanococcus* fruit extract was utilized to create silver nanoparticles. The GO-Ag nanocomposites were created by incorporating these biosynthesized nanoparticles into the graphene matrix, switching from an insulating to a semiconducting optical band gap. This approach is suggested as a green endeavor for creating semiconducting nanocomposite materials with increased applicability (Joy et al. 2021). Bionanomaterials are created using biomolecules that have been isolated from microbes, plants, agricultural runoff, bugs, aquatic life, and some animals. Due to their enhanced biocompatibility, bioavailability, and bio-reactivity, these bionanomaterials exhibit minimal or insignificant toxicity toward people, other living things, and the environment. When creating bio/nanomaterials for nanomedicine, two of the primary problems to take into account are the stability and safety of (nonviral) nanocarriers. The development of membranes that replicate biological habitats and tumor microenvironments, the reduction of off-target deposition and trafficking, and promoting behavior only in response to the right cues are further obstacles. Immunogenicity and systemic toxicity are another. The many methods depend on crucial elements, including pinpointing the exact effectiveness goal, tissue for exposure, patient or study model safety, and directly applying benchtop work to a possible clinical or commercial environment. The use of functional materials has handled the bulk of these important issues, i.e., surveillance, diagnosis, and treatment, either via releasing control at the proper location or a hand-in-glove strategy for treatment, which might make it easier to monitor the impact in real time. All of these, in addition to just being facilitated by dynamic fictionalization, complex chemical responses with scheduled process parameters that allow tracking of the intricate metabolic (or deterioration) product lines, nevertheless combine to raise costs, production times, and regulatory hurdles that ultimately affect the price of the final market output.

6. Challenges

Ecologically friendly, durable, and greener solutions have drawn a lot of attention as preferred and commonly utilized technology when ramping up and transporting into commercial or industrial uses (Saleh et al. 2018). These solutions include developed plants, microorganisms, and contemporary biopharmaceuticals (Paliwal et al. 2014). NP synthesis has undergone breakthroughs. A nanomedicine item is further developed with numerous phases, scaling up processes (dependent on precursors or specific uses), and approvals before the ultimate market release, as noted in a number of research studies (Paliwal et al. 2014) on comment optimization. Massive information is substituted when promoting multidisciplinary scientific studies in therapeutic settings (as in the case of nano-drugs); however, commercial scale-up is still constrained by the requirement for large-scale financing, exceptional accuracy, and cost-effectiveness.

7. Future Perspective of Green Nanomaterials

This chapter will contribute to the creation of unique, enhanced methods for producing nanoparticles from biomaterials as well as fresh ideas for reaching the requisite yield in a cost-effective, environmentally friendly, and secure way. Biomaterial quantification enhances the product's reproducibility and predictability. The hazards present before, throughout, and after the synthesis of nanoparticles can be addressed before any negative effects. As a result, the search for efficient, economical, and long-lasting illness treatment options will have a greater impact on the biomedical and pharmaceutical sectors.

8. Conclusion

The green synthesis is a more dependable, environmentally responsible, simple, and cost-effective source of nanoparticles than chemical and physical methods. There are several factors that might affect the kind, quantity, characteristics, and usage of nanoparticles. Depending on the kind, location, and makeup of the leaf extracts used to ionize metallic ions and produce the required kind of nanoparticles, they perform in various ways. By modifying and enhancing the characteristics of nanoparticles, repeatability and predictability are increased. It also helps with risk evaluation and management. Due to their varied bionanomaterials and distinctive features, nanoparticles are valuable in a variety of sectors.

References

Abdelghany, T. M., Al-Rajhi, A. M. H. and Al Abboud, M. A. (2018). Recent advances in green synthesis of silver nanoparticles and their applications: about future directions. A review. *Bionanoscience*, 8: 5–16.

Ahmad, N., Sharma, S., Alam, M. K., Singh, V. N., Shamsi, S. F., Mehta, B. R. and Fatma, A. (2010). Rapid synthesis of silver nanoparticles using dried medicinal plant of basil. *Colloids and Surfaces B: Biointerfaces*, 81(1): 81–86.

Arjmand, F., Golshani, Z., Fatemi, S. J., Maghsoudi, S., Naeimi, A. and Hosseini, S. M. (2022). The lead-free perovskite solar cells with the green synthesized BiI3 and AgI nanoparticles using Vitex agnus-castus plant extract for HTM-free and carbon-based solar cells. *Journal of Materials Research and Technology*, 1; 18: 1922–33.

Barbillon, G., Hamouda, F. and Bartenlian, B. (2014). Large surface nano structuring by lithographic techniques for bioplasmonic applications. In *Manufacturing Nanostructures; One Central Press: Altrincham*, UK, pp. 244–262.

Botes, M. and Cloete, T. E. (2010). The potential of nanofibers and nanobiocides in water purification. Crit. Rev. Microbiol., 36: 68–81.

Chakka, V. M., Altuncevahir, B., Jin, Z. Q., Li, Y. and Liua, J. P. (2006). Magnetic nanoparticles produced by surfactant-assisted ball milling. *J. Appl. Phys.*, 99: 08E912.

Chandra, C. and Khan, F. (2020). Nano-scale zerovalent copper: Green synthesis, characterization and efficient removal of uranium. *J. Radioanal. Nucl. Chem.* https://doi.org/10.1007/s1.

Chandran, S. P., Chaudhary, M., Pasricha, R., Ahmad, A. and Sastry, M. (2006). Synthesis of gold nanotriangles and silver nanoparticles using Aloe vera plant extract. *Biotechnol. Prog.*, 22: 577–583.

Dhandapani, P., Maruthamuthu, S. and Rajagopal, G. (2012). Bio-mediated synthesis of TiO_2 nanoparticles and its photocatalytic effect on aquatic biofilm. *J. Photochem. Photobiol. B*, 110: 43–49.

Doan, V. -D., Thieu, A. T., Nguyen, T. -D. et al (2020). Biosynthesis of gold nanoparticles using Litsea cubeba fruit extract for catalytic reduction of 4-Nitrophenol. *J. Nanomater.*, 2020: 4548790. https ://doi.org/10.1155/2020/4548790.

Ebrahiminezhad, A., Taghizadeh, S., Ghasemi, Y. and Berenjian, A. (2018). Green synthesized nanoclusters of ultra-small zero valent iron nanoparticles as a novel dye removing material. *Science of the Total Environment*, 621: 1527–1532.

Ghosh, C., Bera, D. and Roy, L. (2019). Role of nanomaterials in food preservation. *Microbial Nanobionics*, 181–211.

Gokulakrishnan, R., Ravikumar, S. and Raj, J. A. (2012). *In vitro* antibacterial potential of metal oxide nanoparticles against antibiotic resistant bacteria pathogens. *Asian Pacific Journal of Tropical Disease*, 2(5): 411–413.

Gopinath, K., Shanmugam V. K., Gowri, S., Senthilkumar, V., Kumaresan, S. and Arumugam, A. (2014). Antibacterial activity of ruthenium nanoparticles synthesized using Gloriosa superba L. leaf extract. *J. Nanostruct. Chem.*, 4: 83.

Goudarzi, M., Salavati-Niasari, M., Yazdian, F. and Amiri, M. (2019). Sonochemical assisted thermal decomposition method for green synthesis of $CuCO_2O_4$/CuO ceramic nanocomposite using dactylopius coccus for anti-tumor investigations. *J. Alloy. Compd.*, 788: 944–953.

Govindaraju, K., Basha S. K., Kumar, V. G. and Singaravelu, G. (2008). Silver, gold and bimetallic nanoparticles production using single-cell protein (Spirulina platensis). *J. Mater. Sci.*, 43: 5115–5122.

Habte, L., Shiferaw, N., Mulatu, D., Thenepalli, T., Chilakala, R. and Ahn, J. (2019). Synthesis of nano-calcium oxide from waste eggshell by sol-gel method. *Sustainability*, 11(11): 3196.

Hamideh, F. and Akbar, A. (2018). Application of eggshell wastes as valuable and utilizable products: A review. Res. Agric. Eng., 64(2): 104–114.

Huang, J., Li, Q., Sun, D., Lu, Y., Su, Y. and Yang, X. (2007). Biosynthesis of silver and gold nanoparticles by novel sun dried Cinnamomum camphora leaf. *Nanotechnology*, 18: 105104–105115.

Hussain, I., Singh, N. B. and Singh, A. (2016). Green synthesis of nanoparticles and its potential application. *Biotechnol. Lett.*, 38: 545–560.

Iravani, S. (2011). Green synthesis of metal nanoparticles using plants. *Green Chem.*, 13: 2638–2650.

Islam, M. and Islam, M. S. (2013). Electro-deposition method for platinum nano-particles synthesis. Saidul, electro-deposition method platin. *Nano-Particles Synth. Eng. Int.*, 1: 2.

Islam, M. A., Sarkar, D. K., Shahinuzzaman, M., Wahab, Y. A., Khandaker, M. U., Tamam, N., Sulieman, A., Amin, N. and Akhtaruzzaman, M. (2022). Green synthesis of lead sulphide nanoparticles for high-efficiency perovskite solar cell applications. *Nanomaterials*, 5; 12(11): 1933.

Jeyarani, S., Vinita, N. M., Puja, P., Senthamilselvi, S., Devan, J., Velangani, A. J., Biruntha, M., Pugazhendhi, A. and Ponnuchamy, K. (2020). Biomimetic gold nanoparticles for its cytotoxicity and biocompatibility evidenced by fuorescence-based assays in cancer (MDA-MB-231) and noncancerous (HEK-293) cells. *J. Photochem. Photobiol. B Biol.* https://doi.org/10.1016/j.jphotobiol.2019.111715.

Joy, J., Gurumurthy, M. S., Thomas, R. and Balachandran, M. (2021). Biosynthesized Ag nanoparticles: A promising pathway for bandgap tailoring. *Biointerface Res. Appl. Chem.*, 11(2): 8875–8883.

Jung, J. P., Hu, D., Domian, I. J. and Ogle, B. M. (2015). An integrated statistical model for enhanced murine cardiomyocyte differentiation via optimized engagement of 3D extracellular matrices. *Sci. Rep.*, 5: 18705.

Keshari, A. K., Srivastava, R., Singh, P., Yadav, V. B. and Nath, G. (2020). Antioxidant and antibacterial activity of silver nanoparticles synthesized by *Cestrum nocturnum*. *J. Ayurveda Integr. Med.*, 1: 37–44. https://doi. org/10.1016/j.jaim.2017.11.003.

Kong, Y. (2014). Medium optimization for the production of anti-cyanobacterial substances by Streptomyces sp. HJC-D1 using response surface methodology. *Environ. Sci. Pollut. Res.*, 21: 5983–5990.

Krishnappa, K., Pandiyan, J., Paramanandham, J. and Elumalai, K. (2003). Antibacterial and mosquitocidal potentials of selected Indian medicinal plants extracts and synthesized silver nanoparticles. Int. J. Zool Studies, 3: 15–5.

Kumar, C. R. R., Betageri, V. S., Nagaraju, G., Suma, B. P., Kiran, M. S., Pujar, G. H. and Latha, M. S. (2020). One-pot synthesis of ZnO nanoparticles for nitrite sensing, photocatalytic and antibacterial studies. *J. Inorg. Organometall. Polym. Mater.* https://doi. org/10.1007/s10904-020-01544-3.

Li, X., Xu, H., Chen, Z. S. and Chen, G. (2011). Biosynthesis of nanoparticles by microorganisms and their applications. *J. Nanomate.*, 2011: 1–16.

López, Y. C. and Antuch, M. (2020). Morphology control in the plant-mediated synthesis of magnetite nanoparticles. *Current Opinion in Green and Sustainable Chemistry.* doi:10.1016/j.cogsc.2020.02.001.

Mafuné, F., Kohno, J. Y., Takeda, Y. and Kondow, T. (2002). Full physical preparation of size-selected gold nanoparticles in solution: laser ablation and laser-induced size control. *J. Phys. Chem. B*, 106: 7575–7577.

Mahiuddin, M. and Ochiai, B. (2021). Green synthesis of crystalline bismuth nanoparticles using lemon juice. *RSC Advances*, 11(43): 26683–26686.

Mathur, P., Jha, S. and Ramteke, S. (2017). Pharmaceutical aspects of silver nanoparticles. *Artif. Cells Nanomedicine Biotechnol.*, 46: 1–12.

Mittal, A. K., Chisti, Y. and Banerjee, U. C. (2013). Synthesis of metallic nanoparticles using plant extracts. Biotechnol. Adv., 31: 346–356.

Nabi, G., Ain, Q. -U., Tahir, M. B., Khalid, N. R., Iqbal, T., Rafique, M., Hussain, S., Raza, W., Aslam, I. and Rizwan, M. (2020). Green synthesis of TiO_2 nanoparticles using lemon peel extract: their optical and photocatalytic properties. *Int. J. Environ. Anal. Chem.* https://doi. org/10.1080/030673 19.2020.1722816.

Nadagouda, M. N. and Varma, R. S. (2008). Green synthesis of silver and palladium nanoparticles at room temperature using coffee and tea extract. *Green Chem.*, 10: 859–862.

Narasaiah, B. P. and Mandal, B. K. (2020). Remediation of azo-dyes based toxicity by agro-waste cotton boll peels mediated palladium nanoparticles. *J. Saudi Chem. Soc.*, 24: 267–281. https://doi. org/10.1016/j.jscs.2019.11.003.

Nasrollahi, A., Paurshamsian, K. H. and Mansourkiaee, P. (2011). Antifungal activity of silver nanoparticles on some of fungi. *International Journal of Nano Dimension*, 1(3): 233–239.

Orooji, Y. (2022). Valorisation of nuts biowaste: Prospects in sustainable bio(nano)catalysts and environmental applications. *J. Clean Prod.*, 347: 131220.

Paliwal, R., Jayachandra Babu, R. and Palakurthi, S. (2014). Nanomedicine scale-up technologies: Feasibilities and challenges. *AAPS Pharm. Sci. Tech.*, 15: 1527–34.

Pérez-Beltrán, C. H., García-Guzmán, J. J., Ferreira, B., Estevez-Hernandez, O., Lopez-Iglesias, D., Cubillana-Aguilera, L., Link, W., Stănică, N., da Costa, A. M. R. and Palacios-Santander, J. M. (2021). One-minute and green synthesis of magnetic iron oxide nanoparticles assisted by design of experiments and high energy ultrasound: Application to bio sensing and immunoprecipitation. *Mater. Sci. Eng. C*, 123: 112023.

Qi, C., Musetti, S., Fu, L. -H., Zhu, Y. -J. and Huang, L. (2019). Biomolecule-assisted green synthesis of nanostructured calcium phosphates and their biomedical applications. *Chemical Society Reviews*. doi:10.1039/c8cs00489g.

Rafique, M., Sadaf, I. and Rafique, M. S. (2017). A review on green synthesis of silver nanoparticles and their applications. *Artif Cells Nanomedicine Biotechnol.*, 45: 1272–1291.

Rather, G. A., Nanda, A., Chakravorty, A., Hamid, S., Khan, J., Rather, M. A., Bhattacharya, T., Rahman, M. H. (2021). Biogenic green synthesis of nanoparticles from living sources with special emphasis on their biomedical applications. *Res. Sq.*

Sadeghi, B. and Gholamhoseinpoor, F. (2015). A study on the stability and green synthesis of silver nanoparticles using Ziziphora tenuior (Zt) extract at room temperature. *Spectrochimica Acta Part A: Molecular and Biomolecular Spectroscopy*, 134: 310–315. doi:10.1016/j.saa.2014.06.046.

Saleh, N. and Yousaf, Z. (2018a). Tools and techniques for the optimized synthesis, reproducibility and scale up of desired nanoparticles from plant derived material and their role in pharmaceutical properties. pp. 86–131. *In*: Grumezescu, A. M. (ed.). Nanoscale Fabrication, Optimization, Scale-Up and Biological Aspects of Pharmaceutical Nanotechnology (Amsterdam: Elsevier).

Saleh, Nadia. (2018b). Nanoscale Fabrication, Optimization, Scale-Up and Biological Aspects of Pharmaceutical Nanotechnology Tools and techniques for the optimized synthesis, reproducibility and scale up of desired nanoparticles from plant derived material and their role in pharmaceutical properties, 85–131.

Samuel, M. S., Jose, S., Selvarajan, E., Mathimani, T. and Pugazhendhi, A. (2020). Biosynthesized silver nanoparticles using Bacillus amyloliquefaciens; Application for cytotoxicity effect on A549 cell line and photocatalytic degradation of p-nitrophenol. *J. Photochem. Photobiol. B Biol.*, 202: 111642.

Samuel, M. S., Sumanb, S., Venkateshkannanc Selvaraj, E., Mathimanie, T. and Pugazhendhif, A. (2020). Immobilization of $Cu_3(btc)_2$ on graphene oxide-chitosan hybrid composite for the adsorption and photocatalytic degradation of methylene blue. *J. Photochem. Photobiol. B Biol.*, 204: 111809.

Sekatawa, K., Byarugaba, D. K., Kato, C. D., Wampande, E. M., Ejobi, F., Nakavuma, J. L., Maaza, M., Sackey, J., Nxumalo, E. and Kirabira, J. B. (2021). Green strategy–based synthesis of silver nanoparticles for antibacterial applications. *Frontiers in Nanotechnology*, 3: 697303.

Selvi, A. M., Palanisamy, S., Jeyanthi, S., Vinosha, M., Mohandoss, S., Tabarsa, M., You, S. G., Kannapiran, E. and Prabhu, N. M. (2020). Synthesis of Tragia involucrata mediated platinum nanoparticles for comprehensive therapeutic applications: Antioxidant, antibacterial and mitochondria-associated apoptosis in HeLa cells. *Process Biochem*. https://doi.org/10.1016/j.procbio.2020.07.008.

Silva, L. P., Reis, I. G. and Bonatto, C. C. (2015). Green synthesis of metal nanoparticles by plants: Current trends and challenges. pp. 259–275. *In*: Vladimir, A. B. and Elena V. B. (eds.). Green Processes for Nanotechnology From Inorganic to Bioinspired Nanomaterials. Springer International Publisher, Inc: Switzerland.

Singh, A., Singh, N. B., Hussain, I., Singh, H. and Singh, S. C. (2015a). Plant nanoparticle interaction: An approach to improve agricultural practices and plant productivity. *Int. J. Pharm. Sci. Inven.*, 4: 25–40.

Sintubin, L., Gusseme, D. B., Meeren, V. P., Pycke, B. F. G., Verstraete, W. and Boon, N. (2011). The antibacterial activity of biogenic silver and its mode of action. *Appl. Microbiol. Biotechnol.*, 91: 153–162.

Sintubin, L., Windt, D. W., Dick, J., Mast, J., Vander, H. D., Verstraete, W. and Boon, N. (2009). Lactic acid bacteria as reducing and capping agent for the fast and efficient production of silver nanoparticles. *Appl. Microbiol. Biotechnol.*, 84: 741–749.

Smitha, S. L., Philip, D. and Gopchandran, K. G. (2009). Green synthesis of gold nanoparticles using Cinnamomum zeylanicum leaf broth. *Spectrochim. Acta Part A: Mol. Biomol. Spectrosc.*, 74(3): 735–739.

Sobhani-Nasab, A. and Behpour, M. (2016). Synthesis, characterization, and morphological control of $Eu_2Ti_2O_7$ nanoparticles through green method and its photocatalyst application. *Journal of Materials Science: Materials in Electronics*, 27(11): 11946–11951. doi:10.1007/s10854-016-5341-4.

Van der Meel. (2019). Smart cancer nanomedicine. *Nat. Nanotechnol.*, 14: 1007.

Varma, R. S. (2012). Greener approach to nanomaterials and their sustainable applications. *Curr. Opin. Chem. Eng.*, 1: 123–128.

Ventola, C. L. (2017). Progress in nanomedicine: Approved and investigational nanodrugs P. T 42 742–55 PMID: 29234213; PMCID: PMC5720487.

Vossen, J. L., Kern, W. and Kern, W. (1991). Thin Film Processes II; Gulf Professional Publishing: Houston, TX, USA. Volume 2, ISBN 0127282513.

Wang, T., Jin, X., Chen, Z., Megharaj, M. and Naidu, R. (2014). Green synthesis of Fe nanoparticles using eucalyptus leaf extracts for treatment of eutrophic wastewater. *Science of the Total Environment*, 466: 210–213.

Wang, X. W., Zhang, L., Ma, C. L., Song, R. Y., Hou, H. B. and Li, D. L. (2009). Enrichment and separation of silver from waste solutions by metal ion imprinted membrane. Hydrometal, 100: 82–86.

Winterer, M. and Hahn, H. (2003). Nanoceramics by chemical vapour synthesis. *Zeitschriftfur Metallkunde*, 94: 1084–90.

Zhang, X. -F., Liu, Z. -G. and Shen, W. (2016). Silver nanoparticles: Synthesis, characterization, properties, applications, and therapeutic approaches. IJMS, 17: 1534.

Zhang, Y. (2021). Waste eggshell membrane-assisted synthesis of magnetic CuF_{e2O4} nanomaterials with multifunctional properties (adsorptive, catalytic, antibacterial) for water remediation. *Colloids Surf A*, 612: 125874–17.

CHAPTER 7

Green Synthesis of Metal Oxide-Based Electrode Materials for Supercapacitor Electrode

P. Periasamy,[1,*] Murthy Chavali,[2] Maria Plamenova Nikolova,[3] Mohamed Bououdina,[4] B. Selvakumar[5] and K. Senthil Kannan[6]

1. Introduction

Renewable energy sources have gained attention as promising, economically viable, and environmentally friendly alternatives to address the concerns surrounding climate change and the eventual depletion of non-renewable fossil fuel resources (Figure 1). It is important to note that significant portions of renewable energy sources are intermittent or variable. Therefore, there is a need to develop effective storage techniques to ensure a constant and reliable supply of energy. The most common kind of storage system in use right now is a secondary battery (Dubal et al. 2017). The power density of batteries is modest despite their high-energy density. The chemical nature of energy storage in batteries leads to physicochemical changes in the electrolytes and electrodes during charging and discharging. Unfortunately,

[1] Department of Physics, Nehru Institute of Engineering and Technology, T. M. Palayam, Coimbatore – 641105, Tamil Nadu, India.

[2] Office of the Dean (Research) and Department of Chemistry, Faculty of Science Technology, Alliance University (Central Campus), Chandapura – Anekal Main Road, Bengaluru 562106, Karnataka, India.

[3] Department of Material Science and Technology, Faculty of Mechanical and Manufacturing Engineering, University of Ruse "Angel Kanchev", Ruse, Bulgaria.

[4] Department of Mathematics and Sciences, College of Humanities and Sciences, Prince Sultan University, Riyadh, Kingdom of Saudi Arabia.

[5] Department of Physics, Sri Eshwar College of Engineering, Kinathukadavu, Coimbatore, Tamil Nadu, India.

[6] Department of Physics/R and D, Edayathangudy G.S Pillay Arts and Science College (Autonomous – Affiliated to Bharathidasan University, Trichy 620024), Nagapattinam, Tamil Nadu, 611002, India.

* Corresponding author: periasamy.nano@gmail.com

these changes have a negative impact on both the battery life and cyclability. Because of this, their employment in energy storage is limited. However, supercapacitors are superior to batteries in terms of power density and cycle stability (Latha et al. 2021, Palanisamy et al. 2019). Despite their potential benefits, batteries are often deemed impractical for energy storage due to their low energy density, high effective series resistance, and relatively high cost. According to a comparison of energy density between supercapacitors, lead-acid batteries, and lithium-ion batteries, the energy density of supercapacitors is lower than that of lead-acid and lithium-ion batteries, but they outperform secondary batteries in terms of power density, life cycle, cyclability, and safety. They are getting a lot of buzz for their potential as a novel energy storage (Simon and Gogotsi 2008).

Currents day's researcher's primary goal is to find ways to boost supercapacitors' energy density (Conway 1999). Supercapacitors include three main parts: electrode materials, separators, and electrolytes. The energy density and power density of a supercapacitor can be improved by increasing its specific capacitance (C), expanding its operating voltage (V) window, and reducing its equivalent series resistance (R). These improvements can be achieved through various means, such as optimizing the electrode materials, developing new electrolytes, and improving the overall design of the supercapacitor. Selecting an appropriate electrolyte widens the operating voltage range while optimizing the electrode materials in terms of interconnected porosity, electrical conductivity, and surface area boosts the supercapacitor's specific capacitance (Liu et al. 2010). Supercapacitors rely on two different capacitive characteristics to store energy. Both EDL capacitance and pseudocapacitance may be thought of as electrical double layers (Simon and Gogotsi 2008). Pseudocapacitance, on the other hand, is caused by a faradic phenomenon that involves rapid electrochemical processes, in contrast to real capacitance, which is caused by electrostatic contact. The EDL seems to have the benefit that polarization resistance does not restrict the electrochemical kinetics, and there is no inflammation of the electrode material that typically happens.

The energy is stored in highly porous electrodes, one of which has a current collector attached to it; the energy that has been stored is discharged when the charges that have been collected at the electrodes are discharged (Liu et al. 2010). The electrode's surface determines the amount of charges and energy stored. Because of this, materials with a large amount of surface area are considered excellent options for the supercapacitor electrode. The performance of supercapacitors can be improved through various means, including the discovery of novel electrode materials, understanding the behavior of ions in microscopic pores, and constructing hybrid systems by combining faradic and capacitive electrodes. Developing new electrode materials with high surface area and conductivity and optimizing pore size and distribution can lead to increased specific capacitance and improved energy density. Hybrid systems that combine the benefits of both faradic and capacitive electrodes can also enhance the performance of supercapacitors, allowing them to deliver both high-power and high-energy density. The effectiveness of the cell in a supercapacitor is mostly determined by the material employed as an electrode (Conway 1999). Over the past few decades, there has been significant research and development aimed at

Figure 1. Renewable energy source (Images Credit: Freepik.com).

improving the energy density of supercapacitors. This has resulted in the discovery of new materials, improved designs, and better understanding of the underlying mechanisms that govern supercapacitor performance. As a result, supercapacitors have become increasingly viable as an energy storage technology, with potential applications in areas such as electric vehicles, renewable energy, and portable electronics. It sounds like the objective of the chapter is to provide a comprehensive review of the research efforts that have been made to develop novel materials and methodologies for supercapacitor electrodes, with a focus on the progress that has been made in terms of energy storage capacity for practical applications. This review will likely cover a range of topics, including the different types of materials that have been investigated, the methods used to fabricate electrodes, and the performance metrics that have been used to evaluate supercapacitor performance. The chapter may also highlight some of the key challenges that remain in the development of supercapacitor technology, as well as potential avenues for future research and development.

2. Green Synthesizing Method

The growing demand for nanoparticles of metals and metal oxides has prompted the manufacturing of these particles on a massive scale, employing high-energy technologies in conjunction with a variety of hazardous solvents (Pal et al. 2019). Green chemistry is the application of a set of principles to the process of designing and producing a variety of chemical products that are safe for the environment and have a minimal negative impact on both human health and the natural world (Figure 2). These goals are accomplished using the fewest possible quantities of potentially harmful chemicals. The integration of the concepts of green chemistry into the area of nanotechnology marked a significant advancement since it reduced the number of potential threats (Ying et al. 2022). The utilization of hazardous solvents and chemicals is necessitated both for the physical and the chemical

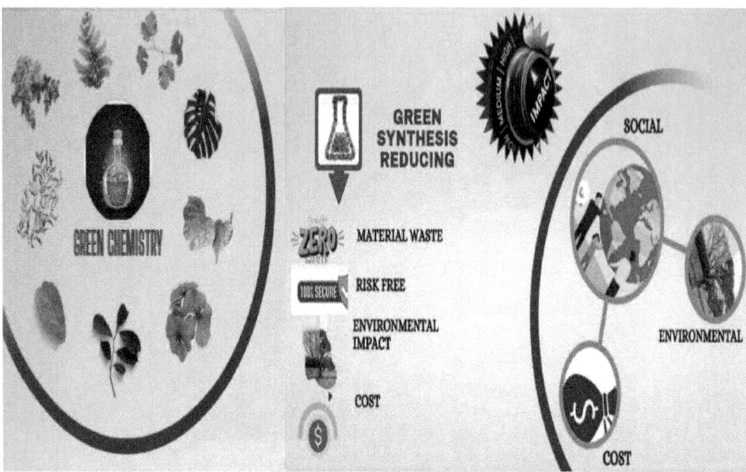

Figure 2. Green synthesizing method and its impact.

approaches. Throughout all of these operations, a significant quantity of energy is used. In contrast hand, green synthesis of nanoparticles is performed with the use of chemicals that are not harmful to humans or living organisms and are more environmentally friendly. It would seem that the green synthesis method yielded more advantageous nanoparticles in terms of their stability and desirability (Pal et al. 2019). Microscopic organisms, crops, algae, bacteria, fermentation, and fungus are the primary organisms used in this kind of procedure.

The synthesis of nanoparticles using environmentally friendly methods is simple, low-cost, and straightforward, with a reduced risk of experiment failure. The primary benefit of using green synthesis is that it leads to the development of nanoparticles that have lower levels of toxicity and are, as a consequence, less detrimental to the environment (Pal et al. 2019). The production of green synthesis is an unforeseen result of environmentally sustainable development that caters to the needs of future generations. This method lowers pollution levels, boosts air and water quality, and uses the vast majority of the earth's naturally existing resources most efficiently while preserving the delicate ecological equilibrium. However, green synthesis offers many advantages (Ying et al. 2022), but still, it has some negative sides, which are as follows:

(a) The implications of nanotechnology for human health have become much more serious;

(b) There have only been a few studies done on the bioaccumulation of nanoparticles and their hazardous effects on the environment;

(c) Because of their tiny size, nanoparticles can quickly enter the body, where they may cause respiratory problems as well as a variety of fatal illnesses;

(d) Green synthesis is still not typically utilized in the industry compared to the nanoparticles that are produced using chemical processes.

3. Green Synthesis of Metal Oxide-Based Electrode Materials for Supercapacitor Electrode

Green synthesis has received widespread attention as a sustainable, efficient, and environmentally friendly methodology for delivering a wide variety of nanostructures, which include nanocomposites, metal/metal oxide nanostructures, and biomimetic materials (Figure 3) (Kumar et al. 2020). Plant flowers include a wide range of secondary chemicals, such as pigments, volatile molecules that contribute to smell, and other phenolics with important ethnobotanical applications, notably treating ailments through floral therapy. Compounds that act as reducing agents are often used in the production of metal oxide nanoparticles. These compounds can donate electrons to the metal ions in a solution, leading to the reduction of the metal ions and the formation of metal nanoparticles. The choice of reducing agent can have a significant impact on the size, shape, and properties of the resulting nanoparticles.

Some common reducing agents in producing metal oxide nanoparticles include sodium borohydride, hydrazine, ascorbic acid, etc. These reducing agents can produce a wide range of metal oxide nanoparticles, including those made from iron, copper, zinc, and titanium. The resulting nanoparticles can have a range of applications, including in catalysis, electronics, and biomedical imaging, among others.

The synthesis and physiochemical analysis of a thin film composed of cobalt oxide (Co_3O_4) was reported by (Gaikar et al. 2022) for supercapacitor application. The precursor powder of Co_3O_4 was synthesized by preparing by using the efficient sol-gel synthesis process and then subjecting it to heat treatment at a temperature that was quite low. A fluorine-doped tin oxide (FTO) substrate was covered with a thin layer of Co_3O_4 utilizing a straightforward approach known as the doctor-blade

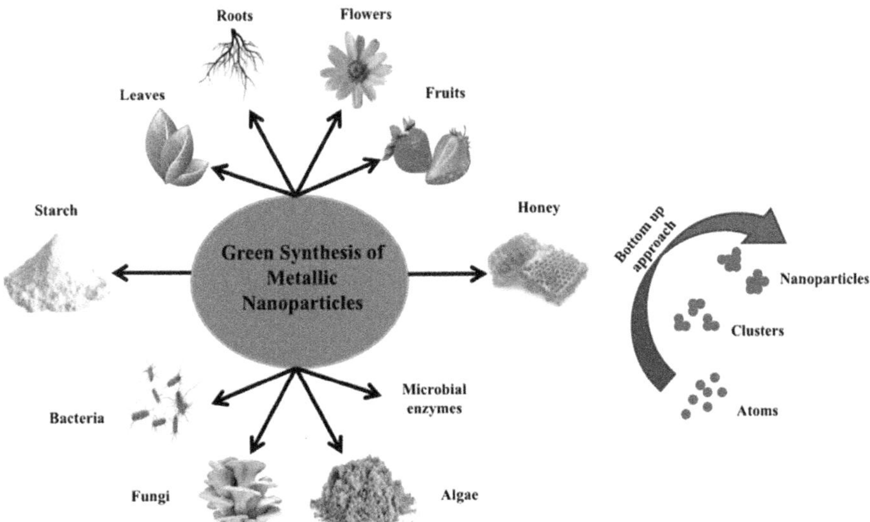

Figure 3. Different types of green synthesis methods for the preparation of metal nanoparticles. [Adopted from (Kumar et al. 2020) under Creative Commons Attribution (CC BY) licenseCopyright@2020, MDPI, Basel, Switzerland].

method. The electrochemical experiments showed that the Co_3O_4 electrode had a capacitive performance of 237 F/g, impressive cycling stability and 77% capacity retention upon 2,000 cycles.

Calcinating $MnCO_3$ precursors resulted in the formation of MnO_2 with a variety of distinct crystal structures and surface morphology (Du et al. 2015). Different $MnCO_3$ precursor morphologies may be synthesized in an environmentally friendly manner using hydrothermal methods employing orange pericarp extraction solution as the reducing agent. Dehydrating the precursor at different temperatures and pressures yields MnOx with different stoichiometric ratios (i.e., MnO, MnO_2, Mn_2O_3, and Mn_3O_4). Temperature, pressure, precursor type, and reaction time are only some of the synthesis parameters that may be modified to alter the stoichiometric ratio of MnOx. Electrochemical investigations show that among manganese oxides, MnO or MnO_2 seem to be more suited as working electrodes for supercapacitors than Mn_2O_3 or Mn_3O_4. They have a substantial specific capacitance (in the range of 296.6 F/g), which makes them excellent candidates for use as electrode materials in supercapacitors. In addition, they have an extraordinary cycle stability.

Graphene-based nanostructures have sparked a positive deal of interest in recent years due to their great results as electrode materials for energy storage applications. The nanocomposite of ternary reduced graphene oxide (rGO) – gold nanoparticles (Au NPs)@polyaniline (PANI) was synthesized in two steps, at a minimal cost, and in an environmentally friendly manner (Çıplak et al. 2020). In the initial phase of the process, graphene oxide (GO) was reduced and stabilized using an extract of the lichen Cetraria Islandica L. Ach. After that, the rGO nanosheets are then uniformly plated with an interconnecting Au@PANI nanoscale film by utilizing HAuCl4 as the oxidizing agent and *in situ* polymerization of aniline monomer. In a two-electrode arrangement, the rGO-Au@PANI nanocomposite demonstrates a high capacitive performance of 212.8 F/g. In addition to this, the nanocomposite electrodes have been shown to have 86.9% of their specific capacitance retention over 5,000 cycles.

The carbon quantum dots/cobalt sulfide nanocomposite, also known as $CQDs/CoS_2$, was synthesized by (Arsalani et al. 2021) via a ball milling process that does not include the use of any solvents and was then subjected to solvothermal treatment. First, the CQDs were synthesized from a simple precursor of glycerol using a microwave-assisted green process, which does not need a catalyst. The quantum yield of this approach was around 16%. The as-prepared $CQDs/CoS_2$ composite materials demonstrate a quite astonishing specific pseudocapacitance, approximately 808 Fg^{-1}, a high rate of capacity retention rate (83.9%), and virtuous cycling stability (98.75%) over 10,000 charge/discharge. These beneficial characteristics take account of a large specific surface area (427.2 m^2g^{-1}).

A polymer nanocomposite film known as PAN-PANI@G that was completely fabricated from a green synthesized graphene quantum dot (GQD, G) doped polyacrylonitrile (PAN), and polyaniline (PANI) systems has been successfully prepared by (Arthisree and Madhuri 2020). In a solution of 0.1 M H_2SO_4, CV, and GCD approaches were used to determine the specific capacitance value of ass-prepared material. With an applied current density of 670 mA g1, the measured supercapacitor values fall between 105 and 587 Fg^{-1} cm^{-2}, which is around 2–1,300 times greater than those published for polypyrrole and polyaniline-based polymer composite film in the existing research. Current efforts in the research to reduce the environmental impact of supercapacitors include replacing effective but poisonous organic electrolytes with safe, bio-friendly electrolytes, including a blend of sodium salt and hydrogel. Biodegradable electrolytes have not yet been shown to be a viable replacement for hazardous organic electrolytes. Pacheco et al. (2020) reported that carbon nanotubes (CNTs) added to a biodegradable hydrogel electrolyte containing sodium salt considerably improved the gel's capacitance as an alternative method.

Nayak et al. (2017) have proposed a straightforward and environmentally friendly solvothermal method for synthesizing graphene-supported tungsten oxide (WO_3) nanowires as an active electrode material. The graphene-WO_3 nanowire nanocomposite, carefully engineered with an optimized weight ratio, exhibited remarkable electrochemical characteristics. It displayed an impressive specific capacitance of 465 F/g, indicating its exceptional ability to store electrical energy. Moreover, this nanocomposite demonstrated a commendable cycling stability of 97.7% throughout 2,000 cycles, signifying its capacity to maintain a consistent level of capacitance retention throughout prolonged usage. The solid-state asymmetric ultracapacitor has achieved an impressive energy density of 26.7 watt-hours per kilogram (Wh/kg) at a power density of 6 kilowatts per kilogram (kW/kg). Remarkably, it demonstrated exceptional stability by maintaining an energy density of 25 Wh/kg under the same power density for a striking 4,000 cycles. This showcases its robust performance and durability under demanding operational conditions.

Functionalization of current polymer moieties in a way that is both efficient and effective for the purpose of achieving particular application targets and significantly boosting electrochemical performance. The structural modification of styrene maleic anhydride (SMA) copolymer was undertaken by incorporating a thiadiazole moiety; this strategic alteration aimed to fabricate a polymer nanocomposite with enhanced properties and functionalities. This was done in conjunction with integrating green-generated ZnO nanoparticles (Chakraborty et al. 2020). Anticipated results suggest that the polymer nanocomposite, slightly larger than both the ZnO nanostructures and the unmodified polymer, is likely to showcase promising behavior as a supercapacitor. Projections indicate a specific capacitance value of 268.5 F/g, underscoring its potential for efficient energy storage capabilities. In addition to this, the materials have excellent cycle stability and retain a maximum amount of capacitance. The newly developed specimen has great promise as a potential environmentally friendly replacement for traditional supercapacitors.

An aqueous solution of NaOH and urea was used to break down the bamboo cellulose fibers for the production of porous nanoplatelets enveloped in carbon aerogel (Yang et al. 2018). The design exhibited beneficial characteristics, such as a significant surface area and an impressive degree of flexibility of 86% when subjected to a stress of 23 kPa. Activated carbon aerogel demonstrated a great specific capacitance of 381 F/g when tested in an aqueous electrolyte containing 6 M KOH. This is a 150% increase in specific capacitance when compared to an unactivated sample. At a high scan rate of 200 mV/s, the admissible micropores and mesopores made a significant contribution to a remarkable retention rate of 90%. A novel type of high-performance asymmetric supercapacitor has been devised using pyrolyzed bacterial cellulose (p-BC)-coated MnO_2 (positive electrode material) and nitrogen-doped p-BC (negative electrode material) employing straightforward, effective, scalable, and environmentally friendly method (Chen et al. 2013). The ideal asymmetric device exhibits remarkable supercapacitive behavior and possesses a power and energy density which is exceptionally high.

Cellulosic trash in the type of sugarcane bagasse may be turned into three-dimensional (3D), interlinked, conductive, and a large surface area carbon nanochannels by the application of a novel methodology (hydrothermal preparation) before pyrolysis (Wahid et al. 2014). The wide buffer gaps in such a porous carbon sample resulted in an outstanding electrochemical capacitance (Csp) of 280 F/g and 275 F/g, with 72% of capacitance retention. The non-hydrothermally processed specimen, in comparison, indicated a specific capacitance value of 180 F/g and 52% capacity retention. Bagasse-derived carbon (BHAC) that undergoes hydrothermal preprocessing exhibits enhanced performance due to the retention of solvents within the buffer spaces formed. This advantageous feature enables the BHAC material to overcome the diffusional limitations caused by pore inaccessibility, particularly when operating at higher scan speeds. The material has an excellent cyclability of 90% after 1,000 cycles and produces a significant energy density of 5 Wh/kg at 3.5 kW/kg.

The activations of a mixture of potassium hydroxide and potassium carbonate are used to produce active carbon materials (ACs), which are produced from falling leaves (FLs) (Li et al. 2015). The findings from the characterizations indicate that employing a single activator, specifically KOH, generates a higher quantity of micropores within a three-dimensional structure. Conversely, utilizing K_2CO_3 as an activator creates a greater number of mesoporous and macrospores within a shallower structure. These observations highlight the influence of the activator choice on the pore characteristics and structural features of the material. Within a two-electrode configuration, the obtained sample exhibits a notably high specific capacitance, measuring approximately 242 F/g. This exceptional value signifies the material's substantial ability to store electrical charge per unit mass and highlights its potential for efficient energy storage applications. After 2,000 cycles, they reach excellent cycling stability with almost no capacitance loss.

To synthesize the activated carbon to use in supercapacitor electrodes, sugar cane bagasse is chemically treated with zinc chloride (Rufford et al. 2010). Sugar cane bagasse carbons (SCCs) demonstrate impressive performance characteristics in two-electrode, sandwich-type supercapacitor cells utilizing 1 M H_2SO_4 electrolyte. These SCCs exhibit specific energy values reaching up to 10 Wh kg^{-1}, indicating their remarkable capacity to store electrical energy per unit mass. Additionally, they showcase a specific capacitance of up to 300 Fg^{-1}, highlighting their exceptional ability to store and deliver electrical charge. Such attributes make SCCs highly promising materials for supercapacitor applications, offering significant energy storage capabilities. The substantial electrochemical properties exhibited by sugar cane bagasse carbons (SCCs) can be attributed to two key factors: their notable large specific surface area and the formation of mesopores, particularly when impregnated with ZnCl2 at a percentage of 1 or higher. The large specific surface area allows for increased contact between the SCCs and the electrolyte, facilitating efficient charge transfer processes. Simultaneously, the presence of mesopores, which are generated through $ZnCl_2$ impregnation, further enhances the accessibility of the electrolyte to the electrode material, promoting improved electrochemical performance. This combination of factors contributes to the enhanced electrochemical properties observed in SCCs. In contrast, the development of carbon with a low specific capacitance resulted from the pyrolysis of bagasse without $ZnCl_2$. The electrochemical performance of sugar cane bagasse carbons (SCCs) was examined under high charge-discharge rates, and it was found that the SCCs produced with a ZnCl2 ratio of 3.5 exhibited the most optimal and stable performance. These SCCs demonstrated superior electrochemical characteristics, including enhanced stability, efficiency, and capacity retention, when subjected to rapid charging and discharging processes. The careful optimization of the $ZnCl_2$ ratio during production proved instrumental in achieving such exceptional and consistent electrochemical performance at high charge-discharge rates.

$NiCo_2O_4$ nanograss (NG)-array coated carbon fiber ($NiCo_2O_4$ NG@CF) is used as the positive electrode in the fabrication of a unique flexible fiber hybrid supercapacitor (HSC), which provides a pseudocapacitance (Senthilkumar et al. 2016). To produce an electrode material at a reduced cost, specifically, the porous carbon is generated from discarded lemon peels. It is interesting to note that the developed HSC displays the highest specific capacitance of 17.5 F gl $(25.03 \text{ mF cm}^{-2})$ and the energy density of 6.61 Wh kg^{-1} when subjected to a current of 1 mA; the obtained values significantly surpass the findings reported in previous studies, indicating a substantial advancement in the observed measurements.

Using a simple one-pot process that involves carbonization and following oxidation of metal ion functionalized biopolymer precursors, a novel method has been developed (Xiang et al. 2019) for the preparation of three-dimensional hierarchical graphitic carbon nanocomposites with dispersed blended Co–Ni oxide nanoparticles (Co–Ni–O/3DG). The nanocomposites have been completely characterized and have shown outstanding electrochemical performance with an excellent specific capacitance of 1,586 F/g, and the capacitance retention after 10,000 cycles was 94.5%, exhibiting excellent cycling stability. Incredibly high power densities of 11358–748.6 W kg^{-1} and high-energy densities of 32.8–54.7 Wh kg^{-1} were achieved.

Carbon dots (CDs) can be used to disperse graphene and serve as a reducing agent for KMnO4 in the preparation of manganese oxide (MnOx)-graphene hybridization nanocomposites for electrochemical devices (Unnikrishnan et al. 2016). The carbon dots (CDs) generated through the degradation of ammonium citrate under dry heating conditions display remarkable solubility in water. This characteristic highlights the ability of these CDs to readily dissolve and disperse in aqueous environments, making them highly suitable for various applications requiring water-based solutions. Flexible solid-state supercapacitors fabricated using as-formed nanostructure showed excellent specific capacitance of 280 F/g, high cyclic stability (> 10,000 cycles), and good capacitance retention. The one-pot synthesis technique for MnOx-CDGs is generally environmentally friendly, easy, and speedy.

In order to use Cladophora glomerata (CG) as electrodes in supercapacitors, an easy-to-implement slow pyrolysis procedure has been developed to produce magnetic biochar (MBC) with olive-shaped pores and specialized biochar (FBC) with a new linked 3D pore network structure (Pourhosseini et al. 2018). FBC was designed and synthesized with HNO_3 and H_2SO_4 for the purpose of prolonging the synthesis of iron oxide by the hydrothermal process in order to produce Fe composite biochar, also known as FCBC. When used for the first time as electrodes for supercapacitors, these electrodes displayed exceptional electrochemical characteristics. The asymmetric supercapacitor is subjected to transient cycling within a broad potential range of 0–1.8 V. Remarkably, this configuration showcased an impressive energy density of 41.5 Wh kg^{-1} when operating at a power density of 900 W kg^{-1}. owing to its favorable attributes, including highly porous nature, large specific capacitance, excellent cycle stability, and an additional potential window in the electrochemical cell. These combined factors contribute to the exceptional performance and energy storage capabilities of the system.

Improved electrochemical properties of WO_3-CdS nanostructures were synthesized through the microwave-assisted process for varying wt.% of CdS (Figure 4). Electrochemical studies have been performed, and substantially improved specific capacitances were determined to be 44 F/g for WO_3 nanostructures and 172 F/g for WO_3-CdS nanocomposites (Periasamy et al. 2019). Microspheres of nickel cobaltite ($NiCo_2O_4$) in a hierarchical floral form have been generated through a quick and template-free microwave-assisted heating (MAH) reflux method, then pyrolyzed from the precursors (Lei et al. 2014). $NiCo_2O_4$ microspheres exhibit a distinctive flower-like morphology composed of nanopetals, each measuring a mere 15 nm in thickness. Despite their slender dimensions, these nanopetals possess a

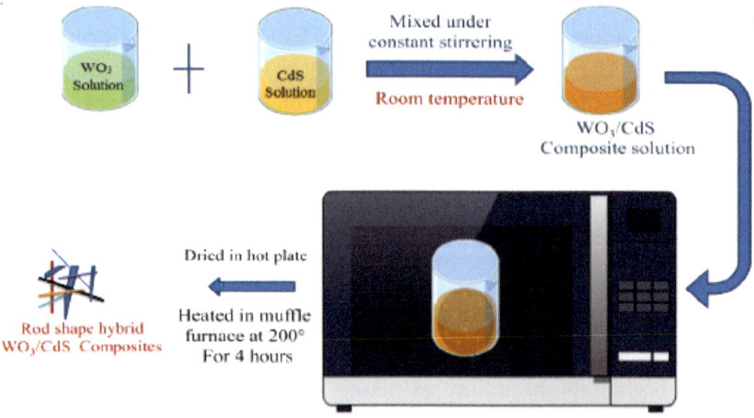

Figure 4. Preparation of hybridized WO_3 nanostructure (Periasamy et al. 2019). (Copyright@2019, Springer Science+Business Media, LLC, part of Springer Nature 2019.)

notable specific surface area of 148.5 m^2 g^{-1}, enabling efficient surface interactions with the surrounding environment. Furthermore, the microspheres demonstrate a narrow distribution of pore sizes ranging from 5 nm to 10 nm, further contributing to their enhanced surface area and optimized pore structure. This combination of features enhances their overall performance and makes them promising candidates for various applications. The fabricated flower-shaped $NiCo_2O_4$ microspheres, known for their exceptional permeability, showcased a remarkable specific capacitance of 1,006 F/g when subjected to a current density of 1 A/g. Furthermore, these microspheres exhibited optimized rate capability, enabling efficient charge/discharge processes even at high current densities. Impressively, they demonstrated superior electrochemical longevity with a retention rate of 93.2% after 1,000 cycles, even under the most demanding condition of a current density of 8 A/g. These findings highlight the outstanding performance and durability of the microspheres, making them highly promising for advanced energy storage applications.

Yeo et al. (2019) reported that a one-step electric field had driven combustion waves to synthesize the extensively moistened hybridization of single-crystalline, spherical Ag micro/nanoparticles, and CNT webs, thereby considerably strengthening the contact surfaces and lowering the grain boundaries. Electric fields within the stacked films of nitrocellulose, silver oxide nanoparticles, and carbon nanotubes (CNTs) play a crucial role in promoting superfast heating-cooling combustion waves. These waves facilitate the migration of reduced and liquid silver species through the fiber-like networks established by the materials. As a result, the process enables the fabrication of single-crystalline silver particles. Interestingly, the CNTs within the structure are generally preserved without undergoing oxidation, maintaining their structural integrity throughout the combustion process. This phenomenon demonstrates the significance of electric fields in governing the unique combustion behavior and preservation of specific components in the stacked film structure. As supercapacitor electrodes, these hybrids demonstrated exceptional performance in terms of both their specific capacitance (1,083 F/g) and their capacitance retention (95% following 10,000 cycles). In addition, the symmetric, solid-state, and flexible supercapacitors that were developed by employing the electrodes showed improved electrochemical properties (458 F/g, 100% following 10,000 cycles), as well as stability under severe mechanical loads, such as stretching and bending (97%).

Shin et al. (2017) reported that structure-guided combustion waves (SGCWs) are stimulated by partial oxidation through the chemical fuel-wrapped materials. It is easily possible to modify the reduced states of manganese oxides and synthesize carbon coatings surrounding them in a single step. The direct production of reduced $Mn_2O_3/Mn_3O_4/MnO$ and MnO was successfully achieved using sequential gas-condensed-water (SGCW) systems. In the presence of air, SGCWs facilitated the mediated oxygen release from MnO_2, resulting in the reduction of $Mn_2O_3/Mn_3O_4/MnO$ to their respective lower oxidation states. Conversely, in an argon (Ar) environment, SGCWs allowed for the controlled reduction of MnO to MnO without oxygen. This innovative approach demonstrates the efficacy of SGCWs in enabling the controlled reduction reactions and highlights their potential for tailored synthesis of manganese oxide materials. Compared with bare MnO_2, these electrodes demonstrated enhanced stability as well as higher specific capacitance (438 F/g for $Mn_2O_3/Mn_3O_4/MnO@C$). The use of SGCWs makes it possible to promptly, inexpensively, and on a wide scale synthesize reduced metallic nanoparticles and organic substances coatings, both of which have the potential to make substantial contributions to electrochemical applications. The synthesis of porous nanowires composed of cobalt oxide ($CoCo_2O_4$) is achieved by a microwave-assisted hydrothermal method (MHP), which is followed by annealing (Ragupathi et al. 2022). Electrochemical analyses demonstrate that the MHP $CoCo_2O_4$ nanowire configurations have a higher specific capacity (743.8 C/g), superior rate efficiency (67.7% retention), and superior cycling efficiency with 89.7% retention. It was determined that the use of microwave assistance was accountable for the increase in the precursor's degree of purity. Furthermore, the CoCo2O4/multi-walled carbon nanotube (m-CNT) hybrid device, designed as a battery-supercapacitor hybrid system, exhibits a notable energy density of 26.8 W h kg^{-1} at a power density of 775.4 W kg^{-1}. This impressive energy density reflects the device's capacity to store significant electrical energy per unit mass. Moreover, the hybrid device demonstrates excellent cycle life, with a remarkable capacity retention of 82.3% even after undergoing 8,000 charge-discharge cycles. This outstanding cycle stability further underscores the device's durability and longevity, making it a highly promising candidate for advanced energy storage applications.

A simplified and scalable approach has been developed to rapidly synthesize reduced graphene oxide (RG-O) by utilizing ionic liquids in conjunction with microwave chemistry, as outlined by Kim et al. (2012). The application of microwave heating to graphite oxide (GO) in the presence of an ionic liquid (IL) enables the rapid reduction of GO within a mere 15 seconds, resulting in the formation of RG-O structures that encapsulate the IL within their porous networks. This process, known as microwave-enhanced graphite oxide (MEGO), offers a highly efficient method for producing reduced graphene oxide. The electrodes prepared from IL-assisted microwave irradiation (mRG-O) exhibit an impressive specific capacitance of approximately 135 F/g. This exceptional capacitance can be attributed to the open architecture of the mRG-O material, which allows for the incorporation of IL moieties within its structure. Subsequently, a supercapacitor was constructed utilizing mRG-O electrodes and an IL electrolyte. This device showcased remarkable power density, measuring 246 kW kg^{-1}, and energy density, reaching 58 W h kg^{-1} while operating at a voltage of 3.5 volts. These findings underscore the potential of mRG-O and IL-based systems for high-performance energy storage applications.

A straightforward two-step technique involving homogeneous precipitation followed by microwave-mediated reduction was employed to fabricate an rGO-Fe2O3 composite material, as outlined in the work of Saraf et al. (2016). To evaluate its potential as an electrode material for supercapacitors, cyclic voltammetry (CV) and galvanostatic charging-discharging (GCD) experiments were conducted. Comparing the composite to pristine rGO, it exhibited superior supercapacitor performance. At a current density of 2 A g^{-1}, the composite displayed a high specific capacitance of 577.5 F g^{-1}. Furthermore, it maintained a specific capacitance of

437.5 F g^{-1} even at a high current density of 10 A g^{-1}, demonstrating good rate capability. These exceptional electrochemical properties observed in the rGO-Fe$_2$O$_3$ composite can be attributed to the beneficial synergistic effects of the interaction between the rGO platelets and the Fe$_2$O$_3$ nanostructures. The composite's impressive performance highlights its potential for advanced energy storage applications.

Achieving a significant areal capacitance is of utmost importance when integrating energy storage systems into microscale electronic devices. This requirement arises from the need to maximize the energy storage capabilities within a limited physical space. These microscale devices can effectively store and deliver electrical energy by attaining a large areal capacitance, enabling their successful integration into various compact electronic applications. The fabrication of supercapacitor devices comprised of porous metal oxides that have a significant areal capacitance remains a difficult task. Synthesis of porous nickel cobaltite microspheres was done using a microwave technique, which is then followed by the decomposition of the as-prepared precursor (Khalid et al. 2016). Microspheres fabricated from porous NiCo$_2$O$_4$ exhibit notable attributes, including a high specific surface area and a diminutive crystallite size. These characteristics contribute to the microspheres' ability to provide a substantial areal capacitance. The asymmetric and symmetric devices demonstrate distinct performance metrics, with the highest recorded values for areal capacitance and energy density. Specifically, the asymmetric devices exhibit a remarkable areal capacitance of 194 mF cm^{-2} and an energy density of 19.1 Wh kg^{-1}. On the other hand, the symmetric devices showcase an even higher areal capacitance of 380 mF cm^{-2}, while maintaining the same energy density of 19.1 Wh kg^{-1}. The successful fabrication of a symmetric device using NiCo$_2$O$_4$ as the negative electrode material underscores its potential as a promising candidate for future asymmetric devices. The combination of enhanced electrical conductivity and an effective porous structure renders 3D superstructure binary metal oxides highly desirable as electrode materials for supercapacitors. This characteristic enables efficient charge storage and delivery, thereby maximizing the overall performance of the supercapacitor system. The findings highlight the suitability of NiCo$_2$O$_4$ as a negative electrode material, paving the way for its potential application in advanced asymmetric supercapacitor devices in the future. A facile and template-free microwave-assisted heating (MAH) reflux method, preceded by the decomposition of the as-prepared precursors, was used to produce 3D hierarchical flower-shaped nickel cobaltite (NiCo$_2$O$_4$) microspheres (Lei et al. 2014). These microspheres are in the form of nickel cobaltite. The flower-shaped NiCo$_2$O$_4$ microspheres are bestowed with a high specific surface area (148.5 m^2 g^{-1}) and a small pore size distribution. The flower-shaped NiCo$_2$O$_4$ microspheres exhibit a unique structure, comprising atomically thin nanopetals with a thickness of approximately 15 nm and a narrow range of 5–10 nm. These porous microspheres, when employed as electrode materials, demonstrate exceptional electrochemical performance. Specifically, they manifest a high specific capacitance of 1,006 F g^{-1} at a current density of 1 A g^{-1}, indicating their remarkable charge storage capacity. Moreover, these microspheres exhibit superior rate capability, enabling efficient charge-discharge processes even at higher current densities. Notably, the microspheres display outstanding electrochemical stability, with an impressive capacity retention of 93.2% after 1,000 cycles, highlighting their robustness and durability over extended usage.

CuO/MnO$_2$ nanocomposites were synthesized (Zhang and Li 2020) using a simple microwave-assisted synthesis method in a typical home microwave oven and studied for electrochemical supercapacitor applications. As demonstrated by the findings of the electrochemical investigation, the capacitance performance of CuO/MnO$_2$ nanocomposites was superior to that of pure CuO material. When immersed in a 6 M KOH electrolyte, the CuO/MnO$_2$ nanocomposites demonstrated a noteworthy specific capacitance of 499.0 F/g when subjected to a current density of 0.5 A/g. Furthermore, an asymmetric supercapacitor was successfully fabricated by integrating activated carbon as the negative electrode and CuO/MnO$_2$ composite as the positive electrode. This unique asymmetric design exhibits a power density of 375.02 W/kg, leading to a remarkable energy density of 32.07 Wh/kg. Furthermore, the cycle stability of this asymmetric device is satisfactory. The Fe-doped NiMnO$_3$ nanostructured electrode materials (Qiao et al. 2019) have been effectively prepared by the microwave-assisted hydrothermal technique, which is simultaneously practical and effective. The investigation revealed that the addition of Fe to the NiMnO$_3$ nanosheet electrode material not only altered the crystal structure but also the shape. In addition, the electrode material displayed remarkable conductivity as well as a high specific surface area. Under the operating conditions of a current density of 1 A/g, the specific capacitance achieved an impressive value of 732.7 F/g. Furthermore, the capacitive retention of the system remained stable at approximately 78.3% over an extensive cycling period of 10,000 cycles when the current density was increased to 3 A/g.

Using dry microwave irradiation (Kumar et al. 2022), nanomaterials with a one-of-a-kind morphology have been successfully prepared to apply supercapacitor electrodes. During the reduction process of graphite oxide, a simultaneous degradation of the FeCl$_3$·6H$_2$O precursor occurred, resulting in the formation of Fe$_3$O$_4$ nanoparticles. This reduction process involves converting the oxidized form of graphite into its reduced form while the FeCl$_3$·6H$_2$O precursor undergoes degradation, leading to the generation of Fe$_3$O$_4$ nanoparticles. Upon analysis within a 1 M KOH electrolyte, the electrode material exhibited an impressive specific capacitance of 771.3 F/g when subjected to a scan rate of 5 mV/s. This finding highlights the high charge storage capacity of the electrode material under the specified testing conditions, suggesting its potential for efficient energy storage in supercapacitor devices. In addition, remarkable stability was reported in the face of continuous cycling, with a capacitance retention of around 95.1% following 5,000 cycles. Microwave-irradiated MnO$_2$ microparticles were coated on two plant-based substrates consisting of Al/lignin and Al/AC/lignin (Mehta et al. 2021). The quasi-solid-state ultracapacitors were built with a PVA/H$_3$PO$_4$ polymeric gel electrolyte (Figure 5). After 700 charge-discharge cycles, electrochemical testing revealed that both supercapacitors retained more than 90% of their capacitance. In contrast to existing technologies, this process offers the advantage of achieving homogenous material structurization while providing customizable properties. This is made possible by adjusting the heating duration and radiation exposure in a tailored

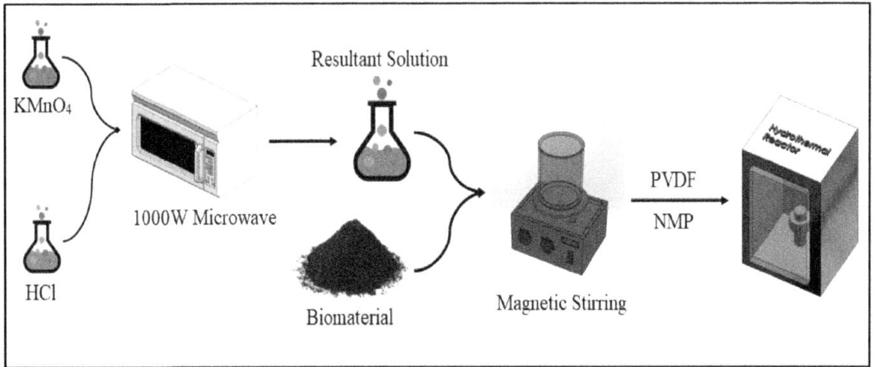

Figure 5. Preparation of the electroactive materials for supercapacitor. (Adopted from (Mehta et al. 2021) under creative common license 4.0, © Siddhi Mehta et al 2021, MDPI, Basel, Switzerland.)

manner. Moreover, this process minimizes waste generation, contributing to a more environmentally friendly approach. The ability to control and optimize the material properties through precise parameter manipulation makes this process a promising avenue for advanced material synthesis and fabrication.

The preparation of a number of other hetero-metal oxides/CNT nanocomposites, including CuO-NiO/CNTs, is perhaps produced with the help of conventional hydrothermal and microwave techniques (Veeman and Karuppuchamy 2022). The present study aimed to explore the synergistic effects resulting from the incorporation of hetero-metal oxides and metal oxide/carbon nanotube (CNT) composites within composite materials. To investigate the impact on surface properties and electrochemical characteristics, the ratio of CNT to metal oxide was systematically varied in three different proportions: 1:1, 0.5:1, and 0.1:1. By altering these ratios, we sought to understand how the combination of these materials influences the overall performance and properties of the resulting composites. A three-electrode setup was used for the purposes of performing electrochemical experiments on the CuO-NiO/CNT composite. The electrochemical experiments conducted in this study revealed that the CuO-NiO/CNT nanocomposites, synthesized using both hydrothermal and microwave techniques, demonstrated superior performance in terms of specific capacitance. Specifically, these nanocomposites exhibited the highest specific capacitance values of 581.3 F/g and 489.6 F/g, respectively, when operated at a current density of 1 A/g. Moreover, the results of a stability test performed on the prepared composite in its role as electrode material demonstrate that all of the electrode materials have remained more than 75% of their starting specific capacitance after being subjected to 2,000 cycles at a current density of 10 A/g. Rosli et al. (2022) effectively produced MoS_2-rGO composites by using the microwave-assisted technique at a temperature of 140°C for 30 minutes with varying amounts of ammonium tetrathiomolybdate. The dissemination of Mo on the rGO sheets was detected by an examination of the morphological characteristics of the composites. At the current density of 0.8 Ag^{-1}, the MoS_2-rGO electrode, which was developed with 0.01 M of $(NH4)\,2MoS_4$, demonstrated a phenomenal specific capacitance of

561.3 Fg^{-1}, and it retained 92% of its capacitance even after 10,000 cycles. When it was brought to use in an asymmetrical supercapacitor device, it achieved an energy density of 39.90 Wh kg1 as well as a power density of 800 W kg1, accordingly.

In order to improve the electrochemical behavior of binary transition-metal oxide nanoparticles to be employed in energy storage applications, it has become desirable to customize those nanoparticles using two-dimensional unique carbon nanomaterials. To be employed as electrodes in supercapacitors (SCs), binary metal oxide of Mn_3O_4-Fe_2O_3/Fe_3O_4 nanoparticles tethered reduced graphene oxide nanosheets were synthesized by microwave-assisted synthesis (Kumar et al. 2020). The electrochemical characteristics of Mn_3O_4-Fe_2O_3/$F_{e3}O_4$@rGO ternary hybrids were investigated in 1.0 M KOH electrolyte solution. The as-prepared ternary hybrids displayed a specific capacity of 590.7 F/g, and its cyclic stability was measured as a capacitance retention of 64.5% over 1,000 cycles. The almost rectangular shape of the cyclic voltammetry contour showed that the electric double-layer capacitance by conductive rGO NSs was dominant in comparison to the pseudocapacitance by binary metal oxide nanoparticles. The microwave-assisted technique was used to synthesize reduced graphene oxide/titanium oxide nanocomposites with varying weight ratios of graphene oxide to titanium oxide (1:1, 1:2, 1:5, and 1:10) (Ates et al. 2017). In order to get the best possible results for nanocomposite materials, the different weight ratio of rGO to TiO_2 was used. The remarkable high specific capacitance of 524.02 F/g and high-energy density of E = 50.07 Wh/kg was observed in the so-prepared symmetrical rGO/TiO_2 nanostructure using the two-electrode setup. In addition, over 1,000 cycles, the symmetrical electrode retains 66.6% of its initial value demonstrating excellent cycling stability.

Copper sulfide mesoporous nanosheet structures are synthesized through an immediate and efficient microwave-assisted technique (Naveed et al. 2019). Supercapacitors developed from CuS nanosheets feature mesoporous structures with a high surface area (169 g1) and a small pore width (< 25 nm). Excellent supercapacitive performance may be achieved due to the large surface area nanosheet structure, which allows for more catalytic activity for electrochemical processes and a shorter ion/electron diffusion route. Under the operating conditions of a current density of 1 A/g in an aqueous electrolyte consisting of 2 M KOH, the mesoporous CuS nanosheet electrodes exhibited a remarkable specific capacitance of 2,535 F/g. This high specific capacitance value indicates the effective charge storage capability of the CuS nanosheets and highlights their potential. The assembled device demonstrated excellent cycling stability, with a retention rate of 88% over 10,000 cycles. Additionally, it exhibited superior capacitance, reaching 177 F/g, and remarkable energy density of 63.2 Wh/kg at a high power density of 400 W/kg. These findings highlight the exceptional performance of the device in terms of both electrochemical stability and energy storage capabilities. This study not only demonstrates the feasibility of utilizing mesoporous CuS electrodes in high-performance hybridization asymmetrical capacitors but also presents a novel approach for synthesizing various transition-metal sulfide architectures through the application of microwave radiation.

The utilization of microwave-assisted hydrothermal processing has been shown to enable efficient synthesis of a $CNT@NiMn_2O_4$ core-shell nanocomposite, as reported by Sun et al. (2020). The resulting nanocomposite exhibits low charge transfer resistance (2.19 Ω), high specific capacitance (up to 915.6 F/g), and excellent cycling stability (93.0% capacity retention after 5,000 cycles). These desirable electrochemical properties can be attributed to several factors, including the microwave-induced weaker crystallinity, the synergistic effect between CNTs and $NiMn_2O_4$ within the core-shell heterojunction, and the enhanced imperfections and vacant positions in CNTs resulting from acidifying pre-treatment. Moreover, an asymmetric supercapacitor device was successfully constructed using the $CNT@NiMn_2O_4$ nanocomposite as the positive electrode. This device demonstrates an outstanding energy density of 36.5 Wh/kg, a high power density of 800 W/kg, and remarkable cycling stability with 82.8% retention over 10,000 cycles.

Lamiel et al. (2017) successfully developed a tailored and binder-free architecture of Ni-Co-Mn oxide on nickel foam using microwave irradiation. The optimized conditions, including composition and holding duration, resulted in hierarchical nanoflakes interconnected by void spaces, creating a continuous and conductive network on the surface of the nickel foam. This unique structure facilitated the achievement of a high specific capacitance of 2,536 F/g in a mixed $KOH/K_3Fe(CN)_6$ electrolyte, owing to the synergistic interactions among the three metal oxides (Ni-Co-Mn). Furthermore, a symmetric device prepared using this architecture exhibited a high capacitance of 298 F/g, along with an impressive energy density of 41.4 Wh/kg and a high power density of 5.4 kW/kg. The highly architectured Ni-Co-Mn oxide nanoflakes greatly enhanced the pseudocapacitive performance, highlighting their potential as electrode materials for efficient binder-free energy storage devices.

The synthesis of SnO_2 nanoparticles electrode for high-performance supercapacitors was accomplished with the help of a technique that included the irradiation of microwave (Gaber et al. 2023). The highest specific capacitance and the specific capacity of a SnO_2 electrode were accomplished at 407 F/g and 163 C/g, correspondingly. In order to meet the requirements of a specific application, a hybrid electrochemical capacitor cell was developed. The cathode in this study was comprised of the as-synthesized SnO_2, while the anode consisted of activated carbon. The battery demonstrated remarkable performance characteristics, including a specific energy of 34 Wh kg^{-1} at a specific power of 773 W kg^{-1} and excellent cycling stability with a capacity retention of 87% after approximately 3,000 cycles. The mesoporous $NiMoO_4$ nanorods were formed on the exterior of reduced graphene oxide composites using a microwave-solvothermal process that was easy to use, quick, and safe for the environment (Liu et al. 2015). The NiMoO4-rGO composite exhibited promising performance as a potential electrode material for supercapacitors. The acquired specific capacitance was 1274 Fg^{-1}, which is significantly higher compared to that of pure $NiMoO_4$. $NiMoO_4$-rGO has the ability to retain around 81.1% of its original capacitance even after being subjected to 1000 cycles. It is remarkable that composites made of NiMoO4 and rGO can be used in asymmetric electrochemical capacitors with ultra-high-energy densities of 30.3 Wh kg^{-1} and power densities of 187 W kg^{-1}.

4. Conclusion

The primary objective in designing supercapacitors is to address the need for efficient storage of renewable resources. This chapter provides an overview of recent advancements in electrode materials for supercapacitors, focusing on green synthesis methods. The review encompasses the design, construction, and electrochemical properties of these materials. Despite significant progress in green synthesis-based electrode materials, the limited energy density remains a key challenge for further advancements in supercapacitors. To overcome this barrier, we propose several suggestions for developing supercapacitors to achieve reasonable energy and power density. One approach is prioritizing research on obtaining low-cost raw materials for electrodes with minimal environmental pollutants, with biomass carbon compounds serving as a promising example. Another area of interest is the development of flexible supercapacitors, driven by the growing demand for wearable and portable electronic devices. Currently, there are numerous publications on transparent and stretchable electrodes for battery systems. Additionally, considerable attention has been given to capacitors based on battery-type electrodes. Moreover, hybrid supercapacitors have emerged as a potential solution to bridge the gap between conventional metal ion batteries and supercapacitors.

References

Arsalani Nasser, Laleh Saleh Ghadimi, Iraj Ahadzadeh, Amin Goljanian Tabrizi and Thomas Nann. (2021). Green synthesized carbon quantum dots/cobalt sulfide nanocomposite as efficient electrode material for supercapacitors. *Energy and Fuels*, 35(11): 9635–45.

Arthisree, D. and Madhuri, W. (2020). Optically active polymer nanocomposite composed of polyaniline, polyacrylonitrile and green-synthesized graphene quantum dot for supercapacitor application. *International Journal of Hydrogen Energy*, 45(16): 9317–27.

Ates Murat, Yuksel Bayrak, Ozan Yoruk and Sinan Caliskan. (2017). Reduced graphene oxide/titanium oxide nanocomposite synthesis via microwave-assisted method and supercapacitor behaviors. *Journal of Alloys and Compounds*, 728: 541–51.

Chakraborty Sohini, Amal Raj, M. and Mary, N. L. (2020). Biocompatible supercapacitor electrodes using green synthesised ZnO/polymer nanocomposites for efficient energy storage applications. *Journal of Energy Storage*, 28: 101275.

Chen Li Feng, Zhi Hong Huang, Hai Wei Liang, Qing Fang Guan and Shu Hong Yu. (2013). Bacterial-cellulose-derived carbon nanofiber@MnO_2 and nitrogen-doped carbon nanofiber electrode materials: An asymmetric supercapacitor with high energy and power density. *Advanced Materials (Deerfield Beach, Fla.)*, 25(34): 4746–52.

Çıplak Zafer, Atila Yıldız and Nuray Yıldız. (2020). Green preparation of ternary reduced graphene oxide-Au@polyaniline nanocomposite for supercapacitor application. *Journal of Energy Storage*, 32: 101846.

Conway, B. E. (1999). Electrochemical supercapacitor: Scientific fundamentals and technological applications. *Electrochemical Supercapacitors*.

Du Wei, Xiaoqian Xu, Di Zhang, Qingyi Lu and Feng Gao. (2015). Green synthesis of MnO x nanostructures and studies of their supercapacitor performance. *Science eChina Chemistry*, 58(4): 627–33.

Dubal Deepak P., Nilesh R. Chodankar, Pedro Gomez-Romero and Do-Heyoung Kim. (2017). Fundamentals of binary metal oxide–based supercapacitors. In *Metal Oxides in Supercapacitors*.

Gaber Amira, Sayed Y. Attia, Aliaa M. S. Salem, Saad G. Mohamed and Soliman I. El-Hout. (2023). Microwave-assisted fabrication of SnO2 nanostructures as electrode for high-performance pseudocapacitors. *Journal of Energy Storage*, 59: 106358.

Gaikar Paresh S., Ankita P. Angre, Gurumeet Wadhawa, Pankaj V. Ledade, Sami H. Mahmood and Trimurti L. Lambat. (2022). Green synthesis of cobalt oxide thin films as an electrode material for electrochemical capacitor application. *Current Research in Green and Sustainable Chemistry*, 5: 100265.

Khalid Syed, Chuanbao Cao, Lin Wang and Youqi Zhu. (2016). Microwave assisted synthesis of porous NiCo2O4 microspheres: Application as high performance asymmetric and symmetric supercapacitors with large areal capacitance. *Scientific Reports*, 6(1): 1–13.

Kim Tae Young, Hyun Chang Kang, Tran Thanh Tung, Jung Don Lee, Hyeongkeun Kim, Woo Seok Yang, Ho Gyu Yoon and Kwang S. Suh. (2012). Ionic liquid-assisted microwave reduction of graphite oxide for supercapacitors. *RSC Advances*, 2(23): 8808–12.

Kumar Harsh, Kanchan Bhardwaj, Kamil Kuča, Anu Kalia, Eugenie Nepovimova, Rachna Verma and Dinesh Kumar. (2020). Flower-based green synthesis of metallic nanoparticles: Applications beyond fragrance. *Nanomaterials*, 10(4): 766.

Kumar Rajesh, Sally M. Youssry, Ednan Joanni, Sumanta Sahoo, Go Kawamura and Atsunori Matsuda. (2022). Microwave-assisted synthesis of iron oxide homogeneously dispersed on reduced graphene oxide for high-performance supercapacitor electrodes. *Journal of Energy Storage*, 56: 105896.

Kumar Rajesh, Sally M. Youssry, Kyaw Zay Ya, Wai Kian Tan, Go Kawamura and Atsunori Matsuda. (2020). Microwave-assisted synthesis of Mn3O4-Fe2O3/Fe3O4@rGO ternary hybrids and electrochemical performance for supercapacitor electrode. *Diamond and Related Materials*, 101: 107622.

Lamiel Charmaine, Van Hoa Nguyen, Deivasigamani Ranjith Kumar and Jae Jin Shim. (2017). Microwave-assisted binder-free synthesis of 3D Ni-Co-Mn Oxide Nanoflakes@Ni foam electrode for supercapacitor applications. *Chemical Engineering Journal*, 316: 1091–1102.

Latha, K., Anbuselvi, S., Periasamy, P., Sudha, R. and Velmurugan, D. (2021). Structural and electrochemical investigation of novel hybridized MnO2/V2O5 nanocomposites prepared by one-step microwave-assisted method for electrochemical supercapacitor application. *Journal of Materials Science: Materials in Electronics*, 32(18): 23293–308.

Lei Ying, Jing Li, Yanyan Wang, Li Gu, Yuefan Chang, Hongyan Yuan and Dan Xiao. (2014). Rapid microwave-assisted green synthesis of 3D hierarchical flower-shaped NiCo2O4 microsphere for high-performance supercapacitor. *ACS Applied Materials and Interfaces*, 6(3): 1773–80.

Li Yin Tao, Yu Tong Pi, Li Ming Lu, Shun Hua Xu and Tie Zhen Ren. (2015). Hierarchical porous active carbon from fallen leaves by synergy of K2CO3 and their supercapacitor performance. *Journal of Power Sources*, 299: 519–28.

Liu Chang, Feng Li, Ma Lai-Peng and Hui Mmg Cheng. (2010). Advanced materials for energy storage. *Adv. Mater.*, 22(8): E28–62.

Liu Ting, Hui Chai, Dianzeng Jia, Ying Su, Tao Wang and Wanyong Zhou. (2015). Rapid microwave-assisted synthesis of mesoporous NiMoO4 nanorod/reduced graphene oxide composites for high-performance supercapacitors. *Electrochimica Acta*, 180: 998–1006.

Mehta Siddhi, Swarn Jha, Dali Huang, Kailash Arole and Hong Liang. (2021). Microwave synthesis of MnO2-lignin composite electrodes for supercapacitors. *Journal of Composites Science*, 5(8): 216.

Naveed Muhammad, Younas, W., Zhu, Y., Souleymen Rafai, Quanqing Zhao, Muhammad Tahir, Nouraiz Mushtaq and Chuanbao Cao. (2019). Template free and facile microwave-assisted synthesis method to prepare mesoporous copper sulfide nanosheets for high-performance hybrid supercapacitor. *Electrochimica Acta*, 319: 49–60.

Nayak Arpan Kumar, Ashok Kumar Das and Debabrata Pradhan. (2017). High performance solid-state asymmetric supercapacitor using green synthesized graphene-WO3 nanowires nanocomposite. *ACS Sustainable Chemistry and Engineering*, 5(11): 10128–38.

Pacheco Marquidia, Maria Fernanda Monroy, Alfredo Santana-Diaz, Joel Pacheco, Ricardo Valdivia-Barrientos, Xin Tu, Alberto Gonzalez-Pedroza and Maria Teresa Ramirez-Palma. (2020). Enhancement of a green supercapacitor with a hydrogel/carbon nanotubes-based electrolyte. *IEEE Transactions on Nanotechnology*, 19: 711–18.

Pal Gaurav, Priya Rai and Anjana Pandey. (2019). Green synthesis of nanoparticles: A greener approach for a cleaner future. *Green Synthesis, Characterization and Applications of Nanoparticles*, 1–26.

Palanisamy Periasamy, Krishnakumar Thangavel, Sandhiya Murugesan, Sathish Marappan, Murthy Chavali, Prem Felix Siril and Devarajan Vaiyapuri Perumal. (2019). Investigating the synergistic effect of hybridized WO3-ZnS nanocomposite prepared by microwave-assisted wet chemical method for supercapacitor application. *Journal of Electroanalytical Chemistry*, 833(November 2018): 93–104.

Periasamy, P., Krishnakumar, T., Sandhiya, M., Sathish, M., Murthy Chavali, Prem Felix Siril and Devarajan, V. P. (2019). Electrochemical investigation of hybridized WO3–CdS semiconducting nanostructures prepared by microwave-assisted wet chemical route for supercapacitor application. *Journal of Materials Science: Materials in Electronics*, 30(10): 9231–44.

Pourhosseini, S. E. M., Omid Norouzi, Pejman Salimi and Hamid Reza Naderi. (2018). Synthesis of a novel interconnected 3D pore network algal biochar constituting iron nanoparticles derived from a harmful marine biomass as high-performance asymmetric supercapacitor electrodes. *ACS Sustainable Chemistry and Engineering*, 6(4): 4746–58.

Qiao Shaoming, Naibao Huang, Junjie Zhang, Yuanyuan Zhang, Yin Sun and Zhengyuan Gao. (2019). Microwave-assisted synthesis of fe-doped $NiMnO_3$ as electrode material for high-performance supercapacitors. *Journal of Solid State Electrochemistry*, 23(1): 63–72.

Ragupathi Hariventhan, Antony Arockiaraj M. and Youngson Choe. (2022). A novel β-MnO_2 and carbon nanotube composite with potent electrochemical properties synthesized using a microwave-assisted method for use in supercapacitor electrodes. *New Journal of Chemistry*, 46(32): 15358–66.

Rosli Nurul Hazwani Aminuddin, Kam Sheng Lau, Tan Winie, Siew Xian Chin, Sarani Zakaria and Chin Hua Chia. (2022). Rapid microwave synthesis of molybdenum disulfide-decorated reduced-graphene oxide nanosheets for use in high electrochemical performance supercapacitors. *Journal of Energy Storage*, 52: 104991.

Rufford Thomas E., Denisa Hulicova-Jurcakova, Kiran Khosla, Zhonghua Zhu and Gao Qing Lu. (2010). Microstructure and electrochemical double-layer capacitance of carbon electrodes prepared by zinc chloride activation of sugar cane bagasse. *Journal of Power Sources*, 195(3): 912–18.

Saraf Mohit, Kaushik Natarajan and Shaikh M. Mobin. (2016). Microwave assisted fabrication of a nanostructured reduced graphene oxide (RGO)/Fe2O3 composite as a promising next generation energy storage material. *RSC Advances*, 7(1): 309–17.

Senthilkumar, S. T., Nianqing Fu, Yan Liu, Yu Wang, Limin Zhou and Haitao Huang. (2016). Flexible fiber hybrid supercapacitor with NiCo2O4 Nanograss@carbon fiber and bio-waste derived high surface area porous carbon. *Electrochimica Acta*, 211: 411–19.

Shin Jungho, Dongjoon Shin, Hayoung Hwang, Taehan Yeo, Seonghyun Park and Wonjoon Choi. (2017). One-step transformation of MnO2 into MnO2−x@carbon nanostructures for high-performance supercapacitors using structure-guided combustion waves. *Journal of Materials Chemistry A*, 5(26): 13488–98.

Simon Patrice and Yury Gogotsi. (2008). Materials for electrochemical capacitors. *Nature Materials*, 7(11): 845–54.

Sun Yin, Xiaomei Du, Junjie Zhang, Naibao Huang, Liu Yang and Xiannian Sun. (2020). Microwave-assisted preparation and improvement mechanism of carbon nanotube@NiMn2O4 core-shell nanocomposite for high performance asymmetric supercapacitors. *Journal of Power Sources*, 473: 228609.

Unnikrishnan Binesh, Chien Wei Wu, Wen Peter Chen, I., Huan Tsung Chang, Chia Hua Lin and Chih Ching Huang. (2016). Carbon dot-mediated synthesis of manganese oxide decorated graphene nanosheets for supercapacitor application. *ACS Sustainable Chemistry and Engineering*, 4(6): 3008–16.

Veeman Sannasi and Karuppuchamy, S. (2022). H2O2 assisted hydrothermal and microwave synthesis of CuO–NiO hybrid MWCNT composite electrode materials for supercapacitor applications. *Ceramics International*, 48(18): 26806–17.

Wahid Malik, Dhanya Puthusseri, Deodatta Phase and Satishchandra Ogale. (2014). Enhanced Capacitance retention in a supercapacitor made of carbon from sugarcane bagasse by hydrothermal pretreatment *Energy and Fuels*, 28(6): 4233–40.

Xiang Cuili, Yin Liu, Ying Yin, Pengru Huang, Yongjin Zou, Marcus Fehse, Zhe She, Fen Xu, Dipanjan Banerjee, Daniel Hermida Merino, Alessandro Longo, Heinz Bernhard Kraatz, Dermot F. Brougham, Bing Wu and Lixian Sun. (2019). Facile green route to Ni/Co oxide nanoparticle embedded 3D graphitic carbon nanosheets for high performance hybrid supercapacitor devices. *ACS Applied Energy Materials*, 2(5): 3389–99.

Yang Xi, Benhua Fei, Jianfeng Ma, Xinge Liu, Shumin Yang, Genlin Tian and Zehui Jiang. (2018). Porous nanoplatelets wrapped carbon aerogels by pyrolysis of regenerated bamboo cellulose aerogels as supercapacitor electrodes. *Carbohydrate Polymers*, 180: 385–92.

Yeo Taehan, Jaeho Lee, Dongjoon Shin, Seonghyun Park, Hayoung Hwang and Wonjoon Choi. (2019). One-step fabrication of silver nanosphere-wetted carbon nanotube electrodes via electric-field-driven combustion waves for high-performance flexible supercapacitors. *Journal of Materials Chemistry A*, 7(15): 9004–18.

Ying Shuaixuan, Zhenru Guan, Polycarp C. Ofoegbu, Preston Clubb, Cyren Rico, Feng He and Jie Hong. (2022). Green synthesis of nanoparticles: Current developments and limitations. *Environmental Technology & Innovation*, 26: 102336.

Zhang Pengjiao and Wei Li. (2020). Microwave-assisted synthesis of CuO/MnO2 nanocomposites for supercapacitor application. *Micro & Nano Letters*, 15(13): 938–42.

CHAPTER 8

Green Synthesis of MOF for Supercapacitor Electrode

Vadivel Siva,[1,2,*] *Sadasivam Kannan,*[3] *Abdul Samad Shameem,*[2,4]
Murugesan Raja,[5] *Anbazhagan Murugan,*[6] *Subramanian Thangarasu*[7]
and Arumugam Raja[8]

1. Introduction

The requirement for renewable and environmentally friendly energy solutions has increased along with the consumption of fossil fuels. Advanced technology must not only at least match the overall functionality of existing energy storage conversion and storage systems but must also be applied and distributed in a way that is both practical and cost-effective. Solar cells, capacitors, batteries, fuel cells, and other topics offer much scope for advancement and scientific focus (Choudhury et al. 2013). A family of porous materials known as metal-organic frameworks (MOFs) is made up of metal ions or clusters that are bonded together by organic ligands. They are extremely crystalline substances with a wide range of structural characteristics and persistent porosity. These characteristics, combined with the flexibility to fine-tune the structure using different ligands and metal centres, result in a wide range

[1] Department of Physics, Karpagam Academy of Higher Education, Coimbatore 641021, Tamil Nadu, India.

[2] Centre for Energy and Environment, Karpagam Academy of Higher Education, Coimbatore 641 021, Tamil Nadu, India.

[3] Department of Physics, Government Arts College for Men, Krishnagiri, 635001, Tamil Nadu, India.

[4] Department of Science and Humanities, Karpagam Academy of Higher Education, Coimbatore 641021, Tamil Nadu, India.

[5] Faculty of Pharmacy, Karpagam Academy of Higher Education, Coimbatore 641021, Tamil Nadu, India.

[6] Department of Science and Humanities, Karpagam College of Engineering, Coimbatore 641032, India.

[7] Department of Physics, Kalasalingam Academy of Higher Education, Krishnankoil-626126, Tamilnadu, India.

[8] CNR-SPIN, University of Salerno, Salerno, Italy.

* Corresponding author: siva.vadivel@kahedu.edu.in, siva33phy@gmail.com

of applications in this fast evolving field, such as gas capture and storage (Suh et al. 2012), separations (Li et al. 2012), catalysis (Leus et al. 2014), magnetism applications (Zhang et al. 2013), fluorescence and sensing applications (Kreno et al. 2012), and more.

The study and creation of new materials for applications involving energy storage have grown significantly during the past several years. Supercapacitors (SCs), a type of energy storage system, have received much attention because of their excellent power density, acceptable energy density, significant rate performance, and long cycle life (Li et al. 2020). Metal-containing molecules with organic substituents (i.e., linkers) are bonded together to form open crystalline building blocks known as MOFs, which are promising electrode materials for energy storage applications. These MOFs have appealing properties like a large specific surface area and capabilities (Kim et al. 2019). For the direct formation of MOFs onto electrically conductive materials, a variety of chemical-based techniques, including hydrothermal, solvothermal, sonochemical, and microwave aided procedures, have been explored to date. Due to its straightforward processes, low-cost equipment, and ability to produce large quantities, electrochemical synthesis provides green synthetic approaches as an alternative (Worrall et al. 2016).

The study of MOF composite has become increasingly popular in the area of electrochemical capacitors and batteries. Due to their unique characteristics, including their enormous surface areas, organised units, ordered porous architectures, and great thermal stability, MOFs have gained increasing interest (Li et al. 2020, Wang et al. 2020). Based on the aforementioned factors, an optimal POAP/Zn-BTC composite has been created and manufactured using an electrochemical approach. Two benefits of the MOF green synthesis approach are its straightforward process and inexpensive cost. The findings show that the POAP/Zn-BTC composite, with its advantageous electrochemical characteristics, can be a useful electrode material for the fabrication of high-performance supercapacitors (Mostaanzadeh et al. 2021).

The category of high-porosity crystalline materials known as MOFs comprises inorganic clusters and organic linkers. Due to their large specific surface areas, variable pore diameter and activity, and adaptable structural topologies, MOFs have previously been used for a variety of applications, including gas storage and separation, catalyst, drug release, sensors, and supercapacitors and conversion (Furukawa et al. 2013). MOFs have poor electronic conductivity by nature, but because of their hybrid, extremely porous, and crystalline character, they are good materials for energy storage and conversion, such as in lithium-ion batteries and supercapacitors (Han et al. 2018). It has been shown to be a promising and effective method to increase the uses of MOFs in this field by using MOFs as sacrificial templates to produce porous structure and/or architecturally complicated carbon or metal oxides (Dang et al. 2017, Guan et al. 2017). Especially when it comes to porous carbons formed from MOFs, these offer prospects for great performance in supercapacitors, notably in EDLCs, as they frequently contain high surface areas, micro-/mesopores, and distinctive nanostructures (Sun et al. 2014, Xia et al. 2015).

Numerous benefits of MOFs, including the quantum size effect, conjugated bonds, exceptional electrolyte penetrability, and minimal steric hindrance, cause electrolyte diffusion and transfer of electrons to occur quickly, which is one of the primary reasons MOFs are so widely used in energy storage systems (Gupta et al. 2013). MOF supercapacitors (Yu et al. 2018) have recently received much interest because of their excellent power density, minimal cyclic retention, and extremely high capacitance (Zhang et al. 2013). However, research on the direct application of MOFs as electrode materials is still in its infancy, and only a small number of classes of ligand materials are known to exist. This is mostly because of their nonconductive qualities (Campagnol et al. 2014). Low conductivities and low cycle rates are characteristics of the MOFs constructed on benzenedicarboxylic acid (Givaja et al. 2012). As a result, conductive ligand substances like benzenetricarboxylic acid and hexaiminotriphenylene are highly sought after and the subject of extensive research as organic MOF components (Choi et al. 2012). The common drawbacks of those based on Maxenes (Ramachandran et al. 2018), polymers (Song et al. 2014), metal oxides (Zhang et al. 2016), and some other energy materials (Veeramani et al. 2017), which are restricted durability and short life cycles, are probably solved by a supercapacitor utilising MOFs. In the meanwhile, numerous investigations have shown that the synthesis procedures are extremely complex and energy consumable. Extensive conditions, including high pressure, high temperature, and high vacuum, are used throughout some procedures (Quah et al. 2015, Gao et al. 2017). A simple and energy-efficient technique is therefore highly desired.

2. Metal-Organic Frameworks

2.1 Physical and Chemical Properties

Low-dimensional metal-organic frameworks (LD MOFs) successfully combine the remarkable chemical and physical characteristics of LD nanomaterials, such as high aspect ratio, abundant accessible active sites, and flexibility, with the special characteristics of MOFs, such as huge surface area, tailorable structure, and uniform cavity, have gained increasing interest in recent years. In the past few years, significant advancements have been made in the morphological and structural regulation of LD MOFs. Further research into the dimensional-dependent characteristics and synthesis principles of LD MOFs is still extremely important (Liu et al. 2019). Similar to the size-dependent features of nanomaterials, shrinking the size of MOF crystals may result in increased in the specific surface area and density of active sites for MOFs, which would enhance performance. Numerous MOFs with 1-D structures, including nanorods, nanotubes, and nanofibers, have been developed thus far and have distinguished physicochemical characteristics (Wang et al. 2018).

However, MOFs have a few flaws that prevent their full potential from being utilised, such as low chemical stability. It is essential to add new functions and further improve the qualities of MOFs in order to fulfil their realistic usage. The development of high-performance composites with complex topologies is made possible by research on MOF composites. One MOF and one or more separate constituent materials, including other MOFs, are combined to form MOF composites or hybrids, which have properties significantly different from those of the individual parts. By effectively combining the benefits of MOFs (structural adaptability and flexibility, high porosity with ordered crystalline pores) and different types of functional materials (superior optical, electrical, magnetic, and catalytic characteristics), composite materials can explore various physical and chemical characteristics and improved performance that are not possible with the individual components (Lykourinou et al. 2011, Kreno et al. 2010). Each MOF-based composite represents a novel material with particular functional features thanks to its flawless command of compositional component, porosity, functionality, and shape. The gas adsorption and separation field, particularly CO_2 capture, H_2 and CH_4 purifications, and CH_4 storage, has been very interested in MOF-based materials with a high surface area, huge void spaces, and controlled surface characteristics (Zhang et al. 2016).

2.2 High Surface Area and Porosity

Porous coordination polymers (PCPs) are another name for MOFs, which have a high level of activity and a wide surface area (Furukawa et al. 2013, Zhai et al. 2011). They have been used in a variety of industries, including sensing (Dolgopolova et al. 2018), adsorption/separation (Dhaka et al. 2019), catalysis (Zheng et al. 2019), and medication delivery (Lázaro et al. 2020). MOFs have seen a rise in electrochemical applications in recent years, particularly electrochemical energy storage (Qiu et al. 2019). Furthermore, the majority of conventional MOFs have poor electrical conductivity, which significantly limits their uses, particularly as electrode materials (Zhang et al. 2021). Although calcination is useful for improving MOFs' conductivity, it may also result in framework collapse and channel damage (Zhong et al. 2021). Conductive MOFs (cMOFs) have gotten a lot of study attention because of ion transfer and improved conductivity. Excellent electrical conductivity is produced by the organic 2D structure that is atomically thin and has in-plane charge delocalisation (Ko et al. 2018). According to Sheberla et al. (2017), the MOF Ni_3 HHTP2 has a capacitance retention of > 90% when used as an electrode material for supercapacitors in an ionic liquid electrolyte of Et4N+ BF4−.

Porous coordination polymers, also known as MOFs, are a new kind of material with expanded pores and distinctive physical and chemical characteristics. MOFs create a special three-dimensional structure by the coordination or intramolecular bonding of metal ions or metal ion clusters with massive organic ligand molecules (Kreno et al. 2012). By either using MOFs directly as an electrode material or as a precursor for the electrode material, supercapacitors have successfully spread their popularity. This is so that the organic molecule can enable effective and reversible charge transfer and the metal ion within the framework functions as an active site for redox processes. By either encouraging redox reactions or maintaining the MOFs' structural integrity during potential cycling, organic ligand molecules play a significant role in the success of MOFs for their supercapacitive behaviour (Millward et al. 2005).

These MOFs are created by using organic ligands to reinforce the bindings between transition metals and oxygen/nitrogen (Chen et al. 2005). These organic ligands are frequently rich in carbon and heteroatoms like phosphorus and nitrogen. Numerous applications, such as chemical sensors, catalysis, batteries, supercapacitors, etc., have drawn attention to MOFs' tunability in terms of pore size, surface area, and surface functionalisation (Zhou et al. 2019).

2.3 Applications

The future development of MOF chemistry and its applicability to environmental domains will depend heavily on improvements in green synthesis techniques. It is possible to broaden the range of applications for MOFs in the environment by easily altering their fundamental features, such as pore size, choice of metal ions, and functional groups. The emphasis has switched to green synthesis methods that generate environmentally benign MOFs for a variety of applications, such as catalysis, adsorption/separation processes, and medicines, in order to reduce the dangers involved with synthesis (e.g., solvothermal approach) (Vikrant et al. 2019, Vikrant et al. 2020). The emphasis has switched toward green synthesis methods that generate environmentally acceptable MOFs for various applications, such as catalysis, adsorption/separation processes, and medicines, in order to reduce the risks involved with synthesis.

Since it can achieve many of the goals of green chemistry and aims to reduce the amount of materials and activation energy needed, catalysis is one of the most active research areas. Additional advantages of catalytic reactions include the possibility to recover the catalyst from the product mixture and reuse it in subsequent cycles of the reaction. By utilising glutamate as a natural ligand, researchers have shown that it is simple to create economically viable Zn-MOF (Kathalikkattil et al. 2016). Due to their potential to sustain high gravimetric and volumetric densities, porous MOFs have become increasingly popular as new adsorbents. Adsorption properties of MOFs are governed by the potency of electrostatic interactions between adsorbent and metal ion sites (Alkordi et al. 2017). Green MOF development has received a lot of research attention, notably for H_2 storage, which is the most environmentally friendly fuel choice.

Due to their production at high pressure, high temperature, and hazardous solvent conditions, traditional MOFs are not suitable for drug delivery. However, green MOFs are currently the most dependable choice for drug delivery applications due to their great biocompatibility and high drug-loading efficiency. Large pores, tunable functionality, high loading capacity, drug adsorption, and release ability, good thermochemical stability, and nontoxicity are just a few of the intriguing properties that zeolite imidazole frameworks (ZIF-8), a significant subclass of Zn-based MOFs, are offering for therapeutics (Kaur et al. 2017).

The early detection and diagnosis of diseases using medical imaging technology today has enhanced patient survival rates. Several imaging modalities, including computed tomography (CT), positron emission tomography (PET), ultrasound (US), and magnetic resonance imaging (MRI), have been developed for this (MRI). Due to their highly sensitive fluorescence, green MOFs offer a significant research potential as contrast agents for the imaging and quantification of biological processes, both at the molecular level and in single cells (Banerjee et al. 2020). Environmentally friendly synthesis of iron(III) carboxylate MOFs was described by Horcajada et al.

(2010) (MIL-101, MIL-100, MIL-89, MIL-88A, MIL-88Bt, and MIL-53). These MOFs (such as MIL-88A), which can operate as contrast agents, can deliver antitumor and retroviral medications effectively and under control, as shown by MRI tests on Wistar female rats. In addition, research is focused on a number of MOF applications, including those in medicine and biology, the separation of harmful compounds from gas and liquid, catalysts, energy storage, and environmental applications.

3. Various Synthesis Methods of MOFs

MOFs are now acknowledged as a sizeable group of porous compounds because of their distinctive operational and structural properties. Metal ion clusters, organic linkers, and metal ions combine to form these frameworks. The chosen primary building components are, without a doubt, another idea that plays a big role in the MOF's ultimate structure and qualities (PBUs). However, there are some additional synthetic procedures and variables that must be taken into account, including solvent, temperature, reaction time, pressure, and pH. Depending on the resulting structures and features, a variety of synthetic techniques, such as slow diffusion, hydrothermal (solvothermal), electrochemical, mechanochemical, microwave-aided, and sonochemistry methods, can be used to make MOFs (Safaei et al. 2019). Its corresponding synthesis methods are given in Figure 1.

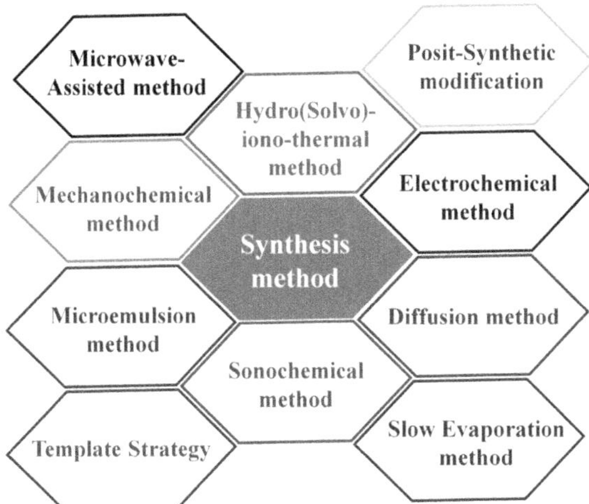

Figure 1. Various synthesis methods of MOFs.

3.1 Diffusion Method

The most common diffusion technique uses liquid-liquid diffusion, in which two layers of solvents—one acting as the precipitant solvent and the other enclosing the product—form due to different densities. In this approach, crystal formation occurs after the precipitant solvent diffuses gradually into a separate layer, or the reactants in two vials of different sizes are diffused gradually by physically separating them. Also, gels are sometimes used as diffusion and crystallisation mediums to slow down diffusion and stop the bulk material from precipitating.

3.2 Hydro (Solvo) Thermal Method

One of the most popular methods is the hydro (solvo) thermal approach, in which soluble precursors self-assemble to produce MOFs. Autogenous pressure is produced in this process inside a sealed environment (autoclave), and the operational temperature typically varies from 80°C to 260°C. When the reaction is finished, the temperature can be lowered at a programmed speed rate. In some circumstances, lengthy reaction times are necessary.

3.3 Microwave Method

The microwave approach is highly valued for the small-scale production of tiny metal and oxide particles. To produce crystals with this process, the solution's temperature may be increased for an hour or longer. Typically, this technique is not used frequently to create crystalline MOFs. More crucially, using this approach allows for high-speed material synthesis with precise control over particle size and form. Typically, the microwave method is unable to create crystals for a single X-ray study. However, it is possible to drastically decrease the cycle of synthesis for multiple activities and regulate the size and shape of the crystals.

3.4 Electrochemical Method

An electrochemical method is used to produce MOF powders on an industrial scale. This method has a number of benefits over solvothermal synthesis, including the avoidance of anions such as nitrates from metal salts, lower reaction temperatures, and exceptionally quick synthesis. It is believed that the reduction of unwanted crystal building during membrane construction is caused by the formation of metal ions *in situ* close to the support surface, which restricts crystallisation in the bulk step. Electrochemical synthesis techniques generate more parameters for fine-tuning than solvothermal techniques since it is simple to change the voltage or apply particular signals.

3.5 Mechanochemical Method

The mechanical chemical transformation, which shows the mechanical disruption of intramolecular bonds, can be used to conduct some chemical reactions and a number of physical processes. In modern chemistry, mechanochemistry is used to create co-crystals in synthetic chemistry, multicomponent (binary and higher) processes, polymer science, inorganic solid-state chemistry, and many other fields. This approach is widely used to synthesise MOFs for a variety of reasons. First, regarding environmental concerns, reactions can be conducted at room temperature without organic solvents. Second, quick reaction times can give quantifiable yields of products (10–16 min).

3.6 Sonochemistry Method

Sonochemistry studies the chemistry that occurs when a reaction mixture is exposed to strong ultrasonic waves. Ultrasound is cyclic mechanical with a frequency of 10

MHz or between 20 kHz and the highest range of human hearing vibration (Vaitsis et al. 2019). These intense vibrations cause activation in the form of microjets on the solid surface, which can erode, activate, and clean the surface by dispersing tiny particle agglomerations. Reactions may occur in the bulk medium or the cavity at the interface in a homogenous liquid with substantial shear pressures. The synthesis of organic and nanomaterials frequently uses sonochemistry. The sonochemical approach in MOF synthesis helps to create a rapid, low-cost, energy-efficient, and user-friendly process at room temperature. In the future, it might help scale up the synthesis of MOF in a short amount of time.

4. Green Synthesis of MOFs

Conventional methods of MOF synthesis are popular tools for a wide range of technological applications. The use of hazardous organic solvents, poisonous metal salts, organic linkers, and the requirement for fabrication at supercritical temperature and pressure make conventional synthesis pathways unsustainable from an ecological standpoint; additionally, the toxicity and health risks associated with the discharge of dangerous byproducts into the environment present difficulties. Green synthesis alternatives for both organic and inorganic materials are increasingly highly sought after as a result. The risks connected with conventional MOFs can be eliminated with the help of a green synthesis platform.

4.1 Significations of Green Synthesis

For the assembly of green MOFs, a number of factors need to be taken into account, including ambient synthesis conditions, benign solvents like water, and organic ligands that are bioinspired or produced from biomass. The stability of MOFs in aqueous media is the main obstacle to their green synthesis. Additionally, one of the constraints may come from the fact that most organic salts and linkers are insoluble in aqueous solvents. Creating green MOFs that offer excellent space-time yields is the main focus of current scientific study. It is crucial to replace conventional solvents with safer alternatives. Other significant obstacles to the continued development of green MOFs are the use of polytopic linkers and fewer hazardous metal salts.

4.2 Green Synthesised MOFs and MOF Composites

The most significant improvements in mechanochemically synthesised MOF applications, with a focus on environmentally friendly uses such as the adsorption of dangerous gases and volatile organic compounds, as well as the eradication of pollutants from contaminated water. Zinc clusters with terephthalic linkers make up the most common MOF archetype, designated as MOF-5. Although Yaghi's group initially created it two decades ago, several studies continue to be published, mostly focusing on its adsorption and storage capabilities. By employing renewable precursors and consuming less energy and solvent, synthesis can be made more environmentally friendly. Inherently green, mechanical synthesis offers quick chemical interactions between solids while avoiding using solvents (or with a small addition of solvents).

In 2013, ZnO was converted into ZIF-8 for the first time dry. Due to its excellent stability and the strong connections that Zn ions and imidazole linkers form, ZIF-8 is unquestionably one of the most researched zeolitic imidazole frameworks. Due to their remarkable catalytic activity, semiconductor characteristics, and very flexible frameworks, Fe-containing MOFs are of great interest. Iron is also a metal that is abundant on earth, making iron-based raw materials affordable and widely available. Some MOFs can change cellular characteristics in response to diverse stimuli, such as temperature, pressure, and solvent addition, without irreparably damaging their structural integrity. Particularly with regard to flexible MOFs like Cr-MIL-88B and Cr-MIL-53, this phenomenon happens (Głowniak et al. 2021).

5. Supercapacitor

Due to rising world population and living standards, the world's energy consumption will continue to rise in the ensuing decades. People are considering non-conventional energy sources as alternatives due to the rapidly declining supply of fossil fuels and increasing levels of greenhouse gas emissions. However, the availability of non-conventional energy sources like solar, wind, and tidal varies on the time of year. Alternative clean and sustainable energy sources should be investigated and used to address this issue. Today, high electric energy may be produced by transforming natural energy sources like the sun, wind, and sea tides into sustainable and renewable ones (Sharma et al. 2019). However, humans have little control over natural events because renewable resources depend on geographic and environmental elements like climate, location, or gravitational forces exerted on earth by the sun or moon. It is more crucial to use these renewable energy sources for continuous energy production rather than only using them to replace fossil fuels. Additionally, storing energy in a variety of forms is a crucial component that can address the issues. In many real-world applications, electrical energy storage systems play a key role. Future energy demands must be met; hence, it is imperative to develop lightweight, flexible, high-performance energy storage technology. Researchers have been motivated by this predicament to look into more advanced energy storage methods, including lithium-ion, sodium-ion, and supercapacitor batteries. Because they embrace the huge potential in a wide variety of applications, SCs have recently gained a lot of attention.

Two extremely porous electrodes, an electrolyte, and a separator that electrically isolate the two electrodes make up a supercapacitor. Figure 2 displays a typical supercapacitor schematic diagram. In comparison to conventional energy storage systems, supercapacitors provide outstanding qualities as energy storage devices due to their quick charge/discharge efficiency, long cycling life ($>$ 500,000 cycles), and high power density ($>$ 10 kW kg^{-1}). The most crucial element is the electrode material, and the electrochemical performance is greatly influenced by the electrode's surface area, electrical conductivity, wetting ability, and permeability of the electrolyte solutions (González et al. 2016, Ramya et al. 2023). Both positive and negative charges collect on the surfaces of the gadget when an external electric field is applied.

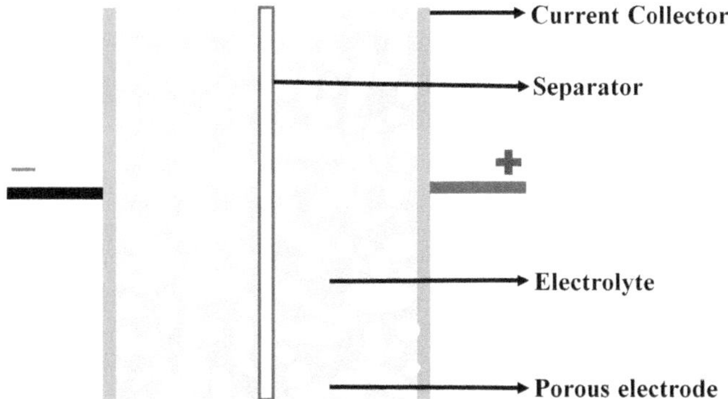

Figure 2. Schematic diagram of supercapacitors.

Ions in an electrolyte solution permeate through the separator and pass into the pores of the electrodes due to the natural attraction between opposite charges. As a result, a double layer of charge is produced at each electrode. Porous electrodes with large surface areas and closer spacing between the electrodes are chosen for supercapacitors to achieve high energy densities and capacitance (González et al. 2016). In hybrid automobiles, telecommunication equipment (remote communication, cell phones, walkie-talkies, satellites, etc.), memory backup systems, pacemakers, portable electronic gadgets, etc., supercapacitors are frequently utilised. Supercapacitors' fundamental strength, meanwhile, is that they have an equivalently lower energy capacity than batteries. Supercapacitors have been proven to be a promising solution to deal with the increased power demand of multi-functional gadgets in the twenty-first century due to their high power.

5.1 Types of Supercapacitors

Electrochemical supercapacitors are divided into a number of categories and subtypes in Figure 3. Supercapacitors are primarily categorised into three groups based on their charge storage mechanisms: electrochemical double-layer capacitors (EDLCs), pseudocapacitors, and hybrid capacitors (Chatterjee et al. 2021). The charge storage mechanism consists of processes that are non-faradic, faradic, and a combination of the two. There is no chemical mechanism involved in the non-Faradaic mechanism. Instead, charges are distributed on surfaces through physical processes rather than by the formation or dissolution of chemical bonds. Charge transfer between the electrolyte and electrode occurs during the Faradaic reactions. This section will provide an overview of these three supercapacitor classes, along with a breakdown of each subclass according to the electrode materials used.

5.2 Electrochemical Double-Layer Capacitors

Electrochemical double-layer capacitors (EDLCs) are constructed from a pair of carbon-based electrodes, an electrolyte, and a separator. Similar to traditional capacitors, EDLCs store charge electrostatically or through a non-Faradaic

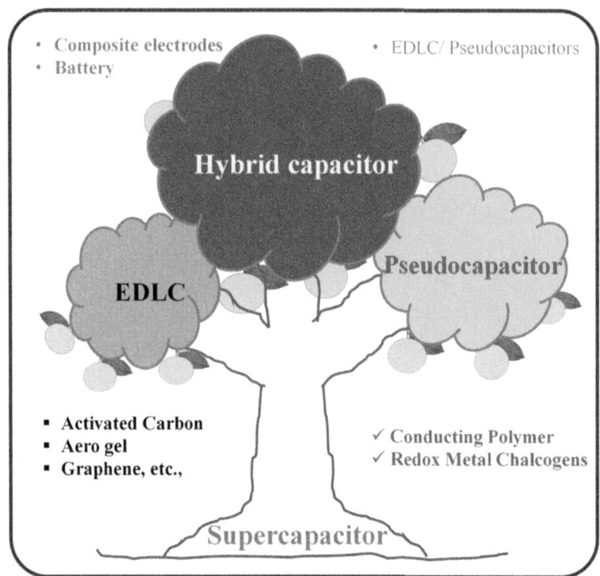

Figure 3. Different types of supercapacitors with examples.

mechanism; there is no charge transfer between the electrode and the electrolyte (Safaei et al. 2019). The energy storage principle used by EDLCs is the electrochemical double layer. There is an accumulation of charge on the electrode surface when voltage is supplied to EDLCs. Ions in the electrolyte solution diffuse over the separator and onto the pores of the opposing charged electrode as a result of the inherent attraction of unlike charges (Najib et al. 2019, Shameem et al. 2020). The electrodes, however, are designed to prevent ion recombination. As a result, at each electrode, a double layer of charge forms. Due to the multiple layers, increased specific surface area, and shorter electrode distances, EDLCs can achieve higher energy densities than traditional capacitors. In contrast, while discharging EDLCs, ions migrate away. Additionally, the charge storage mechanism used in EDLCs, which enables extremely quick energy absorption and distribution through a non-Faradaic process, is a key factor in their improved performance. This enables EDLCs to achieve extremely high cycling stabilities since charge storage in EDLCs is highly reversible. Generally, EDLCs operate with stable performance for millions of charge-discharge cycles than batteries that can be limited to only a few thousand cycles.

5.3 Pseudocapacitors

The energy storage mechanism of pseudocapacitors, in contrast to EDLC supercapacitors, is based on a transfer of charges through reversible Faradaic interactions between electrode and electrolyte. Pseudocapacitor materials are typically made of metal oxides, sulphides, conducting polymers, and phosphate (Najib et al. 2019, Ramesh et al. 2017). Pseudocapacitors produce ten times more capacitance and have a better energy density than EDLCs because their electrode surface has a higher capacity for storing charges than EDLCs. However, a number of characteristics, such as the electrode's surface area, the materials' conductivity and porosity, the size of the particle, etc., have an impact on pseudocapacitors. Pseudocapacitors also have a shorter lifespan than EDLCs because the active material delaminates due to electrochemical redox processes. Pseudocapacitive materials, on the other hand, are gaining popularity due to their high specific capacitance, high energy density, and quick and reversible redox processes.

5.4 Hybrid Capacitors

In order to enhance the overall performance of a device, the characteristics of EDLCs and pseudocapacitors were merged to create a new type of capacitor known as hybrid capacitors. Hybrid systems use both faradic and non-faradic processes for energy storage, which improves power and energy density without compromising cycle stability and device life (González et al. 2016). In hybrid systems, electrode topologies can be either metal oxide materials, which combine the benefits of physical and chemical charge storage mechanisms in a single electrode, or composite electrodes, which incorporate carbon-based materials and polymers. The coupling of carbon electrodes with pseudocapacitor electrodes has acquired a large relevance due to a significant rise in the device's overall capacitance (Sharma et al. 2019).

5.5 Applications of Supercapacitors

Applications for supercapacitors are numerous and vary from simple electronics like LEDs to defence and medical equipment. Due to their exceptional qualities, supercapacitors are widely employed in the automotive, energy, electronics, aerospace, medical, industrial, and other areas (Figure 4). High energy capacities are provided in large part by supercapacitor battery-based hybrid systems. The automotive applications of start-stop, light, ignition, electric, and hybrid vehicle drive support are some of the segments of the supercapacitors market (Ramesh et al. 2017, Murugan et al. 2021). The market for supercapacitors is significantly dominated by railroads and hybrid buses.

It is gaining from rising consumer and operator demand and the accessibility of reasonably priced vehicles. To ensure excellent performance and dependability, a self-healing electronic system also comprises supercapacitors, photodiodes, transistors, and other parts, including self-healing alloys and self-healing conductive ink. Finally, there are numerous biomedical implantable devices that supercapacitors can significantly improve, including automated drug administration systems, bladder, bone, and gastric pacemakers.

6. MOFs-Based Electrode Materials

MOFs are newly created porous materials made of organic ligands and metal centres. Regular 3D architectures are what distinguish MOFs, and their pore structures may be easily changed by changing the bridging organic ligands. As a result, MOFs are

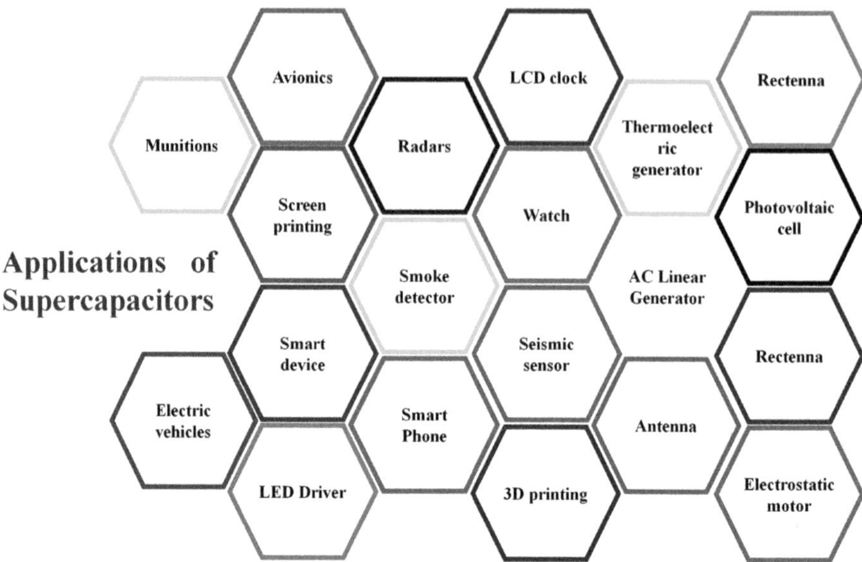

Figure 4. Different technological uses of supercapacitors.

effectively used in a variety of fields, such as gas storage and separation, pollutant adsorption, prolonged medication release, and selective catalysis. They also have potential as supercapacitor electrode materials because of their extraordinarily large specific surface areas and easily tunable pore conditions. Additionally, MOFs are perfect templates or precursors for the production of electrode functional nanomaterials since they have both metal and carbon sources in addition to their typical geometries.

Our research team recently revealed that zeolitic imidazolate frameworks-8 nanocrystals were investigated as an electrode material for supercapacitor application. The enhanced electrode's electrochemical performance was assessed in an aqueous KOH electrolyte. The addition of the nanoporous network improves ion transport in the electrode material and results in a high specific capacity of 122 Fg^{-1} at 5 mVs^{-1} and capacity retention of more than 85.36% after 5,000 continuous charge-discharge cycles. The highest power density that can be given is 3.86 $Wh\ kg^{-1}$, while the maximum power density that can be achieved is 1,250 $W\ kg^{-1}$. Low values of charge transfer resistance provide confirmation of this interface's high specific capacity. Due to the material's high porosity and, consequently, outstanding electrochemical characteristics, ZIF-8 nanostructures are built to create electrode materials that are guaranteed to have practical uses (Siva et al. 2022).

Supercapacitors' electrodes require a lot of cobalt and nickel, but their reserves are scarce and expensive. To create supercapacitors, researchers are attempting to use Zr-MOFs, Fe-MOFs, and other materials. The Zr-MOFs (UIO-66) were created by Tan et al. (2015), using zirconium chloride and hydroquinone at various temperatures (50–110°C) and used for electrochemical energy storage. The specific capacitance of the Zr-MOF is larger than that of the other Zr-MOFs when the scanning speed is

5 mV s1. Their internal resistances (Rs) are 0.42, 0.53, 0.60, and 0.62 Ω, respectively, which shows that the Zr-MOF electrode synthesised at 50°C has low internal resistance. According to a long-cycle test, the Zr-MOF specific capacitance of 654 F/g can still be maintained after 2,000 charging and discharging cycles.

Large specific surface area, strong conductivity, and superior chemical stability are all attributes of carbon material; however, its low specific capacitance and energy density are drawbacks. As a result, we blend MOFs with carbon-based materials to increase MOFs' conductivity and dispersion. Combinations of two materials can result in capacitive properties that are superior to those of the individual materials. These carbon materials offer exceptional qualities like superior conductivity, high contact surface area, and outstanding chemical and thermal stability. They also exhibit distinctive hollow structures and nanoscale mesh patterns that are interconnected. Due to their low specific surface area, low specific capacitance, high self-discharge, and ease of aggregation, CNTs, however, are unable to fulfil actual needs. It can be used in conjunction with MOFs to resolve these problems.

The Ni-MOF was grown directly on the surface of CNTs by Wen et al. (2015) to create Ni-MOF@CNT. CNTs are uniformly covered with Ni-MOF sheets in composites made of Ni-MOF and CNTs. For Ni-MOF/CNTs, a specific capacitance of 1,765 F/g at 0.5 A/g is possible, which is 1.6 times greater than the specific capacitance of 1,080 F g–1 for pure Ni-MOF. Additionally, the Ni-MOF@CNT still has 95% of its original capacity after 5,000 charge and discharge tests. The combined power of Ni-MOF and CNTs may be responsible for this domination. The all-composite capacitor's conductivity can be significantly increased by the CNT, but the Ni-MOF's porous structure can facilitate the entry of electrolytes.

Zhiming Cai et al. (2020) developed a one-pot green synthesis of bismuth and molybdenum metal-organic framework self-assembled dandelion-like bimetallic metal-organic framework with bismuth and molybdenum metal-organic frameworks as linker on nickel foam support (BM-MOF@NF) surface. From the GCD curves, the highest specific capacity of 476 C g–1 was achieved at a current density of 10 A g–1, and at an unusually high current density of 100 A g–1, the BMMOF@NF electrode is still 205 C can provide a specific capacity up to g1, suggesting a high rate capacity of 43%. The BM-MOF@NF electrode displayed a large specific capacity of 476 C g-1 at 10 A g–1 and a capacity retention of 81.2% after 8,000 cycles at 50 A g–1.

Graphene oxide (GO), carbon nanotubes (CNTs), and Fe_3O_4 magnetic nanoparticles (MNPs) were used as platforms for the loading of nanostructured CuBTC MOF and Fe_3O_4 MNPs in a straightforward, environmentally friendly solvothermal approach to create hybrid nanocomposites based on these materials. Pollutant adsorption tests were conducted on MP dye as a model organic pollutant. Pollutant adsorption was found to be significantly higher in the hybrid nanocomposites compared to the parent materials. The adsorption capacity of the composite nanomaterials was observed to have significantly improved for benefits resulting from the synthesis of the parent substance. In this regard, we discovered that the synthesis of MOF in the presence of GO and CNT platforms improved their formation and functionality, including higher dispersion forces, prevention of aggregation, fewer inter-MOF voids, and varied morphology and size. As a result of the *in situ* synthesis of Cu-BTC MOF, the stacking and aggregation of the substrates are also suppressed, and there are also fewer tiny pores created between the MOF and the substrates.

For the creation of MOF/G-based composites, the *in situ* technique is chosen since the connection between the two materials is sufficiently strong. This approach produced nanomaterials with various benefits for electrochemical use, including high electrical conductivity, low resistance, greater stability, and high electrical characteristics. Many researchers have created MOFs/G composites using the *in situ* method for a variety of electrochemical applications. By regulating the preparation temperature, Jabbari et al. (2016) reports G and N-doped porous carbon (HPNCs/rGO-800) nanomaterials are easily carbonised with ZIF-8. They concluded that whereas GO helped decrease aggregation and promote electrical conductivity, ZIF-8 inclusion led to poor surface activity. Results from the CV, CGD, and EIS confirm this assertion. At a constant current density of 1 A/g, the HPNCs/rGO-800 electrode exhibits exceptional supercapacitive characteristics with a high Cs of 245 F/g. Additionally, they created an asymmetric Sc using HPNCs/rGO-800 electrodes that were manufactured, resulting in great power densities of 18.0 Wh/kg and 475 W/kg with good power voltage (0–1.8 V).

Using ZIF-67 as a precursor, carbon polyhedron and carbon nanotube hybrids were created (Xin et al.2017). After producing consistent ZIF-67 polyhedral nanocrystals, carbon nanotubes were next produced using chemical vapour deposition. In HCN, exceptionally long carbon nanotubes are securely bonded to polyhedral porous carbon, displaying a characteristic network topology. By analysing the result, the HCN (HCN-650) electrode's growth temperature was determined to be 650°C. The highest capacitance at the scanning rate is 343 F/g 10 mV/s, and the electrochemical performance was remarkable.

Zhang et al. (2018) reports hydrolysed Ni-MOF in aqueous KOH. The solution is through an 'adaptive change' strategy to prepare $Ni(OH)_2$ with a monolayer structure. $Ni(OH)_2$ produced by soaking in 6 M KOH solution for six hours had the optimum electrochemical performance. At 5 A/g, 830.6 C/g specific capacity can be reached; at 20 A g–1, 461.2 C/g specific capacity can be sustained. The best crystallinity and porosity were found in the Ni(OH)2 material, which is favourable for ion migration and electron transport in the electrode. The performance of electrochemical reactions is significantly influenced by the shape of metal hydroxide.

7. Conclusions and Recommendations

The synthesis of application-oriented MOFs often focuses on computationally assisted framework design selection and screening. Along with materials design, synthesis technique difficulties, including yield, reaction efficiency, and associated environmental impact, must also be taken into account. Aside from using greener synthesis techniques, it is critical to evaluate these materials' suitability for actual field applications while taking environmental and human health considerations into account. The future development of MOF chemistry and its applicability to environmental domains will depend heavily on improvements in green synthesis techniques. It is possible to broaden the range of applications for MOFs in the environment by easily altering their fundamental features, such as pore size, choice of metal ions, and functional groups.

For the green MOF synthesis, the activation, and purification of the MOF, the amount of solvent used should be as minimal as feasible. Water is the recommended solvent for this because of its safety features and the technology that is available for purification and recycling. Organic solvents from renewable sources, such as ethanol or acetic acid, may also be excellent candidates for green synthesis. If the employed solvent is stable under the synthesis conditions and sufficiently recyclable, other organic solvents may also be utilised in green synthesis. This is because green properties are typically seen in connection to one another, and the synthesis process must first be developed to create the desired result. Alternative synthesis techniques, such as mechano-chemical synthesis or the use of ultrasound, can potentially enhance the process's energetic properties. However, a comprehensive comparison of energy characteristics for the synthesis of MOFs is lacking in the state of the research at this time.

Conventional methods of MOF synthesis are popular tools for a wide range of technological applications. The use of hazardous organic solvents, poisonous metal salts, and organic linkers, as well as the requirement for fabrication at supercritical temperature and pressure, make conventional synthesis pathways unsustainable from an ecological standpoint. Additionally, the toxicity and health risks associated with the discharge of dangerous byproducts into the environment present many difficulties. Green synthesis alternatives for both organic and inorganic materials are increasingly highly sought after as a result. The risks connected with conventional MOFs can be eliminated with the help of a green synthesis platform.

The current book chapter offers a thorough analysis of environmentally friendly methods for MOF synthesis. The book chapter addresses many approaches to building sustainable, low-impact MOFs. For the assembly of green MOFs, a number of factors need to be taken into account, including ambient synthesis conditions, benign solvents like water, and organic ligands that are bioinspired or produced from biomass. The stability of MOFs in aqueous media is the main obstacle to their green synthesis. Additionally, one of the constraints may come from the fact that most organic salts and linkers are insoluble in aqueous solvents. The creation of green MOFs that offer excellent space-time yields is the main focus of current scientific study. It is crucial to replace conventional solvents with safer alternatives. Other significant obstacles to the continued development of green MOFs are using polytopic linkers and fewer hazardous metal salts. Improved water purification, medication delivery, and greenhouse gas catalytic transformation should result from a greater understanding of the environmentally friendly procedures used in MOF production. Exploring new pathways for study in the synthesis of green MOFs can advance this discipline and increase the number of societally useful applications.

References

Abánades Lázaro, I., Wells, C. J. and Forgan, R. S. (2020). Multivariate modulation of the Zr MOF UiO-66 for defect-controlled combination anticancer drug delivery. *Angewandte Chemie*, 132(13): 5249–5255. https://doi.org/10.1002/ange.201915848.

Alkordi, M. H., Belmabkhout, Y., Cairns, A. and Eddaoudi, M. (2017). Metal–organic frameworks for H_2 and CH_4 storage: Insights on the pore geometry–sorption energetics relationship. *IUCrJ*, 4: 131. https://doi.org /10.1107/S2052252516019060.

Banerjee, S., Lollar, C. T., Xiao, Z., Fang, Y. and Zhou, H. -C. (2020). Biomedical integration of metal-organic frameworks. *Trends in Chemistry*, 2: 467–479. https://doi.org/10.1016/j.trechm.2020.01.007.

Cai, Z., Deng, L., Song, Y., Li, D., Hong, L. and Shen, Z. (2020). Facile synthesis of hierarchically self-assembled dandelion-like microstructures of bimetallic-MOF as a novel electrode material for high-rate supercapacitors. *Materials Letters*, 281: 128616. https://doi.org/10.1016/j.matlet.2020.128616.

Campagnol, N., Romero-Vara, R., Deleu, W., Stappers, L., Binnemans, K., De Vos, D. E. and Fransaer, J. (2014). A hybrid supercapacitor based on porous carbon and the metal-organic framework MIL-100 (Fe). *ChemElectroChem*, 1(7): 1182–1188. https://doi.org/10.1002/celc.201402022.

Chatterjee, D. P. and Nandi, A. K. (2021). A review of the recent advances in hybrid supercapacitors. *Journal of Materials Chemistry A*, 9(29): 15880–15918. https://doi.org/10.1039/D1TA02505H.

Chen, B., Ockwig, N. W., Millward, A. R., Contreras, D. S. and Yaghi, O. M. (2005). High H_2 adsorption in a microporous metal-organic framework with open metal sites. *Angewandte Chemie International Edition*, 44(30): 4745–4749. https://doi.org/10.1002/anie.200462787.

Cherianá Kathalikkattil, A. and Juneá Cho, S. (2016). A sustainable protocol for the facile synthesis of zinc-glutamate MOF: an efficient catalyst for room temperature CO_2 fixation reactions under wet conditions. *Chemical Communications*, 52(2): 280–283. https://doi.org/10.1039/C5CC07781H.

Choi, B. G., Chang, S. J., Kang, H. W., Park, C. P., Kim, H. J., Hong, W. H., Lee, S. and Huh, Y. S. (2012). High performance of a solid-state flexible asymmetric supercapacitor based on graphene films. *Nanoscale*, 4(16): 4983–4988. https://doi.org/10.1039/C2NR30991B.

Choudhury, A., Chandra, H. and Arora, A. (2013). Application of solid oxide fuel cell technology for power generation—A review. *Renewable and Sustainable Energy Reviews*, 20: 430–442. https://doi.org/10.1016/j.rser.2012.11.031.

Dang, S., Zhu, Q. L. and Xu, Q. (2017). Nanomaterials derived from metal-organic frameworks. *Nature Reviews Materials*, 3(1): 1–14. https://doi.org/10.1038/natrevmats.2017.75.

Dhaka, S., Kumar, R., Deep, A., Kurade, M. B., Ji, S. W. and Jeon, B. H. (2019). Metal-organic frameworks (MOFs) for the removal of emerging contaminants from aquatic environments. *Coordination Chemistry Reviews*, 380: 330–352. https://doi.org/10.1016/j.ccr.2018.10.003.

Dolgopolova, E. A., Rice, A. M., Martin, C. R. and Shustova, N. B. (2018). Photochemistry and photophysics of MOFs: Steps towards MOF-based sensing enhancements. *Chemical Society Reviews*, 47(13): 4710–4728. https://doi.org/10.1039/C7CS00861A.

Furukawa, H., Cordova, K. E., O'Keeffe, M. and Yaghi, O. M. (2013). The chemistry and applications of metal-organic frameworks. *Science*, 341(6149): 1230444. https://doi.org/10.1126/science.1230444.

Furukawa, H., Cordova, K. E., O'Keeffe, M. and Yaghi, O. M. (2013). The chemistry and applications of metal-organic frameworks. *Science*, 341(6149): 1230444. https://doi.org/10.1126/science.1230444.

Gao, W., Chen, D., Quan, H., Zou, R., Wang, W., Luo, X. and Guo, L. (2017). Fabrication of hierarchical porous metal-organic framework electrode for aqueous asymmetric supercapacitor. *ACS Sustainable Chemistry & Engineering*, 5(5): 4144–4153. https://doi.org/10.1021/acssuschemeng.7b00112.

Givaja, G., Amo-Ochoa, P., Gómez-García, C. J. and Zamora, F. (2012). Electrical conductive coordination polymers. *Chemical Society Reviews*, 41(1): 115–147. https://doi.org/10.1039/C1CS15092H.

Głowniak, S., Szczęśniak, B., Choma, J. and Jaroniec, M. (2021). Mechanochemistry: Toward green synthesis of metal-organic frameworks. *Materials Today*, 46: 109–124. https://doi.org/10.1016/j.mattod.2021.01.008.

González, A., Goikolea, E., Barrena, J. A. and Mysyk, R. (2016). Review on supercapacitors: Technologies and materials. *Renewable and Sustainable Energy Reviews*, 58: 1189–1206. https://doi.org/10.1016/j.rser.2015.12.249.

Guan, B. Y., Yu, X. Y., Wu, H. B. and Lou, X. W. (2017). Complex nanostructures from materials based on metal-organic frameworks for electrochemical energy storage and conversion. *Advanced Materials*, 29(47): 1703614. https://doi.org/10.1002/adma.201703614.

Gupta, R., Sanotra, S., Sheikh, H. N. and Kalsotra, B. L. (2013). Room temperature aqueous phase synthesis and characterization of novel nano-sized coordination polymers composed of copper (II), nickel (II), and zinc (II) metal ions with p-phenylenediamine (PPD) as the bridging ligand. *Journal of Nanostructure in Chemistry*, 3(1): 1–9. http://www.jnanochem.com/content/3/1/41.

Han, B., Cheng, G., Zhang, E., Zhang, L. and Wang, X. (2018). Three dimensional hierarchically porous ZIF-8 derived carbon/LDH core-shell composite for high performance supercapacitors. *Electrochimica Acta*, 263: 391–399. https://doi.org/10.1016/j.electacta.2017.12.175.

Horcajada, P., Chalati, T., Serre, C., Gillet, B., Sebrie, C., Baati, T., Eubank, J. F., Heurtaux, D., Clayette, P., Kreuz, C. and Chang, J. S. (2010). Porous metal-organic-framework nanoscale carriers as a potential platform for drug delivery and imaging. *Nature Materials*, 9(2): 172–178. https://doi.org/10.1038/nmat2608.

Jabbari, V., Veleta, J. M., Zarei-Chaleshtori, M., Gardea-Torresdey, J. and Villagrán, D. (2016). Green synthesis of magnetic MOF@GO and MOF@CNT hybrid nanocomposites with high adsorption capacity towards organic pollutants. *Chemical Engineering Journal*, 304: 774–783. https://doi.org/10.1016/j.cej.2016.06.034.

Jia, Z., Hao, S., Wen, J., Suoding Li, Peng, W., Renyao Huang and Xin Xu. (2020). Electrochemical fabrication of metal-organic frameworks membranes and films: A review. *Microporous and Mesoporous Materials*, 305: 110322. https://doi.org/10.1016/j.micromeso.2020.110322.

Kaur, H., Mohanta, G. C., Gupta, V., Kukkar, D. and Tyagi, S. (2017). Synthesis and characterization of ZIF-8 nanoparticles for controlled release of 6-mercaptopurine drug. *Journal of Drug Delivery Science and Technology*, 41: 106–112. https://doi.org/10.1016/j.jddst.2017.07.004.

Kim, H., Sohail, M., Wang, C., Rosillo-Lopez, M., Baek, K., Koo, J., Seo, M. W., Kim, S., Foord, J. S. and Han, S. O. (2019). Facile one-pot synthesis of Bimetallic Co/Mn-MoFs@ Rice Husks, and its Carbonization for supercapacitor electrodes. *Scientific Reports*, 9(1): 1–10. https://doi.org/10.1038/s41598-019-45169-0.

Ko, M., Mendecki, L. and Mirica, K. A. (2018). Conductive two-dimensional metal-organic frameworks as multifunctional materials. *Chemical Communications*, 54(57): 7873–7891. https://doi.org/10.1039/C8CC02871K.

Kreno, L. E., Hupp, J. T. and Van Duyne, R. P. (2010). Metal-organic framework thin film for enhanced localized surface plasmon resonance gas sensing. *Analytical Chemistry*, 82(19): 8042–8046. https://doi.org/10.1021/ac102127p.

Kreno, L. E., Leong, K., Farha, K., Allendorf, M., Van Duyne, R. P. and Hupp, J. T. (2012). Metal-organic framework materials as chemical sensors. *Chemical Reviews*, 112(2): 1105–1125. https://doi.org/10.1021/cr200324t.

Kreno, L. E., Leong, K., Farha, O. K., Allendorf, M., Van Duyne, R. P. and Hupp, J. T. (2012). Metal-organic framework materials as chemical sensors. *Chemical Reviews*, 112(2): 1105–1125. https://doi.org/10.1021/cr200324t.

Leus, K., Liu, Y. Y. and Van Der Voort, P. (2014). Metal-organic frameworks as selective or chiral oxidation catalysts. *Catalysis Reviews*, 56(1): 1–56. https://doi.org/10.1080/01614940.2014.864145.

Li, J. R., Sculley, J. and Zhou, H. C. (2012). Metal-organic frameworks for separations. *Chemical Reviews*, 112(2): 869–932. https://doi.org/10.1021/cr200190s.

Li, J., Luo, S., Wang, C., Tang, Q., Wang, Y., Han, X., Ran, H., Wan, J., Gu, X., Wang, X. and Hu, C. (2020). Low Li ion diffusion barrier on low-crystalline FeOOH nanosheets and high performance of energy storage. *Nano Research*, 13(3): 759–767. https://doi.org/10.1007/s12274-020-2691-2.

Li, X., Yang, X., Xue, H., Pang, H. and Xu, Q. (2020). Metal-organic frameworks as a platform for clean energy applications. *Energy Chem.*, 2(2): 100027. https://doi.org/10.1016/j.enchem.2020.100027.

Liu, W., Yin, R., Xu, X., Zhang, L., Shi, W. and Cao, X. (2019). Structural engineering of low-dimensional metal-organic frameworks: Synthesis, properties, and applications. *Advanced Science*, 6(12): 1802373. https://doi.org/10.1002/advs.201802373.

Lykourinou, V., Chen, Y., Wang, X. S., Meng, L., Hoang, T., Ming, L. J., Musselman, R. L. and Ma, S. (2011). Immobilization of MP-11 into a mesoporous metal-organic framework, MP-11@

mesoMOF: A new platform for enzymatic catalysis. *Journal of the American Chemical Society*, 133(27): 10382–10385. https://doi.org/10.1021/ja2038003.

Millward, A. R. and Yaghi, O. M. (2005). Metal-organic frameworks with exceptionally high capacity for storage of carbon dioxide at room temperature. *Journal of the American Chemical Society*, 127(51): 17998–17999. https://doi.org/10.1021/ja0570032.

Mostaanzadeh, H., Shahmohammadi, M., Ehsani, A., Moharramnejad, M. and Honarmand, E. (2021). Green-synthesized Zn-BTC metal-organic frameworks as a highly efficient material to improving electrochemical pseudocapacitance performance of P-type conductive polymer. *Journal of Materials Science: Materials in Electronics*, 32(22): 26539–26547. https://doi.org/10.1007/s10854-021-07030-x.

Murugan, A., Siva, V., Shameem, A. and Asath Bahadur, S. (2021). Optimization of adsorption and reaction time of SILAR deposited Cu_2ZnSnS_4 thin films: structural, optical and electrochemical performance. *J. Alloys Compd.*, 856: 158055. https://doi.org/10.1016/j.jallcom.2020.158055.

Najib, S. and Erdem, E. (2019). Current progress achieved in novel materials for supercapacitor electrodes: Mini-review. *Nanoscale Advances*, 1(8): 2817–2827. https://doi.org/10.1039/C9NA00345B.

Qiu, M., Sun, P., Cui, G., Tong, Y. and Mai, W. (2019). A flexible microsupercapacitor with integral photocatalytic fuel cell for self-charging. *ACS Nano*, 13(7): 8246–8255. https://doi.org/10.1021/acsnano.9b03603.

Quah, H. S., Chen, W., Schreyer, M. K., Yang, H., Wong, M. W., Ji, W. and Vittal, J. J. (2015). Multiphoton harvesting metal-organic frameworks. *Nature Communications*, 6(1): 7954–7961. https://doi.org/10.1038/ncomms8954.

Ramachandran, R., Rajavel, K., Xuan, W., Lin, D. and Wang, F. (2018). Influence of Ti3C2Tx (MXene) intercalation pseudocapacitance on electrochemical performance of CoMOF binder-free electrode. *Ceram. Int.*, 42: 14425–14431. https://doi.org/10.1016/j.ceramint.2018.05.055.

Ramesh, S., Haldorai, Y., Kim, H. S. and Kim, J. H. (2017). A nanocrystalline Co3O4@ polypyrrole/MWCNT hybrid nanocomposite for high performance electrochemical supercapacitors. https://doi.org/10.1039/c7ra06093a.

Ramya, W., Siva, V., Murugan, A., Shameem, A., Kannan, S. and Venkatachalam, K. (2023). A novel biodegradable polymer-based hybrid nanocomposites for flexible energy storage systems. *Journal of Polymers and the Environment*, 31: 1610–1627. https://doi.org/10.1007/s10924-022-02695-9.

Safaei, M., Foroughi, M. M., Ebrahimpoor, N., Jahani, S., Omidi, A. and Khatami, M. (2019). A review on metal-organic frameworks: Synthesis and applications. *TrAC Trends in Analytical Chemistry*, 118: 401–425. https://doi.org/10.1016/j.trac.2019.06.007.

Shameem, A., Devendran, P., Siva, V., Murugan, A., Sasikumar, S., Nallamuthu, N., Hussain, S. and Bahadur, S. A. (2020). Robust one-step synthesis of bismuth molybdate nanocomposites: A promising negative electrode for high-end ultracapacitors. *Solid State Sciences*, 106: 106303. https://doi.org/10.1016/j.solidstatesciences.2020.106303.

Sharma, K., Arora, A. and Tripathi, S. K. (2019). Review of supercapacitors: Materials and devices, *Journal of Energy Storage*, 21: 801–825. https://doi.org/10.1016/j.est.2019.01.010.

Sheberla, D., Bachman, J. C., Elias, J. S., Sun, C. J., Shao-Horn, Y. and Dincă, M. (2017). Conductive MOF electrodes for stable supercapacitors with high areal capacitance. *Nature Materials*, 16(2): 220–224. https://doi.org/10.1038/nmat4766.

Siva, V., Murugan, A., Shameem, A., Thangarasu, S. and Bahadur, S. A. (2022). A simple synthesis method of Zeolitic Imidazolate Framework-8 (ZIF-8) nanocrystals as superior electrode material for energy storage systems. *Journal of Inorganic and Organometallic Polymers and Materials*, 32(12): 4707–4714. https://doi.org/10.1007/s10904-022-02475-x.

Song, X., Yang, S., He, L., Yana, S. and Liao, F. (2014). Ultra-flyweight hydrophobic poly(m-phenylenediamine) aerogel with micro-spherical shell structures as a high-performance selective adsorbent for oil contamination. *RSC Advances*, 4: 49000–49005. https://doi.org/10.1039/C4RA09080B.

Suh, M. P., Park, H. J., Prasad, T. J. and Lim, D.W. (2012). Hydrogen storage in metal-organic frameworks. *Chemical Reviews*, 112: 782–835. https://doi.org/10.1021/cr200274s.

Sun, J. K. and Xu, Q. (2014). Functional materials derived from open framework templates/precursors: Synthesis and applications. *Energy & Environmental Science*, 7(7): 2071–2100. https://doi.org/10.1039/C4EE00517A.

Tan, Y., Zhang, W., Gao, Y., Wu, J. and Tang, B. (2015). Facile synthesis and supercapacitive properties of Zr-metal organic frameworks (UiO-66). *RSC Adv.*, 5(23): 17601–17605. https://doi.org/10.1039/c4ra11896k.

Vaitsis, C., Sourkouni, G. and Christos Argirusis. (2019). Metal Organic Frameworks (MOFs) and ultrasound: A review. *Ultrasonics Sonochemistry*, 52: 106–119. https://doi.org/10.1016/j.ultsonch.2018.11.004.

Veeramani, V., Sivakumar, M., Chen, S. M., Madhu, R., Alamri, H. R., Alothman, Z. A., Hossain, M. S. A., Chen, C. K., Yamauchi, Y., Miyamoto, N. and Wu, K. C. W. (2017). Lignocellulosic biomass-derived, graphene sheet-like porous activated carbon for electrochemical supercapacitor and catechin sensing. *RSC Advances*, 7: 45668–45675. https://doi.org/10.1039/C7RA07810B.

Vikrant, K., Kim, K. H., Kumar, S. and Boukhvalov, D. W. (2020). Metal-organic frameworks for the adsorptive removal of gaseous aliphatic ketones. ACS Applied Materials & Interfaces, 12(9): 10317–10331. https://doi.org/10.1021/acsami.9b20375.

Vikrant, K., Na, C. J., Younis, S. A., Kim, K. H. and Kumar, S. (2019). Evidence for superiority of conventional adsorbents in the sorptive removal of gaseous benzene under real-world conditions: Test of activated carbon against novel metal-organic frameworks. *Journal of Cleaner Production*, 235: 1090–1102. https://doi.org/10.1016/j.jclepro.2019.07.038.

Wang, K. B., Xun, Q. and Zhang, Q. (2020). Recent progress in metal-organic frameworks as active materials for supercapacitors. *Energy Chem.*, 2(1): 100025. https://doi.org/10.1016/j.enchem.2019.100025.

Wang, L., Zheng, M. and Xie, Z. (2018). Nanoscale metal-organic frameworks for drug delivery: A conventional platform with new promise. *Journal of Materials Chemistry B*, 6(5): 707–717. https://doi.org/10.1039/C7TB02970E.

Wen, P., Gong, P., Sun, J., Wang, J. and Yang, S. (2015). Design and synthesis of Ni-MOF/CNT composites and rGO/carbon nitride composites for an asymmetric supercapacitor with high energy and power density. *J. Mater. Chem. A*, 3(26): 13874–13883. https://doi.org/10.1039/c5ta02461g.

Worrall, S. D., Mann, H., Rogers, A., Bissett, M. A., Attfield, M. P. and Dryfe, R. A. (2016). Electrochemical deposition of zeolitic imidazolate framework electrode coatings for supercapacitor electrodes. *Electrochimica Acta*, 197: 228–240. https://doi.org/10.1016/j.electacta.2016.02.145.

Xia, W., Mahmood, A., Zou, R. and Xu, Q. (2015). Metal-organic frameworks and their derived nanostructures for electrochemical energy storage and conversion. *Energy & Environmental Science*, 8(7): 1837–1866. https://doi.org/10.1039/C5EE00762C.

Xin, L., Liu, Q., Liu, J., Chen, R., Li, R., Li, Z. and Wang, J. (2017). Hierarchical metal-organic framework derived nitrogen-doped porous carbon/graphene composite for high performance supercapacitors. *Electrochim. Acta.*, 248: 215–224. https://doi.org/10.1016/j.electacta.2017.07.138.

Yu, H., Xia, H., Zhang, J., He, J., Guo, S. and Xu, Q. (2018). Fabrication of Fe-doped Co-MOF with mesoporous structure for the optimization of supercapacitor performances. *Chinese Chemical Letters*, 29(6): 834–836. https://doi.org/10.1016/j.cclet.2018.04.008.

Zhai, Y., Dou, Y., Zhao, D., Fulvio, P. F., Mayes, R. T. and Dai, S. (2011). Carbon materials for chemical capacitive energy storage. *Advanced Materials*, 23(42): 4828–4850. https://doi.org/10.1002/adma.201100984.

Zhang, G., Jin, L., Zhang, R., Bai, Y., Zhu, R. and Pang, H. (2021). Recent advances in the development of electronically and ionically conductive metal-organic frameworks. *Coordination Chemistry Reviews*, 439: 213915. https://doi.org/10.1016/j.ccr.2021.213915.

Zhang, H., Xu, B., Xiao, Z., Mei, H., Zhang, L., Han, Y. and Sun, D. (2018). Optimizing crystallinity and porosity of hierarchical $Ni(OH)_2$ through conformal transformation of metal-organic framework template for supercapacitor applications. *CrystEngComm.*, 20(30): 4313–4320. https://doi.org/10.1039/c8ce00741a.

Zhang, Q., Li, B. and Chen, L. (2013). First-principles study of microporous magnets M-MOF-74 (M= Ni, Co, Fe, Mn): The role of metal centers. *Inorganic Chemistry*, 52(16): 9356–9362. https://doi.org/10.1021/ic400927m.

Zhang, W., Ma, C., Fang, J., Cheng, J., Zhang, X., Dong, S. and Zhang, L. (2013). Asymmetric electrochemical capacitors with high energy and power density based on graphene/CoAl-LDH and activated carbon electrodes. *Rsc Advances*, 3(7): 2483–2490. https://doi.org/10.1039/C2RA23283A.

Zhang, Y., Lin, B., Wang, J., Tian, J., Sun, Y., Zhang, X. and Yang, H. (2016). All-solid-state asymmetric supercapacitors based on ZnO quantum dots/carbon/CNT and porous N-doped carbon/CNT electrodes derived from a single ZIF-8/CNT template. *Journal of Materials Chemistry A*, 4: 10282–10293. https://doi.org/10.1039/C6TA03633C.

Zhang, Y., Yuan, S., Feng, X., Li, H., Zhou, J. and Wang, B. (2016). Preparation of nanofibrous metal-organic framework filters for efficient air pollution control. *Journal of the American Chemical Society*, 138(18): 5785–5788. https://doi.org/10.1021/jacs.6b02553.

Zheng, S., Guo, X., Xue, H., Pan, K., Liu, C. and Pang, H. (2019). Facile one-pot generation of metal oxide/hydroxide metal-organic framework composites: highly efficient bifunctional electrocatalysts for overall water splitting. *Chemical Communications*, 55: 10904–10907. https://doi.org/10.1039/C9CC06113D.

Zhong, L., Zhou, H., Li, R., Bian, T., Wang, S. and Yuan, A. (2021). *In situ* confinement pyrolysis of ZIF-67 nanocrystals on hollow carbon spheres towards efficient electrocatalysts for oxygen reduction. *Journal of Colloid and Interface Science*, 584: 439–448. https://doi.org/10.1016/j.jcis.2020.10.020.

Zhou, S., Kong, X., Zheng, B., Huo, F., Strømme, M. and Xu, C. (2019). Cellulose nanofiber@ conductive metal-organic frameworks for high-performance flexible supercapacitors. *ACS Nano*, 13(8): 9578–9586. https://doi.org/10.1021/acsnano.9b04670.

CHAPTER 9

Electrodes Preparation From Biomass

Tatiana Kuleshova

1. Introduction

Nowadays, an increase in demand for energy sources is being observed, which entails an increase in the consumption of major fossil fuels. This causes two main problems—the depletion of fossil fuels and the deterioration of the environmental situation. In this regard, it is important to develop environmentally friendly alternative renewable energy sources that can partially satisfy the growing demand for global energy.

One of the clean energy sources is biomass, in particular, the products of its processing. One of the biomass processing perspectives is the production of activated carbon, which is promising as an electrode material.

The development of progressive renewable and environmentally friendly energy technologies is the solution to a number of numerous environmental problems. Recently, more and more attention has been attracted to supercapacitors—electrochemical devices for energy storage, which simultaneously have high power and energy intensity. Compared to conventional capacitors, they have a higher energy density due to their porous structure and energy storage mechanism—the capacity of supercapacitors can reach thousands of F. Supercapacitors show promise as a replacement for the main electrochemical energy storage (batteries and capacitors), which makes green energy competitive with fossil fuels. Research in this area currently focuses on developing new carbon materials with higher porosity. The increased porosity of the electrodes will provide stronger interactions between the electrode material and adsorbed electrolyte ions and result in higher specific capacitance.

Agrophysical Research Institute, Saint-Petersburg, Grazhdanskiy pr. 14, 195220, Russia.
Email: kuleshova@agrophys.ru

Activated carbon is a substance that contains a huge number of pores. It has a very large specific surface area per unit mass and has a high adsorption capacity. Activated carbon is obtained from various carbonaceous materials of organic origin. Among the currently used electrode materials, activated carbon has outstanding characteristics: excellent electrochemical performance, high specific surface area, high adsorption, fast ion/electron transport, tunable surface chemistry, abundant functional moieties, and low cost (Shaker et al. 2021). All this made them perspective candidates as electrodes for supercapacitors. These advantages can be enhanced if activated carbon is obtained from biomass precursors. In recent days, researchers have focused on biomass because it is renewable and abundant, easy to process, and environmentally friendly.

The purpose of this work was to summarize the available data and to identify the prospects for using biomass as a precursor for activated carbon production for electrode systems of electrochemical devices.

2. Materials for Electrodes

Electrical energy storage devices typically consist of electrodes in contact with electrolyte that stores electricity via electrochemical reaction at a desired voltage (Azwar et al. 2018). Common materials used as electrodes for electrochemical devices include conducting metal oxides, polymers, and porous materials such as activated carbon, carbon nanotubes, and carbon aerogels (Farma et al. 2013b). Metals, synthetic polymers, and graphite have been widely used as active materials for electrochemical device electrodes, but they have their advantages and disadvantages described in Table 1. In most of these systems, carbon in some form is needed to ensure electrical conductivity, considering the high resistivity of many metal oxide materials since easy electron transportation is required in the direction of the electrode-electrolyte interface (Dos Reis et al. 2020).

Activated carbon (AC) is predominantly amorphous in nature and has a high porosity, primarily due to the manufacturing process. Activated carbon is advantageous because of its high surface area and porosity, good thermal and electrical conductivity, good anti-causticity, high stability, low cost, and commercial-scale availability (Pandolfo and Hollenkamp 2006). Several carbon sources, such as carbon black, activated carbon, graphene, carbon nanotubes, carbide-derived carbon, or carbon aerogels, have been employed as sole electrode materials or components for hybrid electrodes (Dos Reis et al. 2020).

Carbon materials have various dimensions (Bi et al. 2019), including one-dimensional (1D) biomass nanostructures (fibrous and tubular structures), two-dimensional (2D) carbonaceous materials with abundant sp2 hybridizations, three-dimensional (3D) biomass carbon material with well-interconnected small and large pores (Table 2).

Activated carbon has excellent performance among currently used materials. Their superior electrochemical performance, high specific surface area, high adsorption, tunable surface chemistry, fast ion/electron transport, abundant functional moieties, low cost, and abundance have made them promising candidates as electrochemical capacitor electrodes (Shaker et al. 2021).

Table 1. Advantages and disadvantages of using different materials as electrodes.

Electrode material	Advantages	Disadvantages
Metals	• High efficiency in energy storage.	• Environmental hazard. • Toxic materials (such as lead and ruthenium) could accumulate in the human body and cause various diseases. Bottleneck for sustainable development and safety concern.
Polymer	• High energy storage.	• Poor life cycle. • Can swell, shrink, and overoxidize during a pseudocapacitive reaction.
Graphite	• Fast electron movement • High electrical conductivity	• Nonrenewable. • Complex, expensive, and energy-consuming synthesis. • Can aggregate during fabrication; as a result, the electrochemical characteristics of the electrodes are reduced.
Biomass	• "Carbon-neutral" • Renewable • Widely available • Does not harm the environment • Good cycling stability • Have a lower material cost	• Low energy density in energy storage application.

Most of the commercial activated carbons are produced from fossil fuel-based precursors (petroleum and coal), which made them expensive and environmentally non-friendly, hence the increasing focus on biomass precursors, which are renewable, readily available, structurally porous, cheaper, and environmentally friendly (Farma et al. 2013b). Agricultural by-products and waste materials are among the low-cost precursors for producing AC (Hesas et al. 2013).

3. Biomass Composition

The International Union of Pure and Applied Chemistry (IUPAC) defines biomass as the material produced during the biological growth of microorganisms, plants, animals, and other living organisms.

AC electrodes can be produced from biomass precursors and have been tested for supercapacitor applications. Biomass precursors are cheaper, readily available, environmentally friendly, renewable, and structurally porous (Farma et al. 2013b).

The preparation techniques, microstructure, and chemical composition of biomass-derived carbons directly impact their final composition and efficiency. Carbon has the highest ratio in biomass composition; oxygen and hydrogen are the main constituents of biomass after carbon; nitrogen and sulfur are minor constituents; ash and small quantities of other elements, including alkali metals, alkaline earth metals, and heavy metals, are other constituents of biomass (Shaker et al. 2021).

A wide variety of waste biomass materials include (Abioye and Ani 2015, Wang et al. 2015) waste coffee beans, cassava peel waste, apricot shell, sugarcane

Table 2. Various dimensions of the biomass carbon materials.

Characteristic	Dimension of biomass carbon materials		
	One-dimensional (1D)	**Two-dimensional (2D)**	**Three-dimensional (3D)**
Advantages	• Excellent mechanical properties. • Reactive hydroxyl group on the surface of the fiber can be chemically modified with other highly active materials. • Fiber geometries with a high aspect ratio can be collected in a bond-free and interconnected network to form a strong film or substrate. • Due to the nanoscale effect, good braidability, high specific surface area, thermal stability, and easy processing of nanofibers. Processing into a thermally stable nonporous membrane with controlled pore structures is easy. • One-dimensional linear channels can provide a direct current path, which can facilitate the transmission of electrons compared to particle electrodes.	• The strong in-plane covalent bond feature makes the unique two-dimensional structure. • High in-plane conductivity, which can accelerate the transmission speed of electrons in the plane. • It can shorten ion transmission distance in the thin dimension. • The large and open flat surface fully exposes the surface atoms on both sides of the two-dimensional carbon nanosheet, which not only provides a high specific surface area but also provides a rich electrochemical active site for electrochemical reactions, thus providing high electrode-electrolyte interfaces and reducing the ionic resistance. • A rich active surface edge and its in-plane defect active sites facilitate charge storage, resulting in higher specific capacitance. • Long-cycle stability is provided by internal free space and damping of volume changes that occur during charge/discharge cycles.	• Continuous electron pathway. • Good electrical contact. • It can contribute to the acceleration of ions transfer by shortening the diffusion pathways, which is essential for designing high-performance electrode materials.
Biomass precursors	• Biomass with fibrous structure (flax, ramie, bacterial cellulose) or tubular structure (cotton, kapok, and willow catkins).	• Puffed rice, bougainvillea flower, silk, biomass wastes, perilla frutescens, eggplant.	• Honeycomb-like porous carbon, sodium lignosulfonate, sugarcane, bagasse, corn straw, garlic skin, sunflower seed, shells, 3D carbon aerogel (cellulose-lignin resorcinol formaldehyde, chitosan, aminated flavonoid tannin), 3D flexible carbon film.

bagasse, rice husk, sunflower seed shell, coffee endocarp, rubber wood sawdust, oil palm empty fruit bunch, camellia oleifera shell, poplar wood, argan seed shell, bamboo species, peanut shell, palm kernel shell, celtuce leaves, tea leaves, fermented rice, waste newspaper, bacteria, fungi, animal's bones and animal's feather, etc., have gained much attention as precursor materials to prepare porous carbons for electrochemical devices due to their abundant availability and low cost.

During the thermal treatment of the biomass precursors, hemicellulose, cellulose, and lignin decompose at different rates and within distinct temperature ranges (Cagnon et al. 2009). While lignin is pyrolyzed over an extensive temperature range and shows the behavior characteristic of solid fuels, hemicellulose, and cellulose decomposition is sharp in a narrow temperature range (Cagnon et al. 2009).

Biochar-derived carbon material includes (Rawat et al. 2022) porous carbon material with pore size below 2 nm, heteroatom-doped biochar (with p-block elements such as N, P, or S), metal-doped biochar, biomass-derived carbon nanotubes with diameters varying from 0.8 to 2 nm for single-walled and 5–20 nm for multi-walled, biomass-derived graphene, biochar derived carbon quantum dots.

Due to porous carbons, high surface area has been widely used as the electrode material for supercapacitors. Mesoporous carbons with a high surface area of 1,527–1,634 m^2/g for supercapacitors were made from biomass such as peanut shells and rice husk by one-step zinc chloride ($ZnCl_2$) activation assisted with microwave heating (He et al. 2013).

Tubular-like porous carbon from natural waste American poplar fruit as a carbon precursor is presented by Kumar et al. (2020). The derived tubular-like porous carbon exhibits an arranged porous structure, an enhanced surface area of 942 m^2/g, and a specific capacitance of 423 F/g at 1 A/g.

4. Carbon Preparation From Biomass

The preparation and production of activated carbon usually consist of carbonization and activation that can be done either separately in a two-stage process or combined in a single-stage process (Abioye and Ani 2015).

The carbonization of a precursor is usually carried out in the absence of oxygen at a temperature between 400°C and 850°C. In the traditional method, the heat source is located outside the carbon layer. The generated heat is transferred to the particles through the mechanisms of convection, conduction, and radiation, so the surface of the sample is heated before the internal parts, and there is a temperature gradient from the surface to the inside of each particle (Thostenson and Chou 1999).

Using a microwave radiation method is a possible way to solve the problems of the thermal gradient and the high cost of AC preparation (Hesas et al. 2013). Carbon materials are very good absorbers of microwaves. The microwaves interact directly with the particles inside the pressed compact material, resulting in fast and uniform heating, so the use of microwave radiation causes higher sintering temperatures, shorter processing times, and therefore higher energy savings (Thakur et al. 2007).

The main characteristics of microwave radiation on resulting AC parameters include:

(1) Microwave power: On the one hand, there is no continuous reaction between the char produced in the carbonization stage and the activator at low microwave power levels; on the other hand, at higher microwave power levels, carbon is burned, and the porous structure is destroyed, resulting in a decrease in adsorption (Hesas et al. 2013).

(2) Radiation time: The optimal value due to the development of porosity increases the adsorption capacity, but its further increase destroys microporous structures and, vice versa, reduces the adsorption capacity.

(3) Impregnation ratio in chemical activation: An increase in the ratio of the agent, on the one hand, can enhance the activation stage. On the other hand, beyond the optimal impregnation value, the pores can be blocked by an excess of the agent and combustion, which will decrease the available area.

(4) Surface Chemistry: Special effects can increase the carbon/oxygen ratio due to the elimination of acidic oxygen-containing functional groups and the production of basic AC.

Microwave radiation offers the following advantages over conventional heating methods: energy transfer instead of heat transfer, smaller steps, lower activation temperature, selective heating, immediate startup and shutdown, improved safety, simplicity, smaller equipment size, and less automation (Hesas et al. 2013).

Farma et al. (2013a) suggested a procedure for preparing electrodes by varying the carbonization temperature during the preparation of activated carbon monoliths from carbon nanotubes and activated carbon from empty fruit bunch fibers based on a combination of chemical (KOH) and physical activation agent (CO_2). They observed that the changes in the density, electrical conductivity, porosity, microstructure and structure of the activated carbon monoliths due to the change in the carbonization temperature notably affected the electrochemical behaviors of the supercapacitor cells, in particular, carbon monoliths prepared at 700–800°C exhibit better specific capacitance, specific energy and specific power than that from the carbon monoliths prepared a 600°C (Farma et al. 2013a). Therefore, it can be concluded that the carbonization temperature value should be closer to or equal to the activation temperature for producing higher-quality activated carbon monolith electrodes (Farma et al. 2013a). In another work, Farma R. et al. used empty fruit bunches as a raw material to produce monolithic nanoporous activated carbon for supercapacitor electrodes – the processing of empty fruit bunches involves pre-carbonization, pulverization, pelletization, chemical treatment, carbonization, and activation (Farma et al. 2013b). He et al. (2013) have shown that mesoporous carbons made by microwave heating show both higher energy and higher power density than the mesoporous carbons made by conventional heating because of bigger average pore size and unique mesopore structure in the mesoporous carbons made by the microwave-assisted technique.

5. Carbon Activation

Activation is a process of converting carbonaceous materials into activated carbon by thermal decomposition in a furnace (convectional heating) or a microwave using a controlled atmosphere and heat (Abioye and Ani 2015). To obtain activated carbon – the methods of physical, chemical, and combined chemical and physical (physiochemical) – microwave activations are used.

Activated carbons prepared by chemical or physical activation methods usually have a wide pore size distribution from 50 nm (ultramicropores). In this case, the pore size should be large enough to access ions, contributing to the capacity of the double layer. Recently, a number of studies have found that microporous carbons with pore sizes ranging from 0.7 nm to 2 nm show a high capacitive performance (Chmiola 2006).

5.1 Physical or Thermal Activation

Physical or thermal activation is the partial gasification of the carbon material, where it is carbonized at high temperatures, 600°C–900°C, in a furnace under an inert atmosphere, such as nitrogen, to eliminate most of the hydrogen and oxygen content and produce char with the desired porosity (Hesas et al. 2013). The prepared char from the carbonization process is then activated by oxidizing gases, such as steam, carbon dioxide, air, or mixtures of these gases, to produce AC (Rodriguez-Reinoso and Molina-Sabio 1992).

5.2 Chemical Activation

Chemical activation involves mixing an acidic or basic solution with the carbon material to influence the pyrolytic decomposition of the starting materials, suppress tar formation, and lower the pyrolysis temperature (Hesas et al. 2013). In the chemical activation process, the precursor material is mixed with activating agents, such as KOH, K_2CO_3, $NaOH$, H_3PO_4, $ZnCl_2$ and $FeCl_3$, that act as dehydrating and oxidizing agents. Chemical activation is a step process in which the carbonization and activation are carried out simultaneously and usually in the temperature range of 300–950°C, thus influencing the pyrolytic decomposition and resulting in the development of a better porous structure and increased carbon yield (Abioye and Ani 2015).

The probable reactions during KOH activation at high temperatures may be exposed in the following equations (Kumar et al. 2020) (1)–(5).

$$6KOH + 2C \rightarrow 2K + 2K_2CO_3 + 3H_2 \tag{1}$$

$$K_2CO_3 \rightarrow K_2O + CO_2 \tag{2}$$

$$CO_2 + C \rightarrow 2CO \tag{3}$$

$$K_2CO_3 + 2C \rightarrow 2K + 3CO \tag{4}$$

$$K_2O + C \rightarrow 2K + CO \tag{5}$$

Inal et al. (2015) showed that the activated carbon derived from the waste tea by K_2CO_3 activation showed much better supercapacitive performance than that produced with H_3PO_4 activation in an aqueous electrolyte despite having a lower specific surface area. K_2CO_3 activation produced a more uniform microporous structure and fewer surface oxygen groups, resulting in lower internal resistance and excellent electrochemical reversibility in aqueous electrolytes (Inal et al. 2015). A well-developed microporous architecture allows ions to move within the porous structure easily. Jiang et al. (2020) showed that maximum specific capacitances of 92.7 F/g at the current density of 100 mA/g were achieved at indirect activation (low-temperature pyrolysis followed by the physical activation using CO_2) routes compared to 80.9 F/g received at direct activation (one-step conversion of the biomass into an activated carbon using CO_2 as the activating agent). At the same time, the work (Yumak et al. 2020) shows that direct KOH and H_3PO_4 activation increased the specific surface area up to 1,272 and 1,373m²/g and capacitive performance up to 140 F/g, which was more than for the MnO_2 loaded or the indirectly activated biochar.

5.3 Self-Activation

The self-activation method does not employ any external activating agent for activation – for this purpose, the gases emitted when biomass is carbonized can be used for activation, which means converting carbon biomass into AC in one step (Shaker et al. 2021). The other self-activation methods take benefit of already existing inorganic materials in biomass structure for its *in situ* activation (chemical self-activation) (Biswal et al. 2013). The disadvantage of the self-activation method is the difficulty in controlling the composition and amount of emitted gases or already existing elements in the biomass structure.

5.4 Hard Templating/Nanocasting

Inorganic materials such as silica are used as a template in the hard template/ nanocasting method. After activation, the hard template can be easily removed by alkaline washing.

Soft templating methods benefit from polymeric reactions between molecule chains to endow particles with special shapes like spheres (Kubo et al. 2011). This technique does not need any template removal step after activation; overall, though employing a template method can lead to controlled porosity and morphology with high accuracy, the cost of applying this method is usually high (Shaker et al. 2021).

5.5 Pyrolysis

Pyrolysis refers to the thermal decomposition of biomass in the temperature range of 300–900°C under an inert atmosphere. Slow pyrolysis operates at a low heating rate (5–7°C/min) and longer residence time > 1 h, while fast pyrolysis is carried out at a greater heating rate (100–200°C/min) with short residence time < 10 s. Slow pyrolysis of lignocellulosic biomass yields more biochar, while a higher amount of bio-oil is obtained from fast pyrolysis (Al Arni 2018).

5.6 Hydrothermal Carbonization

In order to avoid an energy-intensive step of biomass initial drying, hydrothermal carbonization is performed to obtain the hydrochar. Hydrothermal carbonization is performed at a relatively lower temperature range of 180–250°C and a pressure range of 2–10 MPa (Funke and Ziegler 2010). Increasing the temperature to 250–350°C and pressure to 10–30 MPa leads to a decrease in char and an increase in bio-oil yield; however, the process is then referred to as hydrothermal liquefaction (Rawat et al. 2022).

5.7 Torrefaction

Torrefaction involves heat treatment of biomass at 250–300°C for 10–100 min and is often referred to as mild pyrolysis (Inari et al. 2007).

5.8 Flash Carbonization

Flash carbonization is performed at a typically lower temperature of up to 400°C, and once the desired temperature is achieved, the biochar is exposed to air for a short residence time of 10–20 min (Antal et al. 2003). In the process of flash carbonization, the biomass at elevated pressure is placed in a compacted bed and ignited from below against the direction of the airflow.

5.9 Chemical Exfoliation

The graphitized biomass exfoliation method involves peeling the bulk carbon structure by overcoming the van der Waals force, resulting in graphene sheets (Safian et al. 2020). Mechanisms for producing graphene through chemical exfoliation include the reduction of the interlayer forces to produce graphene-intercalated compounds and exfoliating the graphene into multi-layers by applying fast heating (Idris et al. 2022).

5.10 Chemical Vapor Deposition

Chemical vapor deposition (CVD) has been identified as a viable method for producing graphene and has emerged as a promising method for producing large area graphene because it is inexpensive (Idris et al. 2022). Gas species are fed into the reactor and pass through the hot zone, where biomass precursors decompose to carbon radicals at the surface of the transition metal substrate, forming layers of graphene-like carbons (Idris et al. 2022).

6. Methods for Studying the Performance of Activated Carbons as Electrodes

The performance of the supercapacitor cells using activated carbon monoliths as their electrodes is studied using the following methods:

- Electrochemical impedance spectroscopy (EIS)
- Cyclic voltammetry (CV)
- Galvanostatic charge-discharge (GCD)

According to the equation (6), mass-specific capacitances C_{sp} can be calculated using the EIS data (Farma et al. 2013a)

$$C_{sp} = -\frac{1}{\pi f_\ell Z_\ell'' m} \tag{6}$$

where f_1 is the lowest frequency, Z_1'' is the imaginary impedance at f_1, and m is the weight of the electrode. The EIS data as a function of the frequency is analyzed using equations (7), (8), and (9):

$$C(\omega) = C'(\omega) - jC''(\omega) \tag{7}$$

$$C''(\omega) = Z'(\omega)/\omega \, |Z(\omega)|^2 \tag{8}$$

$$C'(\omega) = -Z''(\omega)/\omega \, |Z(\omega)|^2 \tag{9}$$

Here, $Z(\omega)$ is equal to $1/j\omega C(\omega)$, $C'(\omega)$ is the real capacitance, $C''(\omega)$ is the imaginary capacitance, $Z'(\omega)$ is the real impedance, and $Z''(\omega)$ is the imaginary impedance (Portet et al. 2005).

C_{sp} can be determined from the voltammograms using equation (10):

$$C_{sp} = \frac{2I}{Sm} \tag{10}$$

Here, I – electric current, S – scan rate, and m – weight of electrode.

C_{sp} of the electrodes can be calculated using the GCD data (charge-discharge curve) according to the equation (11):

$$C_{sp} = \frac{2I}{\left(\dfrac{\Delta V}{\Delta t}\right)m} \tag{11}$$

where I is the discharge current, ΔV is the voltage, and Δt is the discharge time (Beidaghi et al. 2011), $\Delta V/\Delta t$ is calculated from the slope obtained by fitting a straight line to the discharge curve from the end of the voltage drop to the end of the discharge process (Wei et al. 2015).

The effective series resistance is calculated using the voltage drop at the beginning of the discharge, V_{drop}, at a certain constant current I (Zhu et al. 2011), with the formula (12)

$$R_{ESR} = \frac{V_{drop}}{I} \qquad (12)$$

The Coulomb efficiency is determined using the following formula (13):

$$\eta = \frac{T_d}{T_c} 100\% \qquad (13)$$

where T_d and Tc are discharge time and charge time, respectively (Peng et al. 2013).

The values of the specific power (P) and specific energy (E) are calculated from the GCD data using the equations below, respectively:

$$P = \frac{VI}{m} \qquad (14)$$

$$E = \frac{VIt}{m} \qquad (15)$$

where I is the discharge current, V is the voltage (excluding the iR drop at the beginning of the discharge), and t is time (Prabaharan et al. 2006).

7. Parameters for Evaluating the Efficiency of Electrodes

AC demonstrated significant adsorption in gas and liquid phases due to its large specific surface area (S_{BET}), high micropore volume (V_{micro}), favorable pore size distribution, capability for rapid adsorption, low acid/base reactivity, and thermal stability (Li et al. 2009). Some parameters describing the properties and structure of the ACs are shown in Table 3.

Table 3. Physical properties of activated carbons.

Parameter	Unit	Characteristic
S_{BET}	m²/g	Specific surface area per unit mass of a material or substance is calculated by the Brunauer–Emmett–Teller (BET) method.
S_{micro}	m²/g	Micropore surface area is the resultant difference between BET area and external surface areas calculated by the t-plot method.
$S_{external}$	m²/g	External surface area – geometric surface area defining the external boundary of a solid object or a quantity of powder or granules.
V_{tot}	cm³/g	Total pore volume.
V_{micro}	cm³/g	Micropore volume calculated by t-plot method.
V_{meso}	cm³/g	Mesopore volume.
Pore size	Å	Average pore size.
Yield	%	Carbon yield, defined as the weight of dried AC to the weight of dry raw material: $Y = (W_1/W_0)*100$, where W_0 – precursor weight, W_1 – resulting AC mass.

8. Activated Carbon Modifications

In order to enhance the surface wettability and electron conductivity of carbon, various heteroatoms (like B, N, and S) can be introduced into the carbon skeleton (Cao et al. 2021). Surface-functional groups, including nitrogen, oxygen, and phosphorous, can considerably increase the electrochemical effects as well as improve the wettability of porous carbon with electrolytes, increasing the electrochemical performances of the electrodes (Dos Reis et al. 2020).

However, an elevated amount of functionalities on the carbon electrode surface is reflected in a high quantity of irreversible redox products that remain inside ACs' pores, resulting in a faster capacitance fade concerning the ACs with lesser functional groups (Dos Reis et al. 2020). Cao et al. (2018) related that the amount of O-O functional groups play a significant influence on the electrochemical supercapacitors' efficiency by having carbon electrodes as electrodes. Widmaeir et al. (2016) concluded that the surface functionality could strongly affect the potential shift of the carbon electrodes through time.

Wei et al. (2015) produced porous nitrogen- and oxygen-doped carbons by hydrothermal treatment of broussonetia papyrifera stem bark in the presence of KOH, followed by simultaneous pyrolysis and KOH activation. The as-synthesized porous heteroatom-doped carbon possesses a partial graphitic structure with a very high specific surface area of 1,212 m^2/g, a high pore volume of 0.81 cm^3/g, and electrodes based on it display a high specific capacitance of 320 F/g at 0.5A/g (Wei et al. 2015).

Some research (Li et al. 2019) report on cost-effective, environmentally friendly SiO_2 to regulate the pore structure of a hollow carbon shell and have boosted the electrochemical properties, such as the capacitance, rate, and lifetime.

Zhang et al. (2020) suggested a biomass liquefaction technology and doping with silicon sources to synthesize ordered pore-structured biomass-based carbon composites with micropores and mesopores. The morphological evolution from tubular-like to a multilayer lamellar structure and the pore structure transformation from completely disordered to ordered were recorded systematically for the un-doped and doped silicon sources into the liquefaction before carbonization samples (Zhang et al. 2020). The changes in morphologies and pore textures could further be explained by the release of certain primary gases (CO_2, CO, CH_4, and H_2) through the degradation, repolymerization, cyclization, and aromatization steps of the carbon precursor (Zhang et al. 2020).

Cao et al. (2021) showed a simple and cost-effective one-step method using different nitrogen-containing compounds (NH_4Cl, $(NH_4)_2CO_3$, and urea) as both activation agent and dopant to prepare nitrogen-doped biomass-derived hierarchical porous carbon materials. The prepared hierarchical porous carbon urea electrode delivers a specific capacitance of 300 F/g at 1 A/g and an energy density of 14.3 Wh/kg (Cao et al. 2021).

Ding et al. (2019) related that an appropriate number of functional groups could enhance the electrochemical stability window. It is also related that the presence of certain functionalities (e.g., O–C=O or C=O) induced higher capacitance to carbon materials and that the proper functional groups hinder the potential shift of the carbon electrodes (Ding et al. 2019). Elmouwahidi et al. (2012) concluded that O-rich ACs exhibited the lowest special capacitance (259 F/g at 125 mA/g) and capacity retention (52% at 1 A/g) because of the presence of surface carboxyl groups that hindered electrolyte diffusion into the carbon electrode pores. They also concluded that N-rich ACs presented the highest special capacitance (355 F/g at 125 mA/g) as well as the highest retention (93% at 1 A/g) because of the pseudo-capacitance effects of N functionalities (Elmouwahidi et al. 2012). Yang et al. (2018) applied carbons with different oxygen and nitrogen contents as well as different porosities. It was found that the electrochemical activity increased in the order of oxygen contents, and the carbon with the largest surface area but lower oxygen content had much lower specific capacitance than samples with higher oxygen contents, so it suggests the importance of oxygen functional groups on electrochemical properties of the carbon electrodes because the oxygen groups can enhance electrolyte wettability and reactions in aqueous electrolytes (Dos Reis et al. 2020).

9. Biomass Sources

Biomass from forest plants and residues, agricultural products and agricultural wastes, and industrial and domestic wastes is used to produce activated carbon. Characteristics of biomass precursors from forest crops and waste selected for the manufacture of electrode materials in capacitors are presented in Figure 1. Activating agents used for them (Shaker et al. 2021) include H_3PO_4 or NaOH for rubber wood, KOH for bamboo, wood tar, firewood, willow wood, metaplexis japonica and steam for firewood.

Characteristics of biomass precursors from agricultural products and agricultural wastes selected for the manufacture of electrode materials in capacitors are presented in Figure 2. Activating agents used for them were (Shaker et al. 2021) include KOH

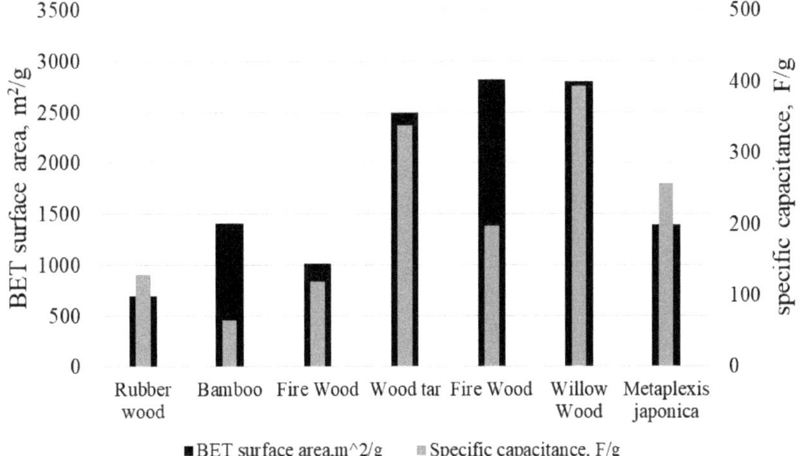

Figure 1. Specific capacitance and surface area of activated carbons from forest plants and residues biomass (on the basis of data from Shaker et al. 2021).

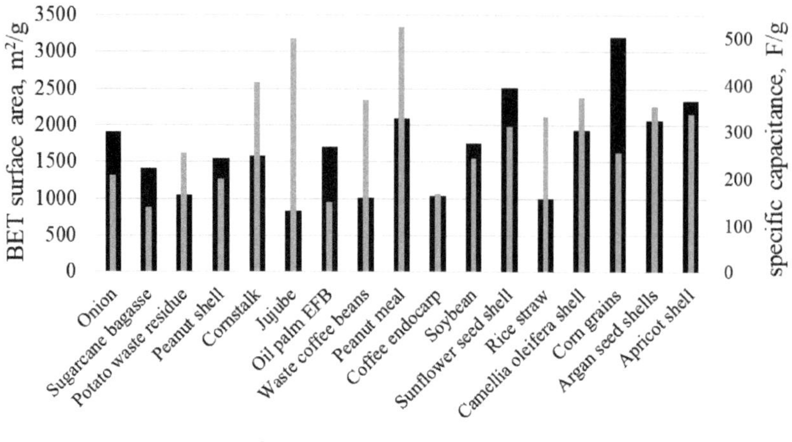

Figure 2. Specific capacitance and surface area of activated carbons from agricultural products and agricultural wastes biomass (on the basis of data from Shaker et al. 2021).

for onion, jujube, oil palm EFB, soybean, sunflower seed shell, rice straw, corn grains, argan seed shells, microwave-$ZnCl_2$ for sugarcane bagasse, peanut shell, $ZnCl_2$ for potato waste residue, waste coffee beans, peanut meal, camellia oleifera shell, air for cornstalk and NaOH for apricot shell.

Characteristics of biomass precursors from industrial wastes, selected for manufacturing electrode materials in capacitors, are presented in Figure 3. Activating agents used for them (Shaker et al. 2021) include KOH for human hair, recycled waste paper, cow dung, waste tea leaves, tobacco rods, $ZnCl_2$ for sugar cane bagasse, sugarcane bagasse, pyrolysis for waste newspaper, sonication for shrimp shells, HNO_3 for waste paper, CO_2 for rubberwood sawdust, and steam for apple pulp.

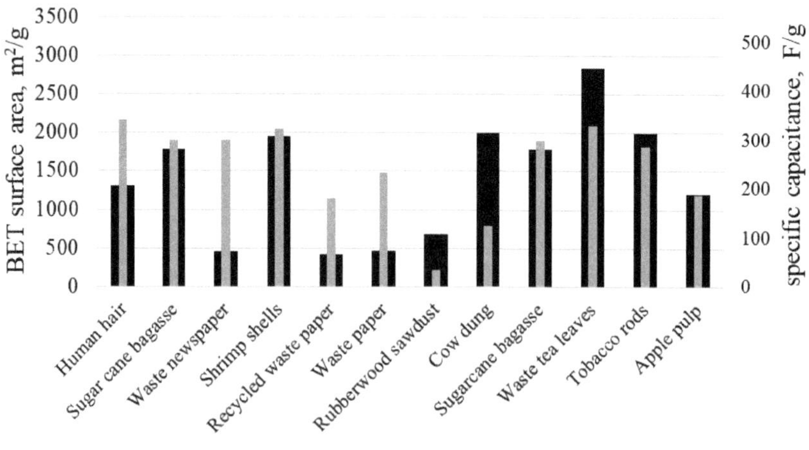

Figure 3. Specific capacitance and surface area of activated carbons from industrial waste biomass (on the basis of data from Shaker et al. 2021).

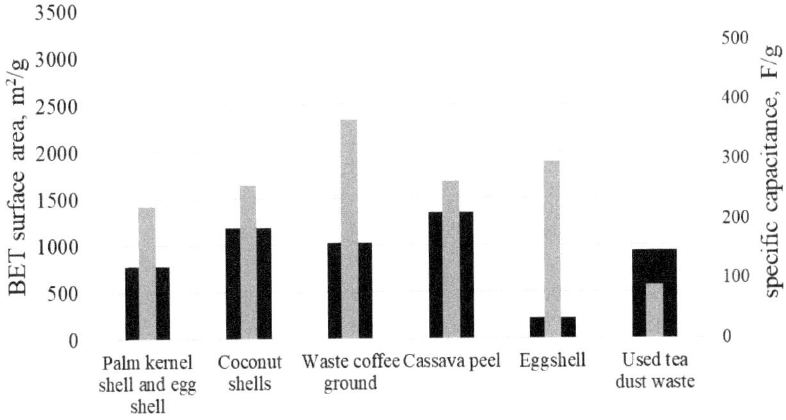

Figure 4. Specific capacitance and surface area of activated carbons from domestic waste biomass (on the basis of data from Shaker et al. 2021).

Characteristics of biomass precursors from domestic wastes selected for the manufacture of electrode materials in capacitors are presented in Figure 4. Activating agents used for them (Shaker et al. 2021) include CaO for palm kernel shell and eggshell, self-activation for coconut shell, $ZnCl_2$ for waste coffee ground, KOH + CO_2 followed by surface modification with H_2SO_4 for cassava peel, pyrolysis (unactivated) for eggshell, and H_3BO_3 for used tea dust waste.

10. Use of Biomass for Various Electrochemical Devices

For the fabrication of the energy storage devices preparation of carbon electrodes, ACs can serve, thereby, either as the sole electrode material (for ion electrosorption via supercapacitors, as a promising electron double-layer capacitive material) or as a tunable substrate to attach heteroatoms (O, N, H, etc.) capable of boosting their electrochemical performances (Dos Reis et al. 2020). ACs are used as electrodes in energy storage devices such as electric double-layer capacitors (EDLCs), pseudocapacitors, asymmetric supercapacitors, and hybrid supercapacitors.

Supercapacitors are electrochemical devices that primarily consist of electrodes, current collectors, an electrolyte, and a spacer and are promising electrical energy storage devices for applications that demand high energy density, maximum power, high reliability, long-term operation stability or long-cycle life, suitable weight and dimensions, fast discharge/charge time, low level of heating, low cost, safety, etc. (Farma et al. 2013b).

It has been found by He et al. (2013) that the supercapacitors made from the mesoporous carbons from biomass such as peanut shell and rice husk have a high energy density of 19.3 Wh/kg at a high power density of 1,007 W/kg in 1 M Et_4NBF_4/PC electrolyte because of bigger average pore size and unique mesopore structure.

Wang et al. (2015) managed to achieve maximal specific capacitances of 340 F/g and high specific surface capacitance of 52.7 mF/cm² at the current density of 0.1 A/g, good rate capability (231 F/g at 10 A/g) and good cycling stability (92% capacitance retention over 3,000 cycles) for supercapacitors with electrodes made of ACs from biomass willow catkins.

Yu et al. (2017) developed a feasible and cost-effective method for obtaining porous carbon from cattail biomass; the as-prepared activated carbon had a high specific surface area of 441.12 m^2/g after CO_2 activation and exhibited exceptional electrochemical performance as supercapacitors, with specific capacitance reaching 126.5 F/g at a current density of 0.5 A/g within a potential range of -1 to 0 V in a 6 M KOH solution.

Taer et al. (2020), using Syzygium oleana leaves biomass, managed to realize like-flower nanosheets structure decorated by nanofiber increases the surface area of activated carbon from 216 m^2/g to 1,218 m^2/g with tuneable well-confirmed mesopores sizes.

It was reported that high specific capacitance of up to 791 F/g can be achieved by supercapacitor containing electrode made by bamboo-derived carbon nanofiber doped with polymer (polyaniline) and carbon nanotube (Yang et al. 2015). Interestingly, the doped carbon nanofiber made from bamboo biomass can store energy (791 F/g) more than the electrode made from non-organic materials such as graphene (117 F/g), ruthenium (200 F/g), cobalt (475 F/g), nickel (153 F/g), and manganese (482 F/g) (Azwar et al. 2018).

Biomass-based electrodes are also used to develop microbial fuel cells (MFC). Yaqoob et al. (2021) studied the use of lignin-graphene-derived anode obtained from oil palm in an MFC. The lignin-based anode displayed a power density of 0.020 mW/m^2 with a current density of 17.54 mA/m^2 (Yaqoob et al. 2021). Hung et al. (2019) showed that coffee waste-activated carbons can be used as material to form anode in an Escherichia coli-based MFC, which achieves a power density of 3,927 mW/m^2. Chen et al. (2018) directly used carbonized waste tires as pepped-up anode material in MFC. The current density of the anode electrode made from waste tires carbonized at 800°C was 23.1 A/m^2 (Chen et al. 2018). Graphene/polyaniline nanocomposite anode derived from cellulosic biomass was utilized to remove toxic metals with electricity production via benthic MFC, as studied by Yaqoob et al. (2020). In the experiment, 87.71 mA/m^2 was observed to be the maximum current density (Yaqoob et al. 2020). Chakraborty et al. (2020) reviewed the applications of waste-derived biochar in MFC. Huggins et al. (2014) synthesized biochar from residues obtained from mills and forestry activities at a carbonization temperature of 1,000°C, which was explored as an anode in MFC. Milling residue and forestry biochar presented the highest power densities of 532 mW/m^2 and 457 mW/m^2, respectively.

11. Conclusion

Further improvement and commercialization of electrochemical devices require cost-effective, renewable, and environmentally friendly electrode materials. These electrode materials must meet a variety of requirements, including high specific surface area and adsorption, fast electron/transfer, and tunable surface chemistry.

Activated carbon has all of the above properties. In order to obtain cost-effective porous carbon, materials with unlimited sources, renewable, low cost, easy processing, and environmentally friendliness must be used. One of the best options to achieve this goal is to use biomass as activated carbon precursors.

In most electrical energy storage devices, carbon in some form is needed to ensure electrical conductivity since easy electron transportation is required in the direction of the electrode-electrolyte interface. Activated carbon is advantageous because of its high porosity and good electrical conductivity. Increasing attention is being paid to renewable, readily available, structurally porous, cheap, and environmentally friendly biomass precursors as a carbon source. Biomass from forest plants and residues, agricultural products and agricultural wastes, and industrial and domestic wastes is used to produce activated carbon. The preparation and production of activated carbon usually consist of carbonization, which takes place at a temperature between 400°C and 850°C and in the absence of oxygen, and activation, which includes physical, chemical, and microwave-induced activations. The performance of the electrochemical devices using activated carbon as their electrode study using the electrochemical impedance spectroscopy, cyclic voltammetry, and galvanostatic charge-discharge. In order to enhance the surface wettability and electron conductivity of carbon, various heteroatoms can be introduced into the carbon skeleton. Activated carbons are used as electrodes in energy storage devices such as electric double-layer capacitors, pseudocapacitors, asymmetric supercapacitors, and hybrid supercapacitors.

Thus, activated carbon obtained from biomass has turned out to be a prospective candidate as an electrode for highly efficient electrochemical devices. As a result, scientists are making great efforts to synthesize advanced carbon materials using biomass from a wide variety of sources. However, these laboratory technologies need to be scaled up so that everyone can enjoy the many benefits of biomass-derived porous carbon.

References

Abioye, A. M. and Ani, F. N. (2015). Recent development in the production of activated carbon electrodes from agricultural waste biomass for supercapacitors: A review. *Renewable and Sustainable Energy Reviews*, 52: 1282–1293.

Al Arni, S. (2018). Comparison of slow and fast pyrolysis for converting biomass into fuel. *Renew. Energy*, 124: 197–201.

Antal, M. J., Mochidzuki, K. and Paredes, L. S. (2003). Flash carbonization of biomass. *Ind. Eng. Chem. Res.*, 42: 3690–3699.

Azwar, E., Mahari, W. A. W., Chuah, J. H., Vo, D. V. N., Ma, N. L., Lam, W. H. and Lam, S. S. (2018). Transformation of biomass into carbon nanofiber for supercapacitor application–a review. International Journal of Hydrogen Energy, 43(45): 20811–20821.

Beidaghi, M., Chen, W. and Wang, C. (2011). Electrochemically activated carbon micro-electrode arrays for electrochemical micro-capacitors. *J. Power Sources*, 196: 2403.

Bi, Z., Kong, Q., Cao, Y., Sun, G., Su, F., Wei, X., Li, X., Ahmad, A., Xie, L. and Chen, C. M. (2019). Biomass-derived porous carbon materials with different dimensions for supercapacitor electrodes: A review. *Journal of Materials Chemistry A*, 7(27): 16028–16045.

Biswal, M., Banerjee, A., Deo, M. and Ogale, S. (2013). From dead leaves to high energy density supercapacitors. *Energy & Environmental Science*, 6(4): 1249–1259.

Cagnon, B., Py, X., Guillot, A., Stoeckli, F. and Chambat, G. (2009). Contributions of hemicellulose, cellulose and lignin to the mass and the porous properties of chars and steam activated carbons from various lignocellulosic precursors. *Bioresour. Technol.*, 100: 292–298.

Cao, L., Li, H., Xu, Z., Zhang, H., Ding, L., Wang, S., Zhang, G., Hou, H., Hu, W., Yang, F. and Jiang, S. (2021). Comparison of the heteroatoms-doped biomass-derived carbon prepared by one-step nitrogen-containing activator for high performance supercapacitor. *Diamond and Related Materials*, 114: 108316.

Chakraborty, I., Sathe, S. M., Dubey, B. K. and Ghangrekar, M. M. (2020). Waste-derived biochar: Applications and future perspective in microbial fuel cells. *Bioresour. Technol.*, 312: 123587.

Chen, W., Feng, H. J., Shen, D. S., Jia, Y. F., Li, N., Ying, X. B., Chen, T., Zhou, Y. Y., Guo, J. Y. and Zhou, M. J. (2018). Carbon materials derived from waste tires as high-performance anodes in microbial fuel cells. *Sci. Total. Environ.*, 618: 804–809.

Chmiola, J. (2006). Anomalous increase in carbon capacitance at pore sizes less than 1 nanometer. *Science*, 313(5794): 1760–1763.

Ding, Z., Trouillet, V. and Dsoke, S. (2019). Are functional groups beneficial or harmful on the electrochemical performance of activated carbon electrodes? *J. Electrochem. Soc.*, 166: A1004–A1014.

Dos Reis, G. S., Larsson, S. H., de Oliveira, H. P., Thyrel, M. and Claudio Lima, E. (2020). Sustainable biomass activated carbons as electrodes for battery and supercapacitors—A mini-review. *Nanomaterials*, 10(7): 1398.

Elmouwahidi, A., Zapata-Benabithe, Z., Carrasco-Marín, F. and Moreno-Castilla, C. (2012). Activated carbons from KOH-activation of argan (Argania spinosa) seed shells as supercapacitor electrodes. *Bioresour. Technol.*, 111: 185–190.

Farma, R., Deraman, Awitdrus, M., Talib, I. A., Omar, R., Manjunatha, J. G., Ishak, M. M., Basri, N. H. and Dolah, B. N. M. (2013a). Physical and electrochemical properties of supercapacitor electrodes derived from carbon nanotube and biomass carbon. *Int. J. Electrochem. Sci.*, 8: 257–273.

Farma, R., Deraman, M., Awitdrus, A., Talib, I. A., Taer, E., Basri, N. H., Manjunatha, J. G., Ishak, M. M., Dollah, B. N. M. and Hashmi, S. A. (2013b). Preparation of highly porous binderless activated carbon electrodes from fibres of oil palm empty fruit bunches for application in supercapacitors. *Bioresource Technology*, 132: 254–261.

Funke, A. and Ziegler, F. (2010). Hydrothermal carbonization of biomass: A summary and discussion of chemical mechanisms for process engineering. *Biofuels, Bioprod. Biorefin.*, 4: 160–177.

He, X., Ling, P., Qiu, J., Yu, M., Zhang, X., Yu, C. and Zheng, M. (2013). Efficient preparation of biomass-based mesoporous carbons for supercapacitors with both high energy density and high power density. *Journal of Power Sources*, 240: 109–113.

Hesas, R. H., Daud, W. M. A. W., Sahu, J. N. and Arami-Niya, A. (2013). The effects of a microwave heating method on the production of activated carbon from agricultural waste: A review. *Journal of Analytical and Applied Pyrolysis*, 100: 1–11.

Huggins, T., Wang, H., Kearns, J., Jenkins, P. and Ren, Z. J. (2014). Biochar as a sustainable electrode material for electricity production in microbial fuel cells. *Bioresour. Technol.*, 157: 114–119.

Hung, Y. H., Liu, T. Y. and Chen, H. Y. (2019). Renewable coffee waste-derived porous carbons as anode materials for high-performance sustainable microbial fuel cells. *ACS Sustain. Chem. Eng.*, 7(20): 16991–16999.

Idris, M. O., Guerrero–Barajas, C., Kim, H. C., Yaqoob, A. A. and Ibrahim, M. N. M. (2022). Scalability of biomass-derived graphene derivative materials as viable anode electrode for a commercialized microbial fuel cell: A systematic review. *Chinese Journal of Chemical Engineering* (in press).

Inal, I. I. G., Holmes, S. M., Banford, A. and Aktas, Z. (2015). The performance of supercapacitor electrodes developed from chemically activated carbon produced from waste tea. *Applied Surface Science*, 357: 696–703.

Inari, G. N., Petrissans, M. and Gerardin, P. (2007). Chemical reactivity of heat-treated wood. *Wood Sci. Technol.*, 41: 157–168.

Jiang, C., Yakaboylu, G. A., Yumak, T., Zondlo, J. W., Sabolsky, E. M. and Wang, J. (2020). Activated carbons prepared by indirect and direct CO_2 activation of lignocellulosic biomass for supercapacitor electrodes. *Renewable Energy*, 155: 38–52.

Kubo, S., White, R. J., Yoshizawa, N., Antonietti, M. and Titirici, M. M. (2011). Ordered carbohydrate-derived porous carbons. *Chemistry of Materials*, 23(22): 4882–4885.

Kumar, T. R., Senthil, R. A., Pan, Z., Pan, J. and Sun, Y. (2020). A tubular-like porous carbon derived from waste American poplar fruit as advanced electrode material for high-performance supercapacitor. *Journal of Energy Storage*, 32: 101903.

Li, W., Peng, J., Zhang, L., Yang, K., Xia, H., Zhang, J. and Guo, S.-H. (2009). Preparation of activated carbon from coconut shell chars in pilot-scale microwave heating equipment at 60 kW. *Waste Management*, 29: 756–760.

Li, Z., Zhang, L., Chen, X., Li, B., Wang, H. and Li, Q. (2019). Three-dimensional graphene-like porous carbon nanosheets derived from molecular precursor for high-performance supercapacitor application. *Electrochim. Acta*, 296: 8–17.

Pandolfo, A. G. and Hollenkamp, A. F. (2006). Carbon properties and their role in supercapacitors. *Journal of Power Sources*, 157: 11–27.

Peng, C., Yan, X., Wang, R., Lang, J., Ou, Y. and Xue, Q. (2013). Promising activated carbons derived from waste tea-leaves and their application in high performance supercapacitors electrodes. *Electrochim. Acta*, 87: 401.

Portet, C., Taberna, O. L., Simon, P., Flahaut, E. and Laberty Robert, C. (2005). High power density electrodes for carbon supercapacitor applications. *Electrochim. Acta*, 50: 4174.

Prabaharan, S. R. S., Vimala, R. and Zainal, Z. (2006). Nanostructured mesoporous carbon as electrodes for supercapacitors. *J. Power Sources*, 161: 730.

Rawat, S., Mishra, R. K. and Bhaskar, T. (2022). Biomass derived functional carbon materials for supercapacitor applications. *Chemosphere*, 286: 131961.

Rodriguez-Reinoso, F. and Molina-Sabio, M. (1992). Activated carbons from lignocellulosic materials by chemical and/or physical activation: An overview. *Carbon*, 30: 1111–1118.

Safian, M. T. U., Haron, U. S. and Mohamad Ibrahim, M. N. (2020). A review on bio-based graphene derived from biomass wastes. *BioResources*, 15(4): 9756–9785.

Shaker, M., Ghazvini, A. A. S., Cao, W., Riahifar, R. and Ge, Q. (2021). Biomass-derived porous carbons as supercapacitor electrodes—A review. *New Carbon Materials*, 36(3): 546–572.

Taer, E., Apriwandi, A., Taslim, R., Agutino, A. and Yusra, D. A. (2020). Conversion Syzygium oleana leaves biomass waste to porous activated carbon nanosheet for boosting supercapacitor performances. *Journal of Materials Research and Technology*, 9(6): 13332–13340.

Thakur, S. K., Kong, T. S. and Gupta, M. (2007). Microwave synthesis and characterization of metastable (Al/Ti) and hybrid (Al/Ti+SiC) composites. *Materials Science and Engineering: A*, 452-453: 61–69.

Thostenson, E. T. and Chou, T. W. (1999). Microwave processing: fundamentals and applications. *Composites Part A Applied Science and Manufacturing*, 30: 1055–1071.

Wang, K., Zhao, N., Lei, S., Yan, R., Tian, X., Wang, J., Song, Y., Xu, D., Guo, Q. and Liu, L. (2015). Promising biomass-based activated carbons derived from willow catkins for high performance supercapacitors. *Electrochimica Acta*, 166: 1–11.

Wei, T., Wei, X., Gao, Y. and Li, H. (2015). Large scale production of biomass-derived nitrogen-doped porous carbon materials for supercapacitors. *Electrochimica Acta*, 169: 186–194.

Widmaier, M., Kruner, B., Jackel, N., Aslan, M., Fleischmann, S., Engel, C. and Presser, V. J. (2016). Carbon as quasi-reference electrode in unconventional lithium-salt containing electrolytes for hybrid battery/supercapacitor devices. *J. Electrochem. Soc.*, 14: 163.

Yang, C., Chen, C., Pan, Y., Li, S., Wang, F., Li, J., Li, N., Li, X., Zhang, Y. and Li, D. (2015). Flexible highly specific capacitance aerogel electrodes based on cellulose nanofibers, carbon nanotubes and polyaniline. *Electrochim Acta*, 82: 264e71.

Yang, W., Li, Y. and Feng, Y. (2018). High electrochemical performance from oxygen functional groups containing porous activated carbon electrode of supercapacitors. *Materials*, 11: 2455.

Yaqoob, A. A., Ibrahim, M. N. M., Yaakop, A. S., Umar, K. and Ahmad, A. (2021). Modified graphene oxide anode: A bioinspired waste material for bioremediation of Pb2+ with energy generation through microbial fuel cells. *Chem. Eng. J*, 417: 128052.

Yu, M., Han, Y., Li, J. and Wang, L. (2017). CO_2-activated porous carbon derived from cattail biomass for removal of malachite green dye and application as supercapacitors. *Chemical Engineering Journal*, 317: 493–502.

Yumak, T., Yakaboylu, G. A., Oginni, O., Singh, K., Ciftyurek, E. and Sabolsky, E. M. (2020). Comparison of the electrochemical properties of engineered switchgrass biomass-derived activated carbon-based EDLCs. *Colloids and Surfaces A: Physicochemical and Engineering Aspects*, 586: 124150.

Zhang, Y., Chen, H., Wang, S., Zhao, X. and Kong, F. (2020). Regulatory pore structure of biomass-based carbon for supercapacitor applications. *Microporous and Mesoporous Materials*, 297: 110032.

Zhu, Y., Murali, S., Stoller, M. D., Ganesh, K. J., Cai, W., Ferreira, P. J., Pirkle, A., Wallace, R. M., Cychosz, K. A., Thommes, M., Su, D., Stach, E. A. and Ruoff, R. S. (2011). Carbon-based supercapacitors produced by activation of graphene. *Science*, 332: 1537.

Chapter 10

Renewable Electrodes Incorporated with Green Synthesis Nanoparticles for the Development of High-Performance Lithium-Ion Batteries

Jeya Preethi Selvam and *Ponmurugan Ponnusamy**

1. Introduction

1.1 Lithium-Ion Batteries

A fundamental factor in developing greener, more efficient alternative energy sources that have replaced conventional and traditional sources is the need to reduce the carbon footprint (Ruiz et al. 2018). The development of new energy conversion technologies has drawn attention in order to address the growing need for energy. Secondary batteries have become quite important for this reason, and Li-ion batteries have received much attention in this regard (Luo et al. 2018). Lithium-ion batteries are frequently utilized to outperform a variety of portable electronic devices due to their higher energy density and sustainability. Despite these advantages, the cycle life and power density still require improvement for usage in electrically powered vehicles (EVs), massive energy-storage systems, and a variety of applications (Zhang et al. 2018). The anode, cathode, separator, and supporting solution are the components of lithium-ion batteries shown in Figure 1. Rechargeable batteries are one of the many energy-storage technologies that offer an effective way to manage power by storing electrical energy as chemical energy. Rechargeable batteries

Department of Botany, Bharathiar University, Coimbatore – 641046, Tamil Nadu, India.
* Corresponding author: drponmurugan@gmail.com

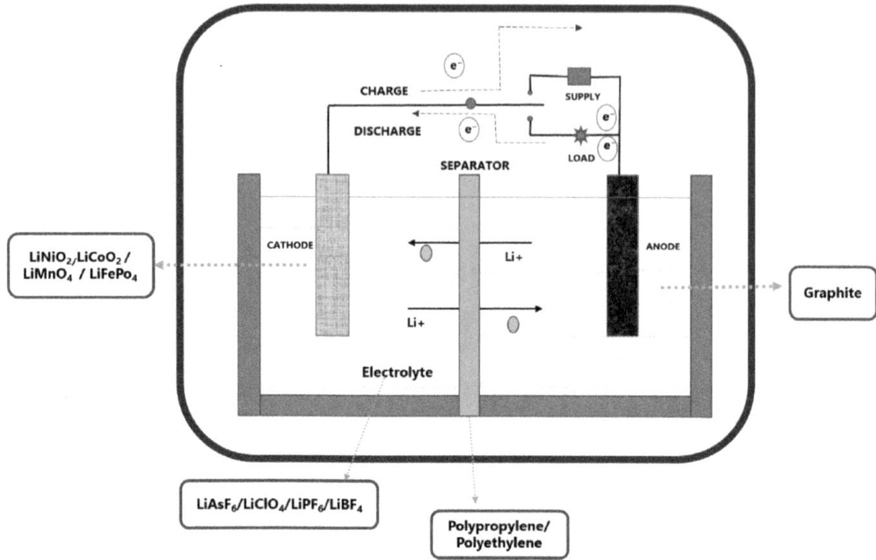

Figure 1. Lithium-ion battery.

like lithium-ion batteries are advancing quickly and use lithium ions as carriers to store electrical charge. Lithium ions travel through the electrolyte from the anode to the cathode when lithium-ion batteries are charged, and they move in reverse when lithium-ion batteries are discharged. Metallic lithium, graphitic carbon, hard carbon, synthetic graphite, lithium titanate, tin-based alloys, and silicon-based materials are the materials that are frequently utilized to fabricate the anode. Lithium manganese oxide, lithium cobalt oxide, iron disulphide, vanadium pentoxide, lithium nickel cobalt manganese oxide, lithium-ion phosphate, and electrical conducting polymers are the materials utilized to make cathode. Several substances, including lithium hexaflurophosphate (LiPF$_6$), lithium perchloride (LiClO$_4$), lithium hexafluoroarsenate (LiAsF$_6$), and lithium triflate (LiCF$_3$SO$_3$) are utilized as electrolytes. Other components, such as a flame retardant, binder, electrolyte solvent, and gel precursor are present in addition to these primary ones. Even though battery technology has made a number of advances, the current batteries still fall well short of the energy demands of electric vehicles. The major cause of this is the battery's own risk due to non-monotonic energy consumption and rapid variations during the battery discharge process. Unsteady high-temperature behavior and a decline in low-temperature charge-discharge performance are two other major problems with Li-ion batteries (Mishra et al. 2018).

Theoretically, graphitic carbons can store a tenth as much energy as lithium metal, and a decade ago, lithium-ion batteries using carbon as the anode material went into commercial production. Additionally, their application is increasingly being expanded into the transportation sector, where they are envisioned as the only power source and in hybrid modes with fuel cells and supercapacitors. Therefore, low-cost electroactive elements with high-rate capability are required for high-power lithium-ion batteries.

1.2 Renewable Electrode

Globally, the supply of nonrenewable fossil fuels is running out, environmental issues are becoming more visible, and the power sector is going through a unique revolution due to the increase in the production of renewable energy from sources including wind, solar, and tidal energy (Wang et al. 2021). By switching to renewable energy sources, electricity generation advances the development of clean energy and aids in lowering air pollution and carbon dioxide emissions. In order to overcome the intermittent nature of renewable energy production, it also lays high demands on dependable energy-storage systems. Therefore, researchers have been interested in the development of renewable electrodes like disordered carbon because of its interesting properties, including (i) a greater lithium intake than the 372 mAhg^{-1} theoretical limit for perfectly graphitic materials, (ii) good cycling properties, and (iii) dependence of their structural properties with their organic precursors, pyrolysis temperature, and soaking time.

Alternative electrode materials are just one of the many technologies and approaches that have been actively promoted to move beyond the current state-of-the-art of Li-ion batteries, with the main goals being to increase energy density, address sustainability concerns, and lower the cost per kW h^{-1} (Hwang et al. 2007).

Many studies have been done on fast-charging electrodes made of carbon-based material for lithium-ion batteries because of their availability, affordability, non-toxicity, and variety of electrochemical reactions. Batteries have traditionally been viewed as possibly hazardous to the environment due to the extensive use of poisonous lead, cadmium, mercury, and other chemicals. These substances are not present in lithium-ion batteries, which utilize 3D metals like nickel or cobalt. Because of their scarcity, rising prices, and potentially harmful extraction processes, their utilization in combination with lithium is controversial. If Li-ion batteries are widely used for grid or automotive applications, it will eventually be necessary to either partially substitute 3D metals or implement an effective technique of battery recovery or recycling (Larcher and Tarascon 2015).

1.3 Nanoparticle for Energy Storage

Nanoparticles have a wide range of uses in many fields, including delivering drugs, bioactivities (such as antifungal, antibacterial, and antioxidant activities), solar energy production, optics, sensors, and water purification techniques; their wide range of applications results from their remarkable properties, including a huge surface area, extremely small particle size, non-toxicity, accessibility to precursor materials, and extremely low synthesis costs as mentioned in Figure 2. As a result, recent research emphasis has been directed to several nanoparticle synthesis pathways (Okpara et al. 2021).

Due to slower kinetics, green synthesis of nanoparticles allows for better manipulation, stability, and control over crystal development. Green synthesis procedures are also quick, eco-friendly, and economical. Due to its potential applications in industry and medicine, metallic nanoparticles made of single metal, metal oxides, or a composite of several metals play an important role in nanotechnology. Noble metal nanoparticles have good biocompatibility, simple

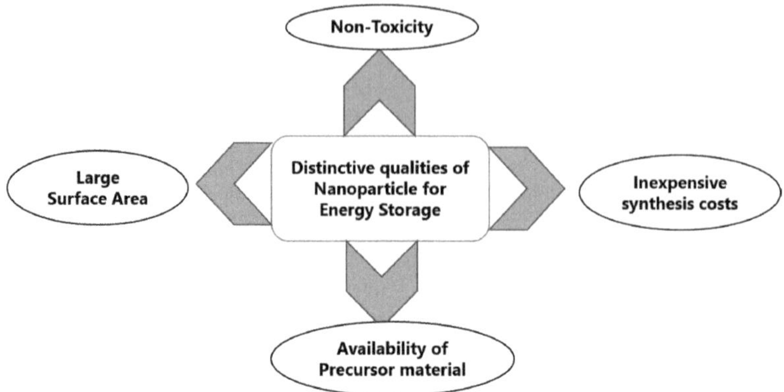

Figure 2. Qualities of nanoparticles for energy storage.

production procedures, low cytotoxicity, and easy surface modification capabilities (Mukherjee et al. 2001, Gayda et al. 2019).

Recent attempts have focused on using nanostructured materials for Li-ion batteries to address the problems with alternate electrode materials. It is generally accepted that those lithium-active materials' nanoscale size-tunable and shape-tunable features can provide supplementary parameters for further improving their electrochemical performances. A potential breakthrough in developing next-generation Li-ion materials is made possible by the interesting size, shape, and associated features of nanostructures. The next generation of Li-ion batteries is expected to include nanostructured electrode materials since they can provide a number of benefits not found in typical bulk materials (Deng 2015). Therefore, in this review chapter, in order to produce lithium-ion batteries for a sustainable environment, we examine the green synthesis of nanoparticles and renewable electrodes.

2. Green Synthesis of Nanoparticles for the Development of Lithium-Ion Battery

The term 'nanoparticle' refers to a subset of solid particles with at least one dimension and a size between 10 and 1,000 nm (Mandal et al. 2018). The most crucial characteristic of nanoparticles is their surface area to volume aspect ratio, which makes it easier for them to interact with other particles (Mandal et al. 2018, Sastry et al. 2005). Organisms have adapted by developing defenses against metals to survive in environments with high metal concentrations. These techniques might entail modifying the chemical makeup of the poisonous metal to render it nontoxic, leading to the development of nanoparticles of the metal in concern.

Nanoparticles exhibit distinct thermal, optical, chemical, physical, magnetic, and electrical properties compared to their bulk material counterparts. Next-generation biosensors, electronics, catalysts, and antimicrobials can benefit from these properties (Sastry et al. 2005). One significant and intensively researched class of materials that exhibits considerable diversity and has a wide range of applications is metallic nanoparticles. Metallic nanoparticles are gaining importance because of the variety of domains in which they may find use. Therefore, it is essential to find a low-price, ecologically friendly technique to synthesize them. Numerous species have the capacity to synthesize nanoparticles, making them potentially exploitable and amenable to modification to make them more effective for this use (Gayda et al. 2019).

Nanoparticles can created in one of two ways: top-down or bottom-up. The physical, chemical, and biological approaches are among them. Chemical precipitation, sonochemistry, one-step wet synthesis, thermal oxidation, sol-gel, solid thermal state decomposition, microwave irradiation, and laser ablation are examples of chemical and physical techniques (Azam et al. 2012).

These physical and chemical techniques have shown to be extremely effective in creating nanoparticles, but they are not without drawbacks. Physical procedures are inherently unsustainable, whereas chemical methods frequently use and produce hazardous byproducts. In addition to the potential for surface structural imperfections, they are frequently quite expensive (Thakkar et al. 2010). On the other hand, ecologically friendly, inexpensive biological resources such as various plant parts, fungi, algae, bacteria, and species of actinomycetes could be used in biological processes, which could be regarded as a competitive alternative (Okpara et al. 2021).

A reliable and environmentally friendly method to create these nanoparticles is becoming necessary. The current chemical and physical processes use harsh substances and high temperatures that are not only harmful to the environment but also expensive. As mentioned below, many groups have concentrated on finding different techniques to create nanoparticles. In an effort to develop a resource-effective, affordable, and sustainable technique of synthesis, researchers have looked into biological systems. For their capacity to produce metallic nanoparticles and withstand the toxic impacts of metal ions, numerous different biological chassis have been explored. They have an advantage over other biological entities since it is simple, inexpensive, and quick to create nanoparticles using extracts from various plant components (Gayda et al. 2019, Mandal et al. 2006).

Green synthesis is the more environmentally friendly method of creating metal nanoparticles from microorganisms (Sharma et al. 2019) and plant material extracts (Kuppusamy et al. 2016), as shown in Figure 3. All of these methods include three processes: agglomeration, stabilization, and metal reduction by bio-reducing agents. Some authors emphasize the benefits of employing extracts from plant derivatives rather than microbes. Plant derivative extracts do not require sterile conditions or the upkeep of cell cultures. Additionally, plant extracts work as natural capping agents for the stability of nanoparticles, simplifying the process of creating nanoparticles (Ahmed et al. 2016). All of these beneficial traits of plant extracts lead to less reagent use and, as a result, to lower costs, which makes it easier to establish large-scale nanoparticle production in an environmentally benign manner. When plant extracts are used as a catalyst for the creation of nanoparticles, a variety of biomolecules, including amino acids, proteins, enzymes, alkaloids, polysaccharides, tannins, terpenoids, phenolic acid, and flavonoids, work together to reduce and stabilize

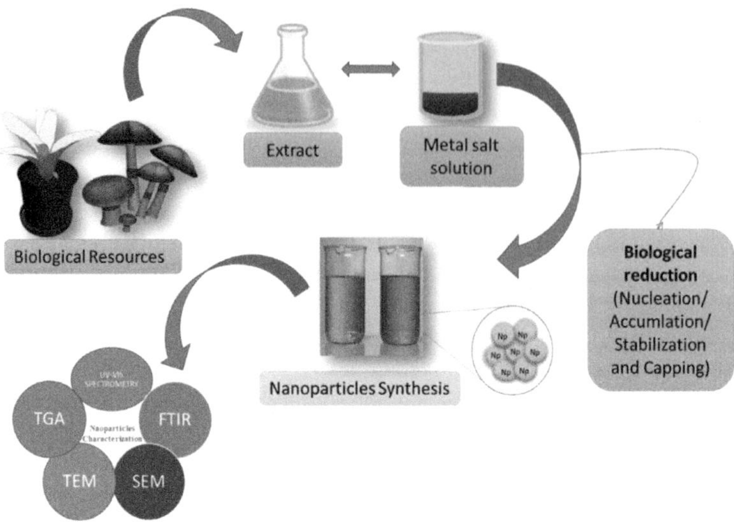

Figure 3. Green synthesis of nanoparticles.

the nanoparticles. It is known that the concentration of these biomolecules, which function as reducing and capping agents and provide synthesized nanoparticles with varying properties (i.e., size, shape, and stability), depends on the source of the extract (Bastos-Arrieta et al. 2018).

Up to now, the precise process of nanoparticle production by biological agents has not been determined. However, it has been hypothesized that a number of biomolecules have a role in this process (Owaid et al. 2019). The nanoparticles in the resulting colloidal solution are prevented from further development and agglomeration by capping and stabilizing agents (polysaccharides, polypeptides, and other bioorganic substances). When microbial cells release reductive enzymes on the surface of the cell wall or outside the cell, nanoparticles are typically synthesized extracellularly (Vetchinkina et al. 2017). It has been detailed how this process is mediated by a variety of enzymes, such as hydrogenases, nitrate reductase, phenol-oxidizing enzymes (laccases, tyrosinases, and Mn-peroxidases), and adhesins. The cell wall of microorganisms appears to be crucial in the intracellular creation of nanoparticles in addition to the lowering of extracellular metabolites (Mukherjee et al. 2001). Metal ions come into contact with it electrostatically and are subsequently trapped. The metal ions are reduced by reducing substances and enzymes. The final step is the production of nanoparticles (Gayda et al. 2019).

2.1 Characterization of Nanoparticles

In many research studies, the size and form of nanoparticles are considered their fundamental properties. Scanning electron microscopy (SEM) analyzes the nanoparticles to identify their size, distribution, and shape. However, the testing nanoparticles properties are changed by the drying and contrasting operations, which causes imaging errors or artifacts (Wang et al. 2017). A high-tech method for characterizing nanoparticles, transmission electron microscopy (TEM) is more potent and effective than SEM. With a high spatial resolution, it may be utilized to analyze the dispersion, particle size, and aggregation in an aqueous environment, revealing the specifics of the crystal structure.

The produced nanoparticles were characterized using a variety of analytical techniques. The appearance of the surface plasmon resonance (SPR) peak of nanoparticles was discovered using UV-Vis spectroscopy. The size distribution of phytochemical-capped nanoparticles and the stability of such particles were determined using the dynamic light scattering technique (DLS). To ascertain the crystalline nature and elemental makeup of the produced nanoparticles, respectively, XRD and EDX tests were carried out. FTIR spectroscopy was used to look into the biomolecules that are involved in metal reduction and capping.

3. Renewable Electrodes for Sustainable Environment

Electronic equipment, including computers, digital cameras, and portable memory devices, have been powered by lithium batteries for more than a decade. Their range of use is gradually being expanded to include the transportation industry. These days, the development of low-emission automobiles is crucial for the evolution of energy-storage devices with great specific power and energy where they are envisioned as the only power source and in hybrid modes with fuel cells and supercapacitors (Broussely 1999, Jansen et al. 1999). Therefore, the development of high-power lithium-ion battery cells for high-rate capability and relatively inexpensive electroactive materials is required. The lithium-ion battery technology's lithiated carbon anode is a crucial component. Today's lithium-ion technology relies heavily on graphitic carbons, which have a theoretical specific capacity of 372 mAh g^{-1}, or around one-tenth that of metallic lithium.

Current Li-ion battery technology depends on using electroactive materials (such as $LiCoO_2$ and $LiNiO_2$) that are not renewable; this uses 30% of the world's cobalt production, of which 3 kg are required for every kWh (Lupi and Pasquali 2003). Furthermore, according to recent estimates from integrated life-cycle assessments, 72 kg kWh^{-1} of CO_2 is released for Li-ion battery manufacturing, materials, and recycling (Tsukuba 2002). These elements are energy-intensive due to their rarity, and their future rarefaction will make this situation worse. Although there is still a significant energy requirement for its manufacture, the recent development of electrode materials is based on the utilization of renewable materials.

Many efforts were made to exploit renewable materials like biomass for usage in energy storage devices as carbon electrodes in order to lessen environmental contamination. Due to their distinctive combination of physical and chemical properties, such as strong conductivity, large surface areas, compatibility with composite materials, and low price, carbon materials are commonly used for energy storage applications as mentioned in Figure 5. Examples of carbon materials used in energy storage are shown in Table 1. It is simple to vary the microstructure and surface of carbon will change the material's electrochemical performance. For energy storage, activated carbons, carbon nanotubes, carbon aerogels, carbon powders, and

Table 1. Biomass precursors for renewable electrodes.

S. No.	Biomass precursors for renewable electrodes	Method	Reference
1.	Cotton wool	Pyrolysis between 700°C and 1,100°C	(Peled et al. 1998)
2.	Almond shells, sugar, oak, starch, filbert shells, walnut, brown sugar, maple, phenolic resin and lignin	Pyrolysis at 1,000°C	(Xing et al. 1996)
3.	Peanut shell	Pyrolysis between 300°C–600°C	(Fey et al. 2003)
4.	Rice husk	Pyrolysis at 900°C	(Fey and Chen 2001)
5.	Coffee shell	Pyrolysis at 900°C	(Hwang et al. 2007)
6.	Rice straw	KOH treatment	(Zhang et al. 2009)
7.	Mushroom	Pyrolysis at 500°C, 800°C, and 900°C	(Tang et al. 2016)
8.	Seaweeds	Pyrolysis between 600°C to 1,000°C	(Divya et al. 2019, Pintor et al. 2013)

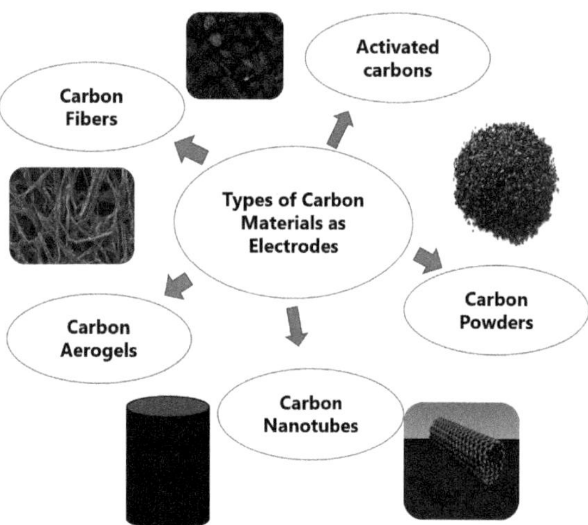

Figure 4. Types of carbon materials as electrodes.

carbon fibers are among the several types of carbon materials that are frequently utilized as electrodes, as shown in Figure 4.

Biomasses have gained a lot of attention recently as a renewable resource due to their promising uses in the production of activated carbons (Karagöz et al. 2008). However, there are not many reports on carbons. The surface areas of the porous carbons generated from biomasses were typically low, and they have been described in the literature as having a hierarchical porous structure. A quick, affordable, and highly productive method for producing porous carbons with a large surface area from biomasses. The porous carbons obtained have a hierarchical porous framework that creates the routes necessary for quick transit of lithium ions and simple accessibility of electrolytes (Zhang et al. 2009).

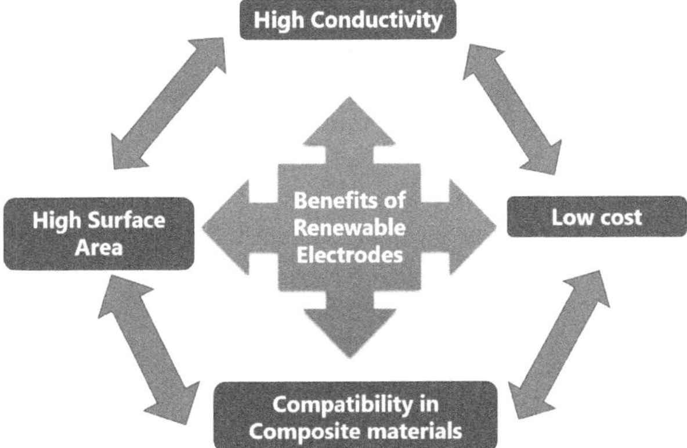

Figure 5. Benefits of renewable electrodes.

3.1 Production of Renewable Electrodes

The urgent need for drastically improved rechargeable lithium batteries has become a global necessity because of these critical environmental issues, the quick development and enthusiasm of the electric automotive industry, the fast development of portable electronics and hybrid electric vehicles, and the desire to increase the use of renewable energies.

Even though they are the least environmentally friendly component of portable electronics, lithium-ion batteries have experienced tremendous commercial success. Under the assumption that improvements in terms of safety and sustainability can be made, lithium systems are contenders for hybrid electric vehicles (HEVs) and plug-in hybrid electric vehicles (PHEVs). The latter is concerned with the availability and abundance of materials, among other things (Kates et al. 2001).

More environmentally friendly developments include the amazing lithium protection from water-based electrolytes provided by a vitro ceramic coating and the enormous capacity of lithium-air batteries, which present the difficulty of indirectly exposing lithium to oxygen (Ogasawara et al. 2006). Although extremely desirable, the hydro/pyrometallurgical methods used to recycle used batteries use quite a lot of energy and chemicals (Lupi and Pasquali 2003). Reduced usage of nonrenewable resources, waste production, and energy use are now necessities. Here, we raise the possibility of a paradigm change in favor of high-performance organic matter derived from biomass, which has a low life-cycle cost, as electrode materials.

The operation of various biochemical processes, such as photosynthesis employing chlorophyll as the organic mediator, enables the chemical processes of life, which are primarily based on organic molecules, to exploit "renewable processes." The main objective is to develop new Li-ion batteries based purely on organic electrodes by searching for electrochemically active organic compounds that may be produced from biomass using environmentally friendly methods and are readily reusable. As the first commercialized Li-ion system was partially organic, based on polyaniline, and never achieved the advertised capacities, we declare that using organic electrodes is clearly nothing new (Chen et al. 2008).

In order to create renewable electrodes, the activated porous carbon, which is used in energy storage devices, many biomass-derived materials must first undergo physical activation, followed by carbonization with oxidizing gases like air, steam, and CO_2, and chemical activation with chemical agents like KOH and $ZnCl_2$ (Habazaki et al. 2004) and chemical vapor deposition (CVD). Using chemicals from organic materials with a high carbon content (Ahmadpour and Do 1996), agricultural waste (Boehm 1989), and animal waste (Tokumitsu et al. 1995) as a source of carbon, it is possible to develop materials with higher surface area and high porosity for use in energy storage devices explained in Figure 6. Jiang et al., created high surface area carbon material from waste seaweeds without any activation (Jiang et al. 2021). Without any activation, Biswal et al. have shown how to create functional microporous conducting carbon from dead Neem leaves (Biswal et al. 2013). Supercapacitors have been studied using biomass-based functional carbon with a high capacitance of 354 F/g produced by direct pyrolysis of *Sargassum Wightii* seaweed (Divya et al. 2019). Research has been done on functional carbon made from *Turbinaria Conoides* seaweed, which is micro- and mesoporous and has a high capacitance of 416 at 1 A/g and a high energy density of 52 Wh/kg (Pintor et al. 2013).

Anode materials with greater specific capacities have received a lot of attention. Low-crystalline carbons, which can hold far more lithium than graphite, are among them. Numerous appealing characteristics of disordered carbons from organic biomass can be tailored by altering the nature of their organic precursors' temperature regimes and good cycling qualities (Dahn et al. 1995); a larger lithium intake than the theoretical limit of 372 mAh g^{-1} for fully graphitic material. Although they lack broad crystallographic organization, disordered carbons primarily produced by pyrolyzing organic precursors have a planar hexagonal network of carbon atoms. Carbons produced through pyrolysis have also been demonstrated to retain up to 30% of the original hydrogen. Pyrolytic carbons' high lithium capacities are linked to both disorder and hydrogen content (Dahn et al. 1995). By pyrolyzing biomass precursors like rice husk (Fey and Chen 2001), lignin, cotton, and wool (Xing et al. 1996), walnut, almond shells, peanut shells (Fey et al. 2003), cotton and wool

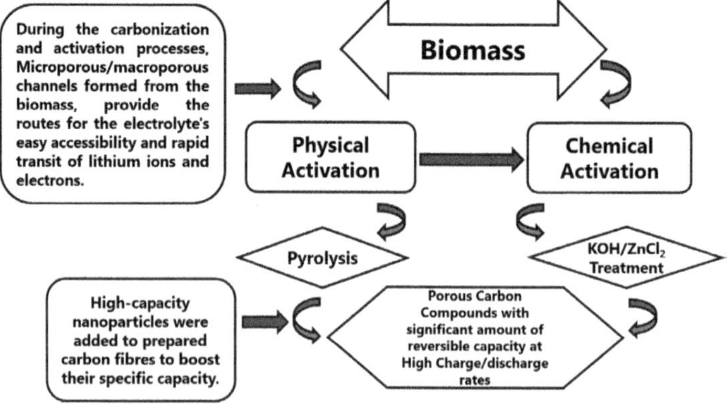

Figure 6. Flow chart for production of renewable electrodes.

(Peled et al. 1998), sugar, starch, and oak (Xing et al. 1996), and banana fibers that have been porogenized using KOH or $ZnCl_2$. A number of these carbons have been produced. The majority of the single carbon layers in these disordered carbons are disorganized, which can serve as locations for accommodating lithium during charging. The "falling cards" concept proposed by Dah states that at the low pyrolytic temperatures where these carbons are generated, thermal energy is inadequate to spin graphene sheets into parallel alignment and stack them (Dahn et al. 1997). Since there are a lot of non-parallel, disordered single layers of hexagonal carbons in low-temperature carbons, the R parameter's low values are a reflection of this fact. The presence of such uncorrelated graphene sheets and fragments significantly influences the ability of a disordered carbon to hold lithium. The "extra" capacity is caused by lithium occupying locations inside nanopores between layers in disordered carbons, where it is frequently challenging to remove the inserted lithium. Large amounts of lithium can be inserted more easily through such holes; however, some of it is lost due to thorough passivation. The passivation byproducts progressively seal the pores and cavities' openings, preventing the lithium trapped inside them from being cycled further. The huge surface area of the carbon is in line with the significant initial-cycle insertion capacity achieved with the carbon made from the $ZnCl_2$ processed fiber.

Large surface areas, though, can also expose a lot of surface area for passivation. Furthermore, the huge pores that are produced in this carbon do not function as cavities that can hold lithium. As an alternative, they give electrolyte access, which can passivate the pore surface and increase the irreversible capacity.

Contrary to the solid electrolyte interphase, when pores become clogged, thick layers of pores must inevitably form, preventing the electrolyte from reaching the active carbon surface. Thus, a decreased recovery of lithium stored in the carbons would result from masking the carbon surface. It follows that having big pores is detrimental to carbon-based anodes. Such pores enable simultaneous plating on the carbon surface and interaction of lithium with surface functional groups. Passivation is then caused by an electrolyte reaction, and the process is then complete. However, smaller sites can still be active lithium-accommodating sites (Stephan et al. 2006).

As anode materials for lithium-ion batteries, carbon-based materials have attracted a lot of interest due to their advantageous properties, which include accessibility, heat and chemical stability, and low price with a variety of textures, morphologies, and crystallinities. Graphite is currently the most widely used anode substance for lithium-ion batteries. However, the demand for electric gadgets cannot be satisfied by graphite's theoretical lithium-storage capacity (theoretical value 372 mA h g^{-1}). In order to increase the capacity of lithium storage, much effort has been put into the study of amorphous porous carbon materials. Studies on carbon-based anode materials show that micropores and macropores have a considerable impact on their electronic structure and performance, including reduced lithium-ion and electron diffusion lengths and enhanced reversible capacity (Cheng et al. 2008).

By employing amphiphilic copolymer surfactants as soft templates and mesoporous silica, zeolites, and tightly packed poly(methyl methacrylate) spheres as hard templates, porous carbons can be created (Ryoo et al. 1999). However, the preparation processes for such templating systems were frequently time-consuming and expensive, which limited their practical applications.

Porous carbon compounds having a significant surface area and a hierarchical porous network have been produced from biomass. Lithium ions and electrons can move rapidly via the macroporous channels made from the biomass, and the micropores formed during the carbonization and activation processes make the electrolyte easily accessible. From biomass, we were able to produce porous carbon compounds with high charge/discharge rates and large reversible capacity.

Large macroporous channels made of biomass offer an accessible pore system that makes it easier for lithium-ion diffusion and electrolyte transit to occur (Fu et al. 2009), while micropores created during the carbonization and activation processes assure a high reversible capacity. Additionally, the channels' thin walls make it easier for electrons to travel into the interior of the porous carbons and for Li-ions to diffuse there. The large reversible capacity and high-rate performance of the hierarchical porous carbons result from all these features.

Surface area increases the reversible capacity. The fact that the reversible capacity increases with surface area may be due to the availability of more active sites for lithium storage in materials with greater surface areas (Zhang et al. 2009).

Hard and soft carbons, which can be made in a variety of methods, are substitutes for graphitic carbons for Li-ion anodes. Sucrose, a naturally occurring organic sugar, is pyrolyzed as part of a classic method of creating hard carbon. Hard carbon anodes produce larger specific capacities than graphitic structures (over 500 mAh/g). However, they frequently have a significant irreversible capacity because of lithium's reactivity with both the electrolyte and the surface functional groups of the carbon structure that develop during pyrolysis. It is necessary to investigate naturally derived carbon precursors in order to commercialize new forms of carbon for Li-ion batteries. The use of activated carbons as scalable, economically efficient Li-ion anodes is also demonstrating amazing possibilities.

When exposed to a concentrated KOH solution or other chemical activating agents, activated carbons are most frequently created. After heat activation, KOH treatment creates more mesopores and/or micropores; the ensuing flaws promote larger capacities by allowing more lithium to be inserted into the carbon. The typical process for activating carbon entails soaking in a KOH bath and then subjecting it to a high-temperature heat treatment (Zheng et al. 2012).

The method for creating hierarchical porous carbons creates new opportunities for using appropriate carbons obtained from biomass as anode materials for lithium batteries with high rates and large capacities.

By using composites, researchers may solve the issue of poor cyclability. In such a composite material, one component, typically carbon, serves as a stress absorber while another, such as silicon/tin or nanoparticles, increases capacity. By using this method, a composite with a capacity greater than carbon and cyclability superior can be produced (Deng 2015).

In order to build up the capacity of the lithium-ion batteries, renewable electrodes obtained from biomass were treated with the nanoparticles produced through the green synthesis method. Using the drop cast process (Bastos-Arrieta et al. 2018), the nanoparticles were incorporated into the electrodes, which is an effective method to produce nanoparticles incorporated in renewable electrodes.

Green synthesis of nanoparticles using extracts from *Solanum macrocarpon* fruits and nano-dimensional crystalline CuO nanoparticles was effectively made using two methods: conventional heating and microwave irradiation. Using the highly electroactive modified with the as-prepared nanoparticles were electrochemically characterized. The redox activities of these electrodes were strongly influenced by the size and aggregation of the nanoparticles. As a result, in the probe and electrolyte media examined, the CuO nanoparticles with smaller diameters and fewer agglomerations displayed greater electroactivity (Okpara et al. 2021).

The green synthesis method is an environmentally safe process for the formation of nanoparticles, in which plants include components that can act as stabilizers and reducing agents in the production of nanoparticles. Thus, the synthesis of silver nanoparticles utilizing an extracted aqueous solution of grape stalk waste as a reducing and capping agent was examined by (Bastos-Arrieta et al. 2018). Finally, the voltammetry measurements were evaluated using screen-printed electrodes enhanced with silver nanoparticles. The results show good repeatability and sensitivity. From this result, we can suggest that high-capacity green synthesized nanoparticles were employed on the prepared carbon fiber in order to increase its specific capacity.

4. Conclusion

Li-ion batteries currently use inorganic insertion compounds. With foreseeable large-scale uses, the quantity and material life-cycle costs of such batteries may pose problems in the long run. A totally different strategy from the traditional method has been implemented to produce novel organic electrode materials to address the problem of sustainability of electrode materials. In contrast, Li-ion cell technology supports sustainable or green chemistry concepts, like battery processing. The ability to secure sustainable energy sources will be humanity's main problem in the twenty-first century.

Due to rising standards of living in the emerging world, the demand for electric energy is expected to rise even more quickly than the global population by the year 2050, which predictions place at roughly 10 billion. Future clean energy economies must completely restructure their economies to promote the use of renewable raw resources in order to completely eradicate CO_2 emissions, which is near to doing so. The development of green and sustainable batteries within the next ten years may be made possible by the inclusion of renewable resources in the design of electrode materials.

References

Ahmadpour, A. and Do, D. D. (1996). The preparation of active carbons from coal by chemical and physical activation. *Carbon*, 34(4): 471–79. https://doi.org/10.1016/0008-6223(95)00204-9.

Ahmed Shakeel, Mudasir Ahmad, Babu Lal Swami and Saiqa Ikram. (2016). A review on plants extract mediated synthesis of silver nanoparticles for antimicrobial applications: A green expertise. *Journal of Advanced Research*, 7(1): 17–28. https://doi.org/10.1016/j.jare.2015.02.007.

Azam Ameer, Arham S. Ahmed, Oves, M., Khan, M. S. and Adnan Memic. (2012). Size-dependent antimicrobial properties of CuO nanoparticles against gram-positive and -negative bacterial strains. *International Journal of Nanomedicine*, 7: 3527–35. https://doi.org/10.2147/IJN.S29020.

Bastos-Arrieta Julio, Antonio Florido, Clara Pérez-Ràfols, Núria Serrano, Núria Fiol, Jordi Poch and Isabel Villaescusa. (2018). Green synthesis of Ag nanoparticles using grape stalk waste extract for the modification of screen-printed electrodes. *Nanomaterials*, 8(11). https://doi.org/10.3390/nano8110946.

Biswal Mandakini, Abhik Banerjee, Meenal Deo and Satishchandra Ogale. (2013). From dead leaves to high energy density supercapacitors. *Energy and Environmental Science*, 6(4): 1249–59. https://doi.org/10.1039/c3ee22325f.

Boehm, H. P. (1989). Surface properties of carbons. *Studies in Surface Science and Catalysis*, 48(C): 145–57. https://doi.org/10.1016/S0167-2991(08)60678-3.

Broussely, M. (1999). Recent developments on lithium ion batteries at SAFT. *Journal of Power Sources*, 81-82: 140–43. https://doi.org/10.1016/S0378-7753(99)00146-9.

Chen Haiyan, Michel Armand, Gilles Demailly, Franck Dolhem, Philippe Poizot and Jean Marie Tarascon. (2008). From biomass to a renewable LiXC6O6 organic electrode for sustainable li-ion batteries. *ChemSusChem*, 1(4): 348–55. https://doi.org/10.1002/cssc.200700161.

Cheng Fangyi, Zhanliang Tao, Jing Liang and Jun Chen. (2008). Template-directed materials for rechargeable lithium-ion batteries. *Chemistry of Materials*, 20(3): 667–81. https://doi.org/10.1021/cm702091q.

Dahn, J. R., Xing, W. and Gao, Y. (1997). The 'falling cards model' for the structure of microporous carbons. *Carbon*, 35(6): 825–30. https://doi.org/10.1016/S0008-6223(97)00037-7.

Dahn Jeff R., Tao Zheng, Yinghu Liu and Xue, J. S. (1995). Mechanisms for lithium insertion in carbonaceous materials. *Science*, 270(5236): 590–93.

Deng Da. (2015). Li-ion batteries: Basics, progress, and challenges. *Energy Science & Engineering*, 3(5): 385–418. https://doi.org/https://doi.org/10.1002/ese3.95.

Divya, P., Prithiba, A. and Rajalakshmi, R. (2019). Biomass derived functional carbon from sargassum wightii seaweed for supercapacitors. *IOP Conference Series: Materials Science and Engineering*, 561(1). https://doi.org/10.1088/1757-899X/561/1/012078.

Fey, G., Ting Kuo, Lee, D. C., Lin, Y. Y. and Prem Kumar, T. (2003). High-capacity disordered carbons derived from peanut shells as lithium-intercalating anode materials. *Synthetic Metals*, 139(1): 71–80. https://doi.org/10.1016/S0379-6779(03)00082-1.

Fey George Ting Kuo and Chung Lai Chen. (2001). High-capacity carbons for lithium-ion batteries prepared from rice husk. *Journal of Power Sources*, 97-98(December 2000): 47–51. https://doi.org/10.1016/S0378-7753(01)00504-3.

Fu, L. J., Yang, L. C., Shi, Y., Wang, B. and Wu, Y. P. (2009). Synthesis of carbon coated nanoporous microcomposite and its rate capability for lithium ion battery. *Microporous and Mesoporous Materials*, 117(1-2): 515–18. https://doi.org/10.1016/j.micromeso.2008.07.008.

Gayda Galina Z., Olha M. Demkiv, Nataliya Ye Stasyuk, Roman Ya Serkiz, Maksym D. Lootsik, Abdelhamid Errachid, Mykhailo V. Gonchar and Marina Nisnevitch. (2019). Metallic nanoparticles obtained via 'green' synthesis as a platform for biosensor construction. *Applied Sciences (Switzerland)*, 9(4). https://doi.org/10.3390/app9040720.

Habazaki, H., Sato, S., Habazaki, H. and Inagaki, M. (2004). No title. *Carbon*, 42: 2756.

Hwang Yun Ju, Soo Kyung Jeong, Kee Suk Nahm, Jae Sun Shin and Manuel Stephan, A. (2007). Pyrolytic carbon derived from coffee shells as anode materials for lithium batteries. *Journal of Physics and Chemistry of Solids*, 68(2): 182–88. https://doi.org/10.1016/j.jpcs.2006.10.007.

Jansen, A. N., Kahaian, A. J., Kepler, K. D., Nelson, P. A., Amine, K., Dees, D. W., Vissers, D. R. and Thackeray, M. M. (1999). Development of a high-power lithium-ion battery. *Journal of Power Sources*, 81-82: 902–5. https://doi.org/10.1016/S0378-7753(99)00268-2.

Jiang Luyun, Seong O. K. Han, Melissa Pirie, Hyun Hee Kim, Young Hoon Seong, Hyunuk Kim and John S. Foord. (2021). Seaweed biomass waste-derived carbon as an electrode material for supercapacitor. *Energy and Environment*, 32(6): 1117–29. https://doi.org/10.1177/0958305X19882398.

Karagöz Selhan, Turgay Tay, Suat Ucar and Murat Erdem. (2008). Activated carbons from waste biomass by sulfuric acid activation and their use on methylene blue adsorption. *Bioresource Technology*, 99(14): 6214–22. https://doi.org/10.1016/j.biortech.2007.12.019.

Kates, R. W., Clark, W. C., Corell, R., Hall, J. M., Jaeger, C. C., Lowe, I., McCarthy, J. J., Schellnhuber, H. J., Bolin, B., Dickson, N. M., Faucheux, S., Gallopin, G. C., Grübler, A., Huntley, B., Jäger, J., Jodha, N. S., Kasperson, R. E., Mabogunje, A., Matson, P., Mooney, H. and Svedlin, U. (2001).

Environment and development. Sustainability science. Science (New York, N.Y.) 292(5517): 641–642. https://doi.org/10.1126/science.1059386.

Kuppusamy Palaniselvam, Mashitah M. Yusoff, Gaanty Pragas Maniam and Natanamurugaraj Govindan. (2016). Biosynthesis of metallic nanoparticles using plant derivatives and their new avenues in pharmacological applications – an updated report. *Saudi Pharmaceutical Journal*, 24(4): 473–84. https://doi.org/10.1016/j.jsps.2014.11.013.

Larcher, D. and Tarascon, J. M. (2015). Towards greener and more sustainable batteries for electrical energy storage. *Nature Chemistry*, 7(1): 19–29. https://doi.org/10.1038/nchem.2085.

Luo Yuqing, Yijian Tang, Shasha Zheng, Yan Yan, Huaiguo Xue and Huan Pang. (2018). Dual anode materials for lithium- and sodium-ion batteries. *Journal of Materials Chemistry A*, 6(10): 4236–59. https://doi.org/10.1039/c8ta00107c.

Lupi Carla and Mauro Pasquali. (2003). Electrolytic nickel recovery from lithium-ion batteries. *Minerals Engineering*, 16(6): 537–42. https://doi.org/10.1016/S0892-6875(03)00080-3.

Mandal Deendayal, Mark E. Bolander, Debabrata Mukhopadhyay, Gobinda Sarkar and Priyabrata Mukherjee. (2006). The use of microorganisms for the formation of metal nanoparticles and their application. *Applied Microbiology and Biotechnology*, 69(5): 485–92. https://doi.org/10.1007/s00253-005-0179-3.

Mandal Dindyal, Sourav Mishra and Rohit Kumar Singh. (2018). Green synthesized nanoparticles as potential nanosensors. *Energy, Environment, and Sustainability*, 137–64. https://doi.org/10.1007/978-981-10-7751-7_7.

Mishra Amit, Akansha Mehta, Soumen Basu, Shweta J. Malode, Nagaraj P. Shetti, Shyam S. Shukla, Mallikarjuna N. Nadagouda and Tejraj M. Aminabhavi. (2018). Materials science for energy technologies electrode materials for lithium-ion batteries. *Materials Science for Energy Technologies*, 1(2): 182–87. https://doi.org/10.1016/j.mset.2018.08.001.

Mukherjee, P., Ahmad, A., Mandal, D., Senapati, S., Sainkar, S. R., Khan, M. I., Ramani, R., Parischa, R., Ajayakumar, P. V., Alam, M., Sastry, M. and Kumar, R. (2001). Bioreduction of AuCl(4)(-) Ions by the Fungus, *Verticillium* sp. and surface trapping of the gold nanoparticles formed D.M. and S.S. thank the Council of Scientific and Industrial Research (CSIR), Government of India, for financial assistance. Angewandte Chemie (International ed. in English), 40(19): 3585–3588. https://doi.org/10.1002/1521-3773(20011001)40:19<3585::aid-anie3585>3.0.co;2-k.

Ogasawara Takeshi, Aurélie Débart, Michael Holzapfel, Petr Novák and Peter G. Bruce. (2006). Rechargeable Li2O2 electrode for lithium batteries. *Journal of the American Chemical Society*, 128(4): 1390–93. https://doi.org/10.1021/ja056811q.

Okpara Enyioma C., Oluwasayo E. Ogunjinmi, Opeyemi A. Oyewo, Omolola E. Fayemi and Damian C. Onwudiwe. (2021). Green synthesis of copper oxide nanoparticles using extracts of solanum macrocarpon fruit and their redox responses on SPAu electrode. *Heliyon*, 7(12): e08571. https://doi.org/10.1016/j.heliyon.2021.e08571.

Owaid Mustafa Nadhim, Tahseen Ali Zaidan, Rasim Farraj Muslim and Mohammed Abdulrahman Hammood. (2019). Biosynthesis, characterization and cytotoxicity of zinc nanoparticles using panax ginseng roots, araliaceae. *Acta Pharmaceutica Sciencia*, 57(1): 19–32. https://doi.org/10.23893/1307-2080.APS.05702.

Peled Emanuel, Victor Eshkenazi and Yuri Rosenberg. (1998). Study of lithium insertion in hard carbon made from cotton wool. *Journal of Power Sources*, 76(2): 153–58. https://doi.org/10.1016/S0378-7753(98)00148-7.

Pintor Marie Julie, Corine Jean-Marius, Valérie Jeanne-Rose, Pierre Louis Taberna, Patrice Simon, Jean Gamby, Roger Gadiou and Sarra Gaspard. (2013). Preparation of Activated carbon from turbinaria turbinata seaweeds and its use as supercapacitor electrode materials. *Comptes Rendus Chimie*, 16(1): 73–79. https://doi.org/10.1016/j.crci.2012.12.016.

Ruiz, V., Pfrang, A., Kriston, A., Omar, N., Van den Bossche, P. and Boon-Brett, L. (2018). A review of international abuse testing standards and regulations for lithium ion batteries in electric and hybrid electric vehicles. *Renewable and Sustainable Energy Reviews*, 81(January): 1427–52. https://doi.org/10.1016/j.rser.2017.05.195.

Ryoo Ryong, Sang Hoon Joo and Shinae Jun. (1999). Synthesis of highly ordered carbon molecular sieves via template-mediated structural transformation. *Journal of Physical Chemistry B*, 103(37): 7743–46. https://doi.org/10.1021/jp991673a.

Sastry Murali, Absar Ahmad, Islam Khan, M. and Rajiv Kumar. (2005). Microbial nanoparticle production. *Nanobiotechnology*, 126–35. https://doi.org/10.1002/3527602453.ch9.

Sharma Deepali, Suvardhan Kanchi and Krishna Bisetty. (2019). Biogenic synthesis of nanoparticles: A review. *Arabian Journal of Chemistry*, 12(8): 3576–3600. https://doi.org/10.1016/j.arabjc.2015.11.002.

Stephan, A. Manuel, Prem Kumar, T., Ramesh, R., Sabu Thomas, Soo Kyung Jeong and Kee Suk Nahm. (2006). Pyrolitic carbon from biomass precursors as anode materials for lithium batteries. *Materials Science and Engineering A*, 430(1-2): 132–37. https://doi.org/10.1016/j.msea.2006.05.131.

Tang Jialiang, Vinodkumar Etacheri and Vilas G. Pol. (2016). Wild fungus derived carbon fibers and hybrids as anodes for lithium-ion batteries. *ACS Sustainable Chemistry and Engineering*, 4(5): 2624–31. https://doi.org/10.1021/acssuschemeng.6b00114.

Thakkar Kaushik N., Snehit S. Mhatre and Rasesh Y. Parikh. (2010). Biological synthesis of metallic nanoparticles. *Nanomedicine: Nanotechnology, Biology, and Medicine*, 6(2): 257–62. https://doi.org/10.1016/j.nano.2009.07.002.

Tokumitsu, K., Mabuchi, A., Fujimoto, H. and Kasuh, T. (1995). Charge/discharge characteristics of synthetic carbon anode for lithium secondary battery. *Journal of Power Sources*, 54(2): 444–47. https://doi.org/10.1016/0378-7753(94)02121-I.

Tsukuba Epochal. (2002). The Fifth International Conference on EcoBalance Practical Tools and Thoughtful Principles for Sustainability.

Vetchinkina Elena P., Ekaterina A. Loshchinina, Ilya R. Vodolazov, Viktor F. Kursky, Lev A. Dykman and Valentina E. Nikitina. (2017). Biosynthesis of nanoparticles of metals and metalloids by basidiomycetes. preparation of gold nanoparticles by using purified fungal phenol oxidases. *Applied Microbiology and Biotechnology*, 101(3): 1047–62. https://doi.org/10.1007/s00253-016-7893-x.

Wang, L., Hu, C. and Shao, L. (2017). The antimicrobial activity of nanoparticles: present situation and prospects for the future. International Journal of Nanomedicine, 12: 1227–1249. https://doi.org/10.2147/IJN.S121956.

Wang Gang, Minghao Yu and Xinliang Feng. (2021). Carbon materials for ion-intercalation involved rechargeable battery technologies. *Chemical Society Reviews*, 50(4): 2388–2443. https://doi.org/10.1039/d0cs00187b.

Xing Weibing, Xue, J. S., Tao Zheng, Gibaud, A. and Dahn, J. R. (1996). Correlation between lithium intercalation capacity and microstructure in hard carbons. *Journal of the Electrochemical Society*, 143(11): 3482–91. https://doi.org/10.1149/1.1837241.

Zhang Di, Shuai Wang, Yang Ma and Shubin Yang. (2018). Two-dimensional nanosheets as building blocks to construct three-dimensional structures for lithium storage. *Journal of Energy Chemistry*, 27(1): 128–45. https://doi.org/10.1016/j.jechem.2017.11.031.

Zhang Feng, Kai Xue Wang, Guo Dong Li and Jie Sheng Chen. (2009). Hierarchical porous carbon derived from rice straw for lithium ion batteries with high-rate performance. *Electrochemistry Communications*, 11(1): 130–33. https://doi.org/10.1016/j.elecom.2008.10.041.

Zheng Honghe, Qunting Qu, Li Zhang, Gao Liu and Vincent S. Battaglia. (2012). Hard carbon: A promising lithium-ion battery anode for high temperature applications with ionic electrolyte. *RSC Advances*, 2(11): 4904–12. https://doi.org/10.1039/c2ra20536j.

CHAPTER 11

Recent Advances and Outlook on Green Synthesis and Applications of ZnO Nanoparticles

Tanuj,[1] Rajesh Kumar,[1,] Santosh Kumar,[1] Kiran Kumar,[1]
Sandeep Chauhan,[1] Dhiraj Singh Rawat,[2] Neerja Kalra[3]
and Subhash Sharma[4,5]*

1. Introduction

In the latest era of technology, the technology of nanoscience is a very swiftly emerging field that promises the synthesis of some innovative nano-sized materials. The objective of nanoscience is to create, depict, and employ material that has a 1–100 nm size. In the modern century, this technology of synthesizing materials of nano-range has grown as an interesting field (Pillai et al. 2020).

Nanotechnology is one of the emerging technologies which can carry scientific fields to a different level. Nano-sized materials have a very broad range of applications because of their morphology and size and have an important contribution in applied and basic sciences like fields. In recent times, a lot of stress has been given to semiconductors having nano-size for their unique properties, due to which they exhibit amazing optoelectronic applications (Fakhari et al. 2019). Nanotechnology

[1] Department of Chemistry, Himachal Pradesh University, Summerhill, Shimla (HP) India-171005.

[2] Department of Bio-Sciences, Himachal Pradesh University, Summerhill, Shimla (HP) India-171005.

[3] Department of Chemistry, Government College, Ateli, Mahendergarh (Haryana) India – 123021.

[4] Centro de Nanociencias y Nanotecnología. Universidad Nacional Autónoma de México, Km 107 Carretera Tijuana-Ensenada AP 14, Ensenada, 22860, B.C., México.

[5] CONAHCyT- Centro de Nanociencias y Nanotecnología. Universidad Nacional Autónoma de México, Km 107 Carretera Tijuana-Ensenada AP 14, Ensenada, 22860, B.C., México.

* Corresponding author: krajeshdhiman@gmail.com

deals with materials that show unique, improved, and amazing chemical, biological, as well as physical properties due to the size of particles in the nanoscale and also work as a capping agent to stabilize them. Metal nanoparticle synthesis is one of the symbolic examples.

Nano-based technology has many uses in different fields like biomedical science, gene delivery, drug delivery, healthcare systems, and foods. A diversity of applications in biomedical as well as in material science and a wide variety of unique properties are achieved by reducing the size of the material (Sivasankarapillai et al. 2019).

Recently, a lot of emphasis has been given to developing methods to synthesize semiconductors having nano-size because of their unique and characteristic properties that style them as favorable constituents for different applications in optical and electronic devices. For use in descendant optoelectronic semiconductor devices, ZnO is alleged to be the most capable material among the different semiconductor materials that have been reported to date. One of the members of semiconductor compounds of group II–V, i.e., ZnO, crystallizes with a wurtzite structure. ZnO has amazing properties, like broadband gap, i.e. (3.36 eV), high room temperature binding energy (60 MeV), very intense absorption band in the UV region, biocompatibility, high photostability, low toxicity, as well as the high value of electrochemical coupling coefficient. Due to these properties, ZnO has been used in many electronic as well as optical devices.

ZnONPs engrossed very electrifying interest in research because of their massive prospect of appliances like targeted drug delivery (Fahimmunisha et al. 2020), smart UV sensors (Sosna-Glebska et al. 2019), anti-oxidant activities (Lingaraju et al. 2016), biosensing (Hwa and Subramani 2014), and environment remediation (Singh et al. 2018). Selectivity of diverse morphology of formed nano-sized particles conferring to biological source used with improved stability is one of the salient features of biosynthesis of nanoparticles.

Chemical as well as physical methods used generally for the synthesis of various nanoparticles are categorized as (i) top-down, and (ii) bottom-up approach (Figure 1). Out of these methods, most of the methods require critical temperature or pressure conditions, are time-consuming, energy consuming, need harmful chemicals that harm the environment by causing pollution, and are very expensive. For the protection of the environment, alternative environment-friendly methods of synthesis are required, which can eradicate the use of harmful chemicals as well. Recently, many researchers have shown interest in a safe approach, i.e., "green synthesis," which is a biosynthetic approach to synthesizing nanoparticles. In comparison to the other conventional methods, green method of nanoparticle synthesis has many benefits like low cost, simple to use, environment-friendly nature, formation of more stable nanoparticles, and no need for additional harmful chemicals. Among various biosynthetic approaches (enzymes, bacteria, and fungi) reported to date, green synthesis of the nano-sized particles using different plant extracts of different plant parts is the most promising green route of synthesizing nanoparticles.

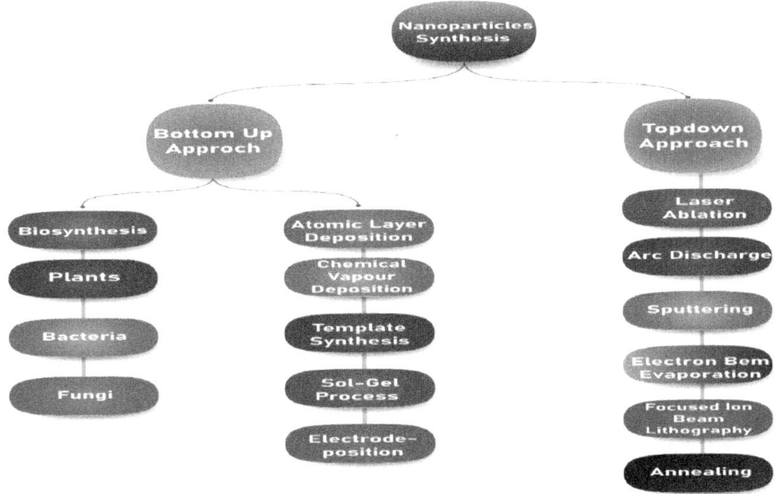

Figure 1. Synthesis of nanoparticles through various physiochemical and biological approaches (Jadoun et al. 2021).

Green chemistry is based on many principles, such as:

(1) Catalysis: It proposes the use of catalysts for improving different progressions like consumption of energy and efficiency.

(2) Accident Prevention: Reduces accident chances and the danger related to it.

(3) Pollution Prevention: Prevents the release of some hazardous, harmful, or toxic substances that mainly pollute the environment.

(4) Safer Solvents: Uses minimum possible chemicals or solvents, which can also lead to the pollution of the environment.

(5) Less Toxic Chemical Synthesis: Plan or propose some safe routes for the synthesis of some new compounds.

(6) Prevention: Minimizes production of waste in every stage of the process.

(7) Atom Economy: Improves the reaction efficiency.

(8) Energy Efficient: Avoids processes that consume large amounts of energy.

(9) Renewable Feedstocks: Using chemicals or solvents made of some renewable sources.

(10) Design for Degradation: Designs biodegradable, non-hazardous, and non-toxic products.

(11) Usage of Safe Chemicals: Minimizes toxicity of processes as well as products.

(12) Reduce the Use of Derivatives: Evades derivative use like protectors and stabilizers (Figure 2) (Bandeira et al. 2020).

In plant parts, like seeds, roots, fruits, and leaves, numerous phytochemicals (those chemicals that are present in the plant parts) are present. Phytochemicals

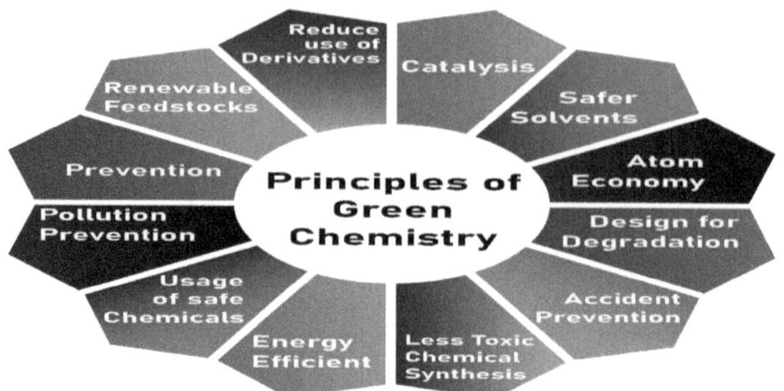

Figure 2. Principles of green chemistry (Bandeira et al. 2020).

include alkaloids, amino acids, flavonoids, proteins, steroids, terpenoids, and tannins, which are found in abundance.

Different phytochemicals include amyrin, anthocyanins, anthraquinones, friedelin, glycosides, hyperin, phlobatannins, phytosterol, quercetin, saponins, ursolic acid, etc. (Figure 3). The green approach includes phytochemical extraction using solvents like water, which is environment-friendly. The extract of phytochemicals collected is then added to the precursor of metal to form the target nanoparticles of precursor metal.

In the green synthesis method, one of the key factors for the formation of nanoparticles is the presence of some bioactive compounds. These compounds have the capacity to form nanoparticles of metal by reducing the ions of the metal to nanoparticles of metal through numerous oxidation methods. Phytochemical compounds can do chelation with metal ions through different functional groups and can also form coordination complexes that thermally decompose to get nanoparticles. Phytochemicals found in the extract of plants perform as a stabilizer by preventing the agglomeration of synthesizing nanoparticles. There are numerous environment-friendly techniques reported to date for ZnONPs synthesis (ultrasonic-assisted synthesis, microwave-assisted synthesis, etc.) (Bayrami et al. 2019). Extracts of plants, microbes, fungi, or algae are used for the formation of nanoparticles of metal or metal oxides by the biological method of synthesis, such as green synthesis (Figure 4). The green approach has many advantages over conventional chemical processes (Jeevanandam et al. 2016). The most significant advantage is the potential application of nanoparticles synthesized by the green approach in living systems directly. This is due to the reason that the nanoparticles prepared from the green approach show a decreased level of toxicity in comparison to other techniques. Biocomponents utilized in the synthesis express stabilizing effects, making it appropriate for the live structures (Singh et al. 2016).

Plant-based synthesis was found to be the best method among different biological approaches because of features like utilization of commonly accessible plants, feasibility, and wide-spread variation in the morphology of the ZnONPs (synthesized by using extract of the plant) (Agarwal et al. 2017), grounded on the previous findings associated with synthesis of nanoparticles and examination of

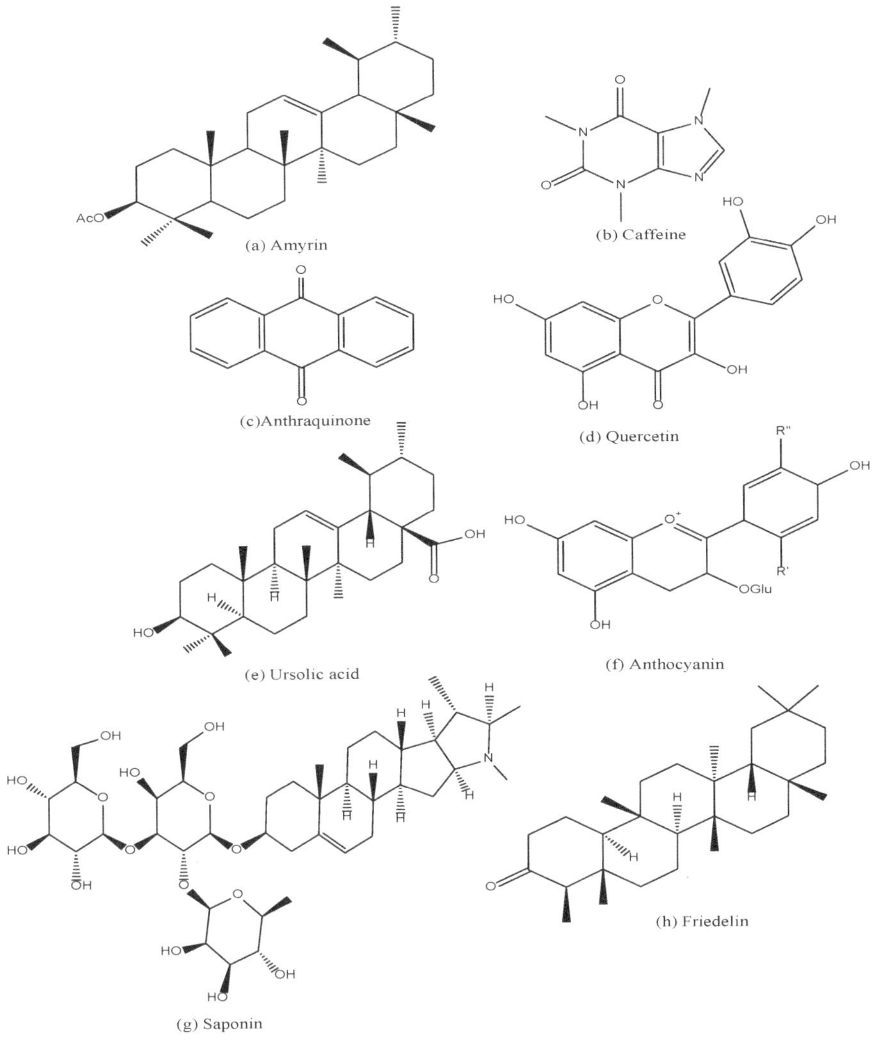

Figure 3. Different phytochemicals that are present in the plants.

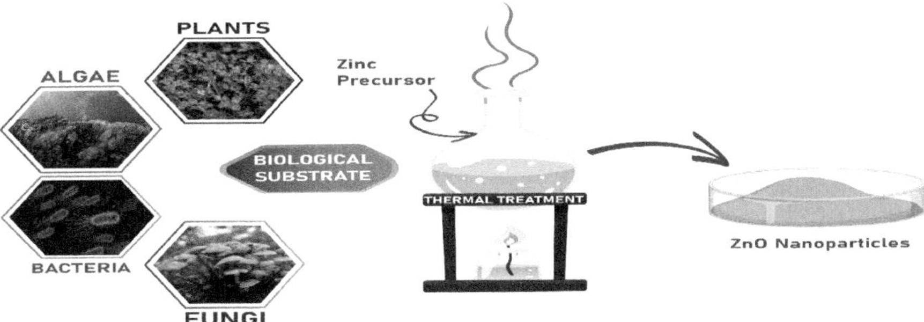

Figure 4. Different approaches to green synthesis of ZnO NPs (Bandeira et al. 2020).

their biological applications (Rahdar et al. 2020). The structural properties of those nanoparticles synthesized by plants acting as reducing agents studied by (XRD) X-ray diffraction, nanoparticles morphology by (SEM) scanning electron microscope, their size by (TEM) Transmission electron microscope, and (PL) Photoluminescence spectra of nanoparticles has been assessed.

ZnO nanoparticles show many properties, such as higher surface area, low toxicity, and biocompatibility. Different industries like the textile, painting, and printing industries cause water pollution as these industries release waste containing organic dyes to water sources. This is a serious issue as it creates a problem for the environment; these organic dyes show a very high resistance to biological degradation. Various dyes like methylene blue (MB), crystal violet (CV), and malachite green (MG) are very common organic dyes found in the wastewater discharged from various industries. The emission of harmful organic dyes into the water causes different health-related issues like irritation in the eye, gastritis, nausea, diarrhea, and toxicity in the central nervous system. If discharged in water, dyes are also dangerous to the aquatic life. For the treatment of wastewater, various direct methods like biodegradation, oxidation, and precipitation have been utilized. All those methods can not mineralize dyes entirely, yield minor pollutants, and are cost ineffective.

In recent times, photocatalysis has developed as a very capable practice for wastewater treatment by the removal of various carbon-based pollutants. Many researchers have shown their attention toward the photocatalysis technique as this technique has many advantages over other conventional methods of wastewater treatment. The photocatalysis technique is cost-effective, highly efficient, simple, and eco-friendly. Different photocatalysts, like ZnO nanoparticles, express the capability of degrading the organic dye aromatic ring completely via an oxidation process into some harmless compounds. Molecules of organic dye were absorbed on the surface of the photocatalyst. Degradation of the dyes occurs due to the formation of very reactive oxygen species (Kanagamani and Muthukrishnan 2019). When compared to other conventional chemical or physical methods, amazing activity is shown by photocatalysts synthesized by the green method. This is all because of the existence of biologically active constituents in plant extract. Coating of the surface of the green synthesized catalyst occurs by the biologically active constituents present in the extract of the plant.

Green synthesized ZnO NPs are used very widely in wastewater treatment applications as an efficient photocatalyst – numerous studies on ZnO NPs biosynthesis by utilizing extract from different natural plants. Results of biosynthesized nanoparticles showed that features of the sample depend on the basis of the extract of the used plant. The properties of various dyes as pollutant has been studied utilizing different methods to estimate their efficiency of degradation illuminated under UV light photo-catalytically (Aldeen et al. 2022). Nanoparticles of metal oxide and metals exhibit a very broad range of functions in various fields like catalysis, agriculture, medicine, textile, pharmacy, and antimicrobial assays (Sonkusare et al. 2018, Akintelu et al. 2019).

ZnO NPs show very remarkable uses in photocatalysis, antimicrobial activity, coating, materials acting as sunscreen, cosmetics, and pesticides (Gupta et al. 2018,

Kadiyala et al. 2019, Agarwal et al. 2019); nowadays, their usage has been increasing (Devatha and Thalla 2018) among different methods and techniques used for ZnO NPs synthesis, which involve use of biomaterial acting as agent for reduction. This is due to their unique properties like easy-to-operate procedures, eco-friendliness, cost-effectiveness, non-hazardous reagents, and low energy consumption (Khatami et al. 2018).

Secondary metabolites and biomolecules are present in extracts of plants, like flavanones, alkaloids, tannins, terpenoids, polyphenols, and saponins, found accountable for the reduction of zinc precursors effectively. Compared to micro-organisms, reduction of precursors of zinc was effective with plant extracts (Rastogi et al. 2018).

Numerous plants, such as *Cinnamomum verum, Cinnamomum tamala, Brassica oleracea var. Italica*, and *Beta vulgaris* (Pillai et al. 2020), *Laurus nobilis L.* (Fakhari et al. 2019), *Phoenix roebelenii* (Aldeen et al. 2022), *Costus igneus* (Nandhini et al. 2018), *Mimosa pudica* (Balogun et al. 2020), *Typha latifolia* (Kumar et al. 2019), *Cayratia pedata* (Jayachandran et al. 2021), *Berberis aristata* (Chandra et al. 2019), *Allium sativum* (Slman et al. 2018), *Piper nigrum* (Thangapandi et al. 2021), *Scutelleria baicalensis* (Chen et al. 2019), onion peel (Canbaz et al. 2020), vegetables (cabbage, tomato, and carrot) (Degefa et al. 2021), *Ocimum tenuiflorum* (Raut et al. 2013), *Aloe barbadensis* (Sangeetha et al. 2011), *Deverrea tortuosa* (Selim et al. 2020), *Hibiscus rosa sinensis* (Devi and Gayathri 2014), fruits (orange, grape, and lemon) (Nava et al. 2017), *Azadirachta indica* (Noorjahan et al. 2015), *Sambucus ebulus* (Alamdari et al. 2020), *Citrus sinensis* (Luque et al. 2018), and *Salvia officinalis* (Alrajhi et al. 2021) are used for synthesis of nanoparticles.

Previous studies have already shown that the nanoparticle's morphological identities influence their various applications. Various green synthesized ZnO NPs are characterized by various techniques such as UV-vis spectrophotometer, Fourier transform infrared spectroscopy (FTIR), SEM, TEM, and Energy dispersive x-ray (EDX) (Degefa et al. 2021).

In the current article, we deliver elaborated material on the latest advances in the ZnO NPs synthesis by utilizing extracts of plants. Likewise, the latest advancements in various applications of biosynthesized ZnO NPs utilizing biological materials, such as plant extracts, are also well presented (Raut et al. 2013).

2. Biogenic Synthesis of ZnO NPs

The most common biological material used for the nanoparticle synthesizing of metal oxide, used nowadays, is plants. Plants as biological substrates have many advantages over other substrates as they are very cost effective, less lethal than bacteria, fungi, and algae and are also very easy to use. Another advantage of using plant parts as substrates is that extract of different parts of the plant can be directly obtained by just exposing that particular plant part to solvent, which is generally either distilled water or ethanol.

Various plant parts have been used for this work, like leaves, seeds, and roots. It is well known that plants have a very high concentration of many active compounds, such as flavonoids, phenolic acids, saponins, and methylxanthines.

All these compounds are capable of neutralizing different chemical species, such as free radicals, chelate metals, and reactive oxygen species (ROS), and they are more commonly regarded as antioxidants. All these antioxidants present in the plants are mainly responsible for metal oxide nanoparticles or metal nanoparticle green synthesis. This is all due to the capability of these antioxidants to chelate metal ions or bio-reduce. They also act as stabilizers of metal oxide nanoparticles produced.

The synthesis of ZnO NPs can be done through a simple process that contains mainly two steps: (1) preparation of plant extract and (2) preparation of nanoparticles of ZnO by using plant extract (Figures 5 and 6). Some of the potential plants proposed for the synthesis of nanoparticles by using their extract are illustrated in Figure 7.

(a) *Costusigneus* (https://alohatropicals.com/product/costus-igneus-fiery-costus/)

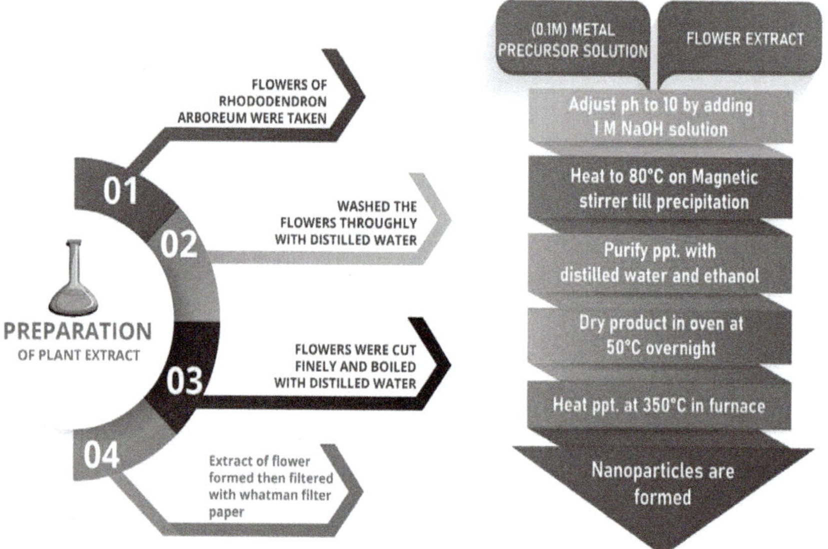

Figure 5. Schematic representation of synthesis of ZnO NPs from plant extract.

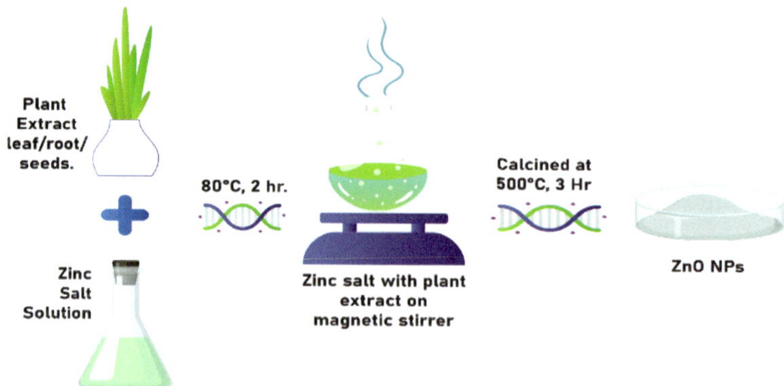

Figure 6. Synthesis of nanoparticles of ZnO from the extract of the plant.

Figure 7. Different plants used as biological substrates for the ZnO NPs synthesis are given below.

(b) *Mimosa pudica* (https://commons.wikimedia.org/wiki/File:Mimosa_pudica_004. JPG)

(c) *Typha latifolia* (https://flickr.com/photos/ryanready/4793337363)

(d) *Cayratiapedata* (https://www.plantslive.in/product/buy-cayratia-pedata-var-pedata-corivalli-tripadi-plant-online-india/)

(e) *Beta vulgaris* (https://www.gardenersworld.com/plants/beta-vulgaris-chioggia/)

(f) *Cinnamomum tamala* (https://herbalistimur.blogspot.com/2019/10/cinnamomum-tamala-buch-ham-nees-eberm.html)

(g) *Cinnamomum verum* (https://www.kamelya.ca/us/cinnamom-essential-oil-cinnamomum-verum.html)

(h) *Brassica oleracea var. Italica* (https://www.biolib.cz/cz/taxonimage/id203688/)

(i) *Berberis aristata* (https://www.medicinalplantsanduses.com/berberis-aristata-medicinal-uses)

(j) *Phoenix roebelenii* (https://jardinage.ooreka.fr/plante/voir/1499/phoenix-roebelenii)

(k) *Allium sativum* (http://www.onlineplantguide.com/Plant-Details/2912/)

(l) *Laurus nobilis* (https://www.boethingtreeland.com/plants/laurus-nobilis/)

(m) *Scutelleria baicalensis* (roots) (http://shlosem.com/en/production/herbs/scutellaria/)

(n) *Piper nigrum* seeds (https://www.exportersindia.com/thakur-agro-industries-pvt-ltd/black-piper-nigrum-seeds-5269670.htm)

(o) *Ocimumtenuiflorum* (http://www.flickr.com/photos/tony_rodd/3243036011/)

(p) *Aloe barbadensis* (https://www.greenex.com/foliage/aloe-barbadensis/)

(q) *Deverreatortuosa* (https://www.plantarium.ru/page/image/id/157624.html)

(r) *Hibiscus rosa sinensis* (https://www.plantmaster.com/gardens/eplant.php?plantnum=306)

(s) *Azadirachta indica* (https://mybageecha.com/products/azadirachta-indica-neem-tree)

(t) *Sambucus ebulus* (http://www.preservons-la-nature.fr/flore/taxon/1038.html)

(u) *Citrussinensis* (http://www.gardensonline.com.au/GardenShed/PlantFinder/Show_929.aspx)

(v) *Salvia officinalis* (http://plantsoftheworldonline.org/taxon/urn:lsid:ipni.org:names:456833-1).

In the literature, different methods of preparation of ZnO nanoparticles using plant extract were reported, as described in Table 1. Various groups illustrated the use of different metal precursors and different calcination temperatures.

3. Some of these Works are Discussed in Detail Below

Nandhini et al. (2018) prepared ZnO nanoparticles from leaves extract of a medicinal plant, i.e., an insulin plant known as *Costus igneus*, in three different solvents, i.e., acetone, methanol, and hot water. By using three different solvent extracts, differences in the morphology and average particle size of nanoparticles were observed. Particles formed from acetone and methanolic extract appeared very large and irregular in their shape and size as compared to the hot water extract, which gave particles in the size range of 65–95 nm. It was also observed that samples of ZnONPs prepared by

Table 1. Different methods of green synthesis of ZnO NPs.

Plant used	Extracting solvent	Precursor	Calcination Temp. (in)	Morphology	Crystallite size (in nm)	Reference
Cinnamomum tamala *Cinnamomum verum* *Beta vulgaris* *Brassica oleracea var. Italica*	Water	Zinc nitrate	400	Spherical Spherical	19 27 14 30	(Pillai et al. 2020)
Laurus nobilis	Water	Zinc acetate Zinc nitrate	NA	Bullets like Flower-like	21.49 25.26	(Fakhari et al. 2019)
Phoenix roebelenii	Water	Zinc nitrate	500	Spherical	11.4	(Aldeen et al. 2022)
Costusigneus	Acetone Methanol Hot water	Zinc acetate	400	NA	41 39 38	(Nandhini et al. 2018)
Mimosa pudica	Ethanol	Zinc acetate	300	Hexagonal	14.7	(Balogun et al. 2020)
Typha latifolia	Ethanol	Zinc chloride	400	Flower-like	NA	(Kumar et al. 2019)
Cayratic pedata	Water	Zinc nitrate	400	Hexagonal	52.24	(Jaychandran et al. 2021)
Berberis aristata	Water	Zinc acetate	NA	Needle shaped	5–25	(Chandra et al. 2019)
Piper nigum	Water	Zinc acetate	300 and 500	Nanorod	NA	(Thangapandi et al. 2021)
Scuttaleria baicalensis	Water	Zinc nitrate	NA	Spherical	50	(Chen et al. 2019)
Onion	Water	Zinc acetate	NA	Spherical	NA	(Canbaz et al. 2020)
Onion Cabbage Carrot Tomato	Water	Zinc acetate	NA	Spherical Nanotube Nanorod Spherical	17 18 24 15	(Degefa et al. 2021)
Ocimum tenuiflorum	Water	Zinc nitrate	130	Hexagonal	13.86	(Raut et al. 2013)
Aloe barbadensis	Water	Zinc nitrate	NA	Spherical	35	(Sangeetha et al. 2011)
Deverrea tortuosa	Water	Zinc nitrate	400	Hexagonal	15.41	Selim et al. 2020)

Table 1 contd. ...

...Table 1 contd.

Plant used	Extracting solvent	Precursor	Calcination Temp. (in)	Morphology	Crystallite size (in nm)	Reference
Hibiscus rosa sinensis	Water	Zinc nitrate	400	Sponge-like	30–35	(Devi and Gayathri 2014)
Tomato Orange Grapefruit Lemon	Water	Zinc nitrate	400	Polyhedral	9.01 12.55 19.66 11.39	(Nava et al. 2017)
Azadirachta indica	Water	Zinc acetate	NA	Spindle-shaped	NA	(Noorjahan et al. 2015)
Sambucus ebulus	Ethanol	Zinc acetate	450	Hexagonal	17	(Alamdari et al. 2020)
Citrus sinensis	Water	Zinc nitrate	400	Hexagonal prism Hexagonal prism Sponge-like	24.3 22.6 12.7	(Luque et al. 2018)
Salvia officinalis	Water	Zinc nitrate	NA	Hexagonal	12.07–14.17	(Alrajhi et al. 2021)

the hot water extract showed a maximum purity of 99.89%, followed by the acetone extract with 99.46% purity. Samples prepared by the methanolic extract were found to be less efficient in forming nanoparticles of ZnO with purity of 99.0%. By using three different solvent extracts for synthesizing nanoparticles of ZnO, Nandhini et al. concluded that out of three solvents, hot water extract was found to be very useful as it gave a higher purity of ZnO. UV-visible spectra of samples prepared by hot water, methanol, and acetone extract of Costusigneus leaves showed a sharp peak having absorption maxima at 356 nm, 379 nm, and 370 nm.

Balogun et al. (2020) prepared ZnO NPs from ethanolic extract of leaves of *Mimosa pudica*. They observed the transmittance of samples by using the spin coating method. In this method, they observed ZnO NPs transmittance at the unlike speed of spin coating. It was found that among various solutions at unlike speed of spin coating, the solution with 500 revolutions per minute has the least transmittance and maximum absorption value. Maximum peaks of absorption occurred at 235 nm, 250 nm, 270 nm, and 300 nm but were found to be lesser in comparison with the size of bulk that rose at a wavelength of 350 nm. Results of FTIR showed peaks at 783.4 cm^{-1} and 682.6 cm^{-1}, representing the peak of the ZnO functional group. XRD of ZnO NPs showed the peaks attributed to ZnO with wurtzite, and the estimated average grain size was 14.7 nm.

Kumar et al. (2019) synthesized ZnONPs by ethanolic extract of *Typha latifolia* leaves. Diffraction peaks in XRD confirmed hexagonal wurtzite structure. It was observed that the nanoparticles formed were in crystalline form. During their studies, it was found that EDX confirmed the presence of oxygen and elemental zinc signal of ZnO NPs with the weight composition of 56.45% copper, 33.12% oxygen, and 9.42% zinc. Nanoflowers of ZnO were observed clearly by SEM images. By EDX peaks of the nanoflowers of ZnO, synthesized ZnO was observed with high purity.

Jayachandran et al. (2021) synthesized ZnONPs using *Cayratia pedata* leaf extract in water. To observe the effect of variation of the concentration in the preparation of the sample, they prepared the extract at two different concentrations; the first sample having 25 g in 50 ml and the other having 25 g in 100 ml of water. An analysis of the UV-visible spectrum showed a specific peak for ZnO nanoparticles at 320 nm. It was also clear by XRD that most of the particles were found to be in horizontal shape. The average size of crystallite of formed nano-sized particles was found to be 52.24 nm. Analyte composition was observed by EDX, which gave very strong peaks of 12.78% for oxygen and 78.32% for zinc.

4. Application of the Green Synthesized ZnO NPs in Wastewater Treatment

ZnONPs are considered to be very useful in various fields, such as the biomedical field or wastewater treatment (Figure 8). Water pollution nowadays is a serious issue for human lives and their better health. Polluted water causes many diseases to living beings and is sometimes fatal, too. Treatment of wastewater is necessary as the amount of water available for drinking purposes is very little for the survival of the increasing population. Industries such as textiles, chemicals, and pharmaceuticals

Figure 8. Different applications of ZnO NPs.

release a massive amount of waste material containing organic dyes, drug wastes, pesticides, etc., into the water bodies.

In the textile industry, organic dyes like MG, CV, MB, methyl orange, indigo carmine, and rhodamine are used in very large amounts for staining clothes. ZnONPs synthesized by various green methods used for the photocatalytic degradation of various dyes such as MG, MB, indigo carmine, methyl orange, and rhodamine dyes are reported to date. Degradation of some pharmaceutical drugs found very commonly in wastewater was also degraded by using ZnO NPs.

5. Wastewater Treatment by Degrading Various Pollutants

The potential application of green synthesized ZnO NPs prepared from different plant extracts has been given in Table 2.

6. Some of These Works Are Discussed in Detail Below

Aldeen et al. (2022) observed the degradation of MB dye photo-catalytically by ZnO NPs prepared by leaves extract of *Phoenix roebelenii*. Results showed a strong band for monomer in UV at 664 nm and 614 nm, a shoulder peak for MB dimer. Degradation of 89% of MB after 105 minutes under UV light as the nanoparticles and high value of the rate constant (k = 0.037) showed increased photocatalytic properties. The amazing catalytic performance of ZnONPs can be credited to their slow energy band gap and small size.

Kumar et al. (2019) described the use of ZnO NPs synthesized by the leaves extract of *Typha latifolia* for the photocatalytic degradation of indigo carmine dye ($C_{16}H_{18}N_2Na_2O_8S_2$). Indigo carmine dye gave its absorption maximum at the wavelength of 610 nm. It was reported that the initial concentration of the indigo carmine degraded to 50% in the first half an hour when exposed to the sunlight. In the absence of a catalyst, degradation of the indigo carmine dye was observed after

Table 2. Photocatalytic degradation of dyes by ZnO NPs prepared from different plant extracts.

Used method	Calcination Temp. (in)	Degraded material	Time (in minutes)	Degradation efficiency (%)	Reference
Phoenix roebelenii	500	MB Dye	105	98	(Aldeen et al. 2022)
Typha latifolia	NA	Indigo Carmine Dye	120	83	(Kumar et al. 2019)
Allium sativum, Allium cepa, Parsley	400	MB Dye	180	NA	(Slman et al. 2018)
Piper nigrum	500 300 60	MB Dye	150	91.49 97.32 86.69	(Thangapandi et al. 2021)
Scutelleriabaicalensis	NA	MB Dye	45 210	50 98.6	(Chen et al. 2019)
Pullulan based	400	AMX PCT	NA	85.7 96.8	(Mohamed et al, 2021)
Eucalyptus globulus	400	MB Methyl Orange	50 60	98.3 96.6	(Reddy et al. 2017)
Sambucus ebulus	450	MB Dye	120	80	(Narath et al. 2021)
Thymus vulgaris	NA	MB Dye	100	96	(Zare et al. 2019)
Lycopersicon esculentum	NA	MB Dye	180	97	(Soto-Robles et al. 2018)

two hours at about 40%, while with the presence of the catalyst, it was reported that the dye was 83% degraded after two hours from its initial value.

Thangapandi et al. (2021) observed photocatalytic degradation of MB by ZnO NPs prepared by seeds extract of *Piper nigrum*. Absorbance of ZnO nanorods in MB was reported at 665 nm. Spectral data revealed that degradation of MB occurred in 30 minutes of periodic gap. Spectrum revealed the degradation of MB dye in a gap of 30 minutes. For ZnO nanorods calcined at 60, 300 and 500, the values of absorbance after 150 minutes of irradiation showed the percentage degradation of dye was nearly 86.69%, 97.32%, and 91.49%, respectively.

Sukri et al. (2020) studied MG dye photocatalytic degradation by biosynthesized ZnO NPs from the fruit peel extract of *Punica granatum*. The maximum peak of the absorbance for the MG dye was reported at 617 nm. Degradation of the MG dye occurred with time normally. While in the presence of the photocatalyst exposed to UV, absorption intensity decreased to zero in nearly 50 minutes, showing complete MG degradation. It was reported that ZnO NPs formed by calcination at 700 showed the fastest MG dye reduction as the peak of absorption reached to zero in 30 minutes. On the other side, ZnO NPs formed by calcining at 600 degraded MG dye in 40 minutes and showed high performance of degradation. The efficiency of removal of dye was at 99% for ZnO NPs formed at calcination temperatures of 600 and 700. The percentage degradation of MG dye by the ZnO NPs formed by calcination at 400 and 500 was reported to be 96%. In spite of having a large size of the particles, ZnONPs calcined at higher temperatures of 600 and 700°C showed amazing results in degrading MG dye compared to the ZnONPs with particles of small size.

Mohamed and Shameli et al. (2021) reported some pharmaceutical drugs degraded photo-catalytically by biosynthesized ZnO NPs via a precipitation technique involving pullulan as biological material. They used two drugs, paracetamol (PCT) and amoxicillin (AMX), as a model for the waste of pharmaceutical companies. PCT drug is used very commonly for fever and pain relief in households, while AMX drug is used extensively in both humans as well as in animals as an antibiotic. They studied the consequence of the catalyst dose, pH, as well as the primary concentration of the drug on the photocatalytic activity of ZnO NPs. It was reported for PCT that the highest degradation occurred at 50 mg of the catalyst, pH 5, and the concentration of the drug was 30 ppm. While for AMX, it was reported to be 50 mg of catalyst dose, pH 9, and drug of 30 ppm concentration.

7. Conclusion

In the last few years, there has been a growing demand for nanotechnology and green chemistry on the way to implementing biomaterial-based methods for nanomaterial formation through plants, bacteria, algae, and fungi. Nanoparticle green synthesis has been growing continuously as the zone of attentive exploration by scientists in the latest time by accepting an eco-friendly method. A lot of work has been reinforced on nanoparticles prepared from the plant extract and their capable applications in different fields because of their unique properties such as easy availability, environment-friendly nature, cost-effectiveness, and non-toxic route. Furthermore, the nanoparticles exhibit a very wide range of applications in different areas, such as wastewater treatment by dye degradation, metal ion uptake, pharmaceutical uptake, pesticide degradation, catalysis, biosensing, bioimaging, biotechnology, optics, bioengineering sciences, and textile engineering. Plants consist of phytochemicals that aid in the formation of nanoparticles and also suggest the synthesis speed. Phytochemicals found in plants act as reducing agents and stabilizers, as capping agents. Plants use for nanoparticles by green method is found to be an exhilarating as well as an emerging share of nanotechnology. It also exhibits very remarkable results on the sustainability of the environment and additional development in the nanoscience field. These biogenic nanoparticles are found to be used as nano-scaled weapons against various pathogens and also in the decontamination of water in different ways for remediation of the environment. From the literature survey, ZnO NPs are found to express a strong photocatalytic activity for dye, pesticides, and pharmaceutical degradation.

Future Prospective

In the future, applications of nanoparticles will exponentially rise, while there is a requirement to take serious note of the growth of nanoparticles in the environment, and the lasting effects of nanoparticles on humans and animals have to be fixed in the future. There might be a future for biosynthesized nanoparticles in the biomedical field as a drug delivery system. Synthesis of nanoparticles that express more efficient dye degradation and prominent antimicrobial properties will be in focus in the upcoming years.

Author Contributions

Rajesh Kumar: writing – original draft and formal analysis. Tanuj: writing – original draft and formal analysis. Santosh Kumar: formal analysis. Kiran Kumar: formal analysis. Sandeep Chauhan: formal analysis. Neerja: formal analysis. Subhash Sharma: formal analysis.

Data Availability

There is no data used in this review article.

List of Abbreviations

AFM	Atomic force microscopy
AMX	Amoxicillin
CV	Crystal violet
DSC	Differential scanning calorimetry
EDX	Energy dispersive X-ray
FTIR	Fourier transform infrared
MB	Methylene blue
MG	Malachite green

PCT   Paracetamol
PL   Photoluminescence
ROS   Reactive oxygen species
SEM   Scanning electron microscope
TEM   Transmission electron microscope
XRD   X-ray diffraction
ZnO   Zinc oxide
ZnO MFs   Zinc oxide microflowers
ZnO NPs   Zinc oxide nanoparticles
ZnO NRs   Zinc oxide nanorods

References

Agarwal, H., Kumar, S. V. and Rajeshkumar, S. (2017). A review on green synthesis of zinc oxide nanoparticles. An eco-friendly approach. *Resource-Efficient Technol.*, 3(4): 406–413. https://doi.org/10.1016/j.reffit.2017.03.002.

Agarwal, H., Nakara, A., Menon, S. and Shanmugam, V. (2019). Eco-friendly synthesis of zinc oxide nanoparticles using *Cinnamomum Tamala* leaf extract and its promising effect towards the antibacterial activity. *J. Drug Deliv. Sci. Technol.*, 53: 101212. https://doi.org/10.1016/j.jddst.2019.101212.

Akintelu, S. A., Folorunso, A. S., Oyebamiji, A. K. and Erazua, E. A. (2019). Antibacterial potency of silver nanoparticles synthesized using *Boerhaaviadiffusa* leaf extract as reductive and stabilizing agent. *Int. J. Pharm. Sci. Res.*, 10(12): 374–380.

Alamdari, S., Ghamsari, M. S., Lee, C., Han, W., Park, H. -H., Tafreshi, M. J., Afarideh, H., Majles Ara, M. H. (2020). Preparation and characterization of zinc oxide nanoparticles using leaf extract of Sambucus ebulus. *Appl. Sci.*, 10: 3620. https://doi:10.3390/app10103620.

Aldeen, T. S., Mohamed, H. E. A. and Maaza, M. (2022). ZnO nanoparticles prepared via a green synthesis approach: Physical properties, photocatalytic and antibacterial activity. *J. Phys. Chem. Solids*, 160: 110313. https://doi.org/10.1016/j.jpcs.2021.110313.

Alrajhi, A. H., Ahmed, N. M., Shafouri, M. A., Almessiere, A. M. and Al-Ghamdi, A. M. (2021). Green synthesis of zinc oxide nanoparticles using salvia officials extract. *Mater. Sci. Semicond. Process.*, 125: 105641. https://doi.org/10.1016/j.mssp.2020.105641.

Balogun, S. W., James, O. O., Sanusi, Y. K. and Olayinka, O. H. (2020). Green synthesis and characterization of zinc oxide nanoparticles using bashful (*Mimosa pudica*), leaf extract: A precursor for organic electronics applications. *SN Appl. Sci.* https://doi.org/10.1007/s42452-020-2127-3.

Bandeira, M., Giovanela, M., Roesch-Ely, M., Devine, D. M. and Crespo, J. S. (2020). Green synthesis of zinc oxide nanoparticles: A review of the synthesis methodology and mechanism of formation. *Sustain. Chem. Pharm.*, 15: 100223. https://doi.org/10.1016/j.scp.2020.100223.

Bayrami, A., Alioghli, S., Pouran, S. R., Habibi-Yangjeh, A., Khataee, A. and Ramesh, S. (2019). A facile ultrasonic-aided biosynthesis of ZnO nanoparticles using Vaccinium arctostaphylos L. leaf extract and its antidiabetic, antibacterial, and oxidative activity evaluation. *Ultrason. Sonochem.*, 55: 57–66. https://doi.org/10.1016/j.ultsonch.2019.03.010.

Canbaz, G. T., Acikel, U. and Acikel, Y. S. (2020). Green synthesis of ZnO nanoparticles from onion peel wastes. EurAsia Waste Management Symposium.

Chandra, H., Patel, D., Kumari, P., Jangwan, J. S. and Yadav, S. (2019). Phyto-mediated synthesis of zinc oxide nanoparticles of *Berberis aristata*: Characterization, antioxidant activity and antibacterial activity with special reference to urinary tract pathogens. *Mater. Sci. Eng. C*, 102: 212–220. https://doi.org/10.1016/j.msec.2019.04.035.

Chen, L., Batjikh, I., Hurh, J., Han, Y., Huo, Y., Ali, H., Li, J. F., Rupa, E. J., Ahn, J. C., Mathiyalagan, R. and Yang, D. C. (2019). Green synthesis of zinc oxide nanoparticles from root extract of *Scutellaria baicalensis* and its photocatalytic degradation activity using methylene blue. *Optik*, 184: 324–329. https://doi.org/10.1016/j.ijleo.2019.03.051.

Degefa, A., Bekele, B., Jule, L. T., Fikadu, B., Ramaswamy, S., Dwarampudi, L. P., Nagaprasad, N. and Ramaswamy, K. (2021). Green synthesis, characterization of zinc oxide nanoparticles, and examination of properties for dye-sensitive solar cells using various vegetable extracts. *J. Nanomater.* https://doi.org/10.1155/2021/3941923.

Devatha, C. P. and Thalla, A. K. (2018). Green synthesis of nanomaterials. *Synthesis of Inorganic Nanomaterials.* Woodhead Publishing, 169–184. https://doi.org/10.1016/B978-0-08-101975-7.00007-5.

Devi, R. S. and Gayathri, R. (2014). Green synthesis of zinc oxide nanoparticles by using Hibiscus rosa-sinensis. *Int. J. Curr. Eng. Technol.*, (4).

Fahimmunisha, B. A., Ishwarya, R., AlSalhi, M. S., Devanesan, S., Govindarajan, M. and Vaseeharan, B. (2020). Green fabrication, characterization and antibacterial potential of zinc oxide nanoparticles using *Aloe socotrina* leaf extract: a novel drug delivery approach. *J. Drug Deliv. Sci. Technol.*, 55: 101465. https://doi.org/10.1016/j.jddst.2019.101465.

Fakhari, S., Jamzad, M. and Fard, H. K. (2019). Green synthesis of zinc oxide nanoparticles: A comparison. *Green Chem. Lett. Rev.*, 12(1): 19–24. https://doi.org/10.1080/17518253.2018.1547925.

Gupta, M., Tomar, R. S., Kaushik, S., Mishra, R. K. and Sharma, D. (2018). Effective antimicrobial activity of green ZnO nano particles of *Catharanthus roseus*. *Front. Microbiol.*, 9: 2030. https://doi.org/10.3389/fmicb.2018.02030.

Hwa, K. -Y. and Subramani, B. (2014). Synthesis of zinc oxide nanoparticles on graphene carbon nanotube hybrid for glucose biosensor applications. *Biosens. Bioelectron.*, 62: 127–133. https://doi.org/10.1016/j.bios.2014.06.023.

Jadoun, S., Arif, R., Jangid, N. K. and Meena, R. K. (2021). Green synthesis of nanoparticles using plant extracts: A review. *Environ. Chem. Lett.*, 19: 355–374. https://doi.org/10.1007/s10311-020-01074-x.

Jayachandran, A., Aswathy, T. R. and Nair, A. S. (2021). Green synthesis and characterization of zinc oxide nanoparticles using *Cayratiapedata* leaf extract. *Biochem. Biophys. Rep.*, 26: 00995. https://doi.org/10.1016/j.bbrep.2021.100995.

Jeevanandam, J., Chan, Y. S. and Danquah, M. K. (2016). Biosynthesis of metal and metal oxide nanoparticles. *ChemBioEng Rev.*, 3(2): 55–67. https://doi.org/10.1002/cben.201500018.

Kadiyala, U., Turali-Emre, E. S., Bahng, J. H., Kotov, N. A. and Van Epps, J. S. (2018). Unexpected insights into antibacterial activity of zinc oxide nanoparticles against methicillin resistant *Staphylococcus aureus*. Nanoscale, 1–40. https://doi.org/10.1039/C7NR08499D.

Kanagamani, K. and Muthukrishnan, P. (2019). Photocatalytic degradation of environmental perilous gentian violet dye using *Leucaena*-mediated zinc oxide nanoparticle and its anticancer activity. *Rare Met.*, 38: 277–286.

Khatami, M., Alijani, H. Q., Heli, H. and Sharifi, I. (2018). Rectangular shaped zinc oxide nanoparticles: green synthesis by Stevia and its biomedical efficiency. *Ceram. Int.* https://doi.org/10.1016/j.ceramint.2018.05.224.

Kumar, B. P., Arthanareeswari, M., Devikala, S., Sridharan, M., Arockiaselvi, J. and Pushpamalini, T. (2019). Green synthesis of zinc oxide nanoparticles using *Typha latifolia*. L leaf extract for photocatalytic applications. *Mater. Today: Proc.*, 14: 332–337. https://doi.org/10.1016/j.matpr.2019.04.155.

Lingaraju, K., Naika, H. R., Manjunath, K., Basavaraj, R. B., Nagabhushana, H., Nagaraju, G. and Suresh, D. (2016). Biogenic synthesis of zinc oxide nanoparticles using *Ruta graveolens* (L.) and their antibacterial and antioxidant activities. *Appl. Nanosci.*, 6(5): 703–710. https://doi.org/10.1007/s13204-015-0487-6.

Luque, P. A., SotoRobles, C. A., Nava, O., GomezGutierrez, C. M., CastroBeltran, A., GarrafaGalvez, H. E., VilchisNestor, A. R. and Olivas, A. (2018). Green synthesis of zinc oxide nanoparticles using Citrus sinensis extract. *J. Mater. Sci.: Mater. Electron.*, 29: 9764–9770. https://doi.org/10.1007/s10854-018-9015-2.

Mohamed Isa, E. D., Shameli, K., Ch'ng, H. J., Che Jusoh, N. W. and Hazan, R. (2021). Photocatalytic degradation of selected pharmaceuticals using green fabricated zinc oxide nanoparticles. *Adv. Powder Technol.*, 32: 2398–2409. https://doi.org/10.1016/j.apt.2021.05.021.

Nandhini, G., Suriyaprabha, R., Pauline, W. M. S., Rajendran, V., Aicher, W. K. and Awitor, O. K. (2018). Influence of solvents on the changes in structure, purity, and *in vitro* characteristics of green-synthesized ZnO nanoparticles from *Costus igneus*. *Appl. Nanosci.* https://doi.org/10.1007/s13204-018-0810-0.

Narath, S., Koroth, S. K., Shankar, S. S., George, B., Mutta, V., Waclawek, S., Cernik, M., Padil, V. V. T. and Varma, R. S. (2021). Cinnamomum tamala leaf extract stabilized zinc oxide nanoparticles: a promising photocatalyst for methylene blue degradation. *Nanomaterials*, 11: 1558. https://doi.org/10.3390/nano11061558.

Nava, O. J., Soto-Robles, C. A., Gomez-Gutierrez, C. M., Vilchis-Nestor, A. R., Castro-Beltran, A., Olivas, A. and Luque, P. A. (2017). Fruit peel extract mediated green synthesis of zinc oxide nanoparticles. *J. Mol. Struct.*, 1147. http://dx.doi.org/10.1016/j.molstruc.2017.06.078.

Noorjahan, C. M., Shahina, S. K. J., Deepika, T. and Summera, R. (2015). Green synthesis and characterization of zinc oxide nanoparticles from neem (Azadirachta indica). *Int. J. Sci. Eng. Res.*

Pillai, M. A., Sivasankarapillai, V. S., Rahdar, A., Joseph, J., Sadeghfar, F., Anuf, A. R., Kumar, R. and Kyzas, G. Z. (2020). Green synthesis and characterization of zinc oxide nanoparticles with antibacterial and antifungal activity. *J. Mol. Struct.*, 1211: 128107. https://doi.org/10.1016/j.molstruc.2020.128107.

Rahdar, A., Beyzaei, H., Saadat, M., Yu, X. and Trant, J. F. (2020). Synthesis, physical characterization, and antifungal and antibacterial activities of oleic acid capped nanomagnetite and cobalt-doped nanomagnetite. *Can. J. Chem.*, 98(1): 34–39. https://doi.org/10.1139/cjc-2019-0268.

Rastogi, A., Singh, P., Haraz, F. A. and Barhoum, A. (2018). Biological synthesis of nanoparticles: An environmentally benign approach. *Fundamentals of Nanoparticles*. https://doi.org/10.1016/B978-0-323-51255-8.00023-9.

Raut, S., Thorat, P. V. and Thakre, R. (2013). Green synthesis of zinc oxide (ZnO) nanoparticles using ocimum tenuiflorum leaves. *Int. J. Sci. Res.*

Reddy, S. B. and Mandal, B. K. (2017). Facile green synthesis of zinc oxide nanoparticles by Eucalyptus globulus and their photocatalytic and antioxidant activity. *Adv. Powder Technol.* http://dx.doi.org/10.1016/j.apt.2016.11.026.

Sangeetha, G., Rajeshwari, S. and Venckatesh, R. (2011). Green synthesis of zinc oxide nanoparticles by aloe barbadensis miller leaf extract: Structure and optical properties. *Mater. Res. Bull.*, 46: 2560–2566. https://doi:10.1016/j.materresbull.2011.07.046.

Selim, Y. A., Azb, M. A., Ragab, I. and Abd El-Azim, M. H. M. (2020). Green synthesis of zinc oxide nanoparticles using aqueous extract of deverratortuosa and their cytotoxic activities. *Sci. Rep.*, 10: 3445. https://doi.org/10.1038/s41598-020-60541-1.

Singh, J., Dutta, T., Kim, K. -H., Rawat, M., Samddar, P. and Kumar, P. (2018). Green synthesis of metals and their oxide nanoparticles: Applications for environmental remediation. *J. Nanobiotechnology*, 16: 84. https://doi.org/10.1186/s12951-018-0408-4.

Singh, P., Kim, Y. -J., Zhang, D. and Yang, D. -C. (2016). Biological synthesis of nanoparticles from plants and microorganisms. *Trends Biotechnol.*, 34(7): 588–599. https://doi.org/10.1016/j.tibtech.2016.02.006.

Sivasankarapillai, V. S., Jose, J., Shanavas, M. S., Marathakam, A., Uddin, Md. S. and Mathew, B. (2019). Silicon quantum dots: Promising theranostic probes for the future. *Curr. Drug Targets*, 20(12): 1255–1263. https://doi.org/10.2174/1389450120666190405152315.

Slman, D. K., Jalil, R. D. A. and Abd, A. N. (2018). Biosynthesis of zinc oxide nanoparticles by hot aqueous extract of *Allium sativum* plants. *J. Pharm. Sci. Res.*, 10(6): 1590–1596.

Sonkusare, V. N., Chaudhary, R. G., Bhusari, G. S., Rai, A. R. and Juneja, H. D. (2018). Microwave-mediated synthesis, photocatalytic degradation and antibacterial activity of α-Bi2O microflowers/novel γ-Bi2O3 microspindles. *Nano-Struct. Nano-Objects*, 13: 121–131. https://doi.org/10.1016/j.nanoso.2018.01.002.

Sosna-Glebska, A., Sibinski, M., Szczecinska, N. and Apostoluk, A. (2019). UV-Visible silicon detectors with zinc oxide nanoparticles acting as wavelength shifters. *Mater. Today: Proc.* https://doi.org/10.1016/j.matpr.2019.08.157.

Soto-Robles, C. A., Nava, O. J., Vilchis-Nestor, A. R., Castro-Beltran, A., Gomez Gutierrez, C. M., Lugo-Medina, E., Olivas, A. and Luque, P. A. (2018). Biosynthesized zinc oxide using Lycopersicon esculentum peel extract for methylene blue degradation. *J. Mater. Sci.: Mater. Electron.*, 29: 3722–3729.

Sukri, S. N. A. M., Isa, E. D. M. and Shameli, K. (2020). Photocatalytic degradation of malachite green dye by plant-mediated biosynthesized zinc oxide nanoparticles. *IOP Conference Series: Mater. Sci. Eng.*, 808: 012034.

Thangapandi, J. R., Chelliah, P., Balakrishnan, S., Muthusamy, N., Joshua, E., Sathiya, J., Thangapandi, B., Kasi, M. and Kanniah, P. (2021). Antibacterial and photocatalytic aspects of zinc oxide nanorods synthesized using *Piper nigrum* seed extract. *J. Nanostructure Chem.* https://doi.org/10.1007/s40097-020-00383-5.

Zare, M., Namratha, K., Thakur, M. S. and Byrappa, K. (2019). Biocompatibility assessment and photocatalytic activity of bio-hydrothermal synthesis of ZnO nanoparticles by Thymus Vulgaris leaf extract. *Mater. Res. Bull.*, 109: 49–59. https://doi.org/10.1016/j.materresbull.2018.09.025.

CHAPTER 12

Synthesis of Reusable Magnetic Nanomaterials Through Green Methods and Its Applications

Juri Kalita and Siddhartha S. Dhar*

1. Introduction

Nowadays, one of the fascinating and developing fields of research is nanoscience and nanotechnology. Recent developments in the domain of nanotechnology have contributed to the advancement and revolutionization of a wide variety of industries. Nanotechnology possesses lots of advantages, and these are expanding rapidly. Materials of a size between 1 nm and 100 nm are referred to as nanoparticles (NPs). Because of their small dimensions, they have larger surface areas than the bulk component forms, increased reactivity, and a variety of features like thermal, electrochemical, and optical properties that can also be tuned (Panigrahi et al. 2004). In between bulk material and molecular or atomic structures, nanoparticles exist as a transitional state, which makes them very interesting. These unique characteristics of nanoparticles have sparked the development of nanoscience and the use of NPs in a variety of industries, including cosmetics, food, electronics, agriculture, energy, paints, biomedicine, environmental remediation, etc. (Shah et al. 2015).

Among these MNPs, i.e., magnetic nanoparticles, nanostructures with distinctive magnetic behavior have been found extensive use in a variety of disciplines, including biomedical, energy, engineering, and environmental fields. Due to the distinctive and specific features of MNPs, which could be used in biomedicine, catalysis, agriculture, and the environment, they have recently attracted a lot of research attention (Hao et al. 2010). Different metal elements and their oxides having magnetic properties are used to synthesize MNPs. Due to its great biocompatibility and low toxicity,

Department of Chemistry, National Institute of Technology, Silchar, Cachar, 788010, Assam, India.
* Corresponding author: jurikalita1512@gmail.com

superparamagnetic magnetite (Fe_3O_4) is the most widely utilized iron oxide. Iron-oxide MNPs have recently attracted much attention as researchers work to develop and understand how they might be used in various fields. The most suitable iron-oxide magnetic nanoparticles for biomedical and biological applications are those with lower diameters. The physicochemical properties of superparamagnetic Fe_3O_4 nanoparticles can be modified by changing their surface chemistry and used in a variety of applications, such as cell separation, magnetic resonance imaging (MRI), immunoassays, and hyperthermia.

MNPs are a relevant area with numerous applications in the biological, environmental, catalysis, drug delivery, and bioimaging fields. They have become a hot topic in recent decades because of their tunable shape and size. MNPs are now an evolving discipline of nanobiotechnology and nanoscience due to recent developments and an extraordinary number of publications. MNPs have different physicochemical characteristics from their parent bulkier substances on the basis of having a greater specific surface area, which increases their superparamagnetic capabilities. Some difficulties, including size control, particle surface effects, and dipolar interactions, are crucial in synthesizing monodisperse magnetic nanostructures. Uncapped magnetic iron oxide nanoparticles are reactive and can oxidize in the presence of oxygen, which can negatively affect magnet behavior and dispersion. Due to these factors, it is essential to develop methods for preventing the destruction of the bare magnetic iron oxide nanoparticles during and after the synthesis processes. One method is to encapsulate the magnetic nanoparticles in layers of organic or inorganic stabilizing agents; the main benefit of this method is that the coating not only provides stability but may also be further employed for functionalization depending on the intended application (Tombácz et al. 2015). Iron-oxide magnetic nanoparticles can be employed as catalysts, biomarkers, and other applications after surface functionalization (Rajkumari et al. 2017).

The chemical, physical, and biological processes used in conventional environmental remediation treatment technologies have some drawbacks, including high capital, high maintenance costs, and leachate toxicity. Therefore, researchers started looking for ways to solve these issues to get over these difficulties. Nanotechnology is considered one of the greatest and most promising treatment methods available for environmental cleanup technology (Kalita et al. 2022a, 2022b, Bharali et al. 2022). Due to their unique features, which make them superior to traditional treatment procedures, the utilization of magnetic nanoparticles in environmental remediation has also been demonstrated to be effective (Peigneux et al. 2020). Over the last few decades, MNPs have been more significant in a variety of disciplines, including biology, chemistry, detection and sensing of drugs, pollutants, and treatment. The employment of MNPs in the sewage treatment industry has a great deal of potential to increase the effectiveness of the purification of materials while also significantly boosting reuse and recycling. Using magnetic nanoparticles in nano-remediation approaches can lower overall costs, minimize treatment duration, increase *in situ* treatment opportunities, and achieve near 100% treatment effectiveness. The major benefit of magnetic NMs and nanotechnology, in general, is that they can improve the performance of traditional treatment methods, such as improved residual control and lower energy requirements for thermal methods.

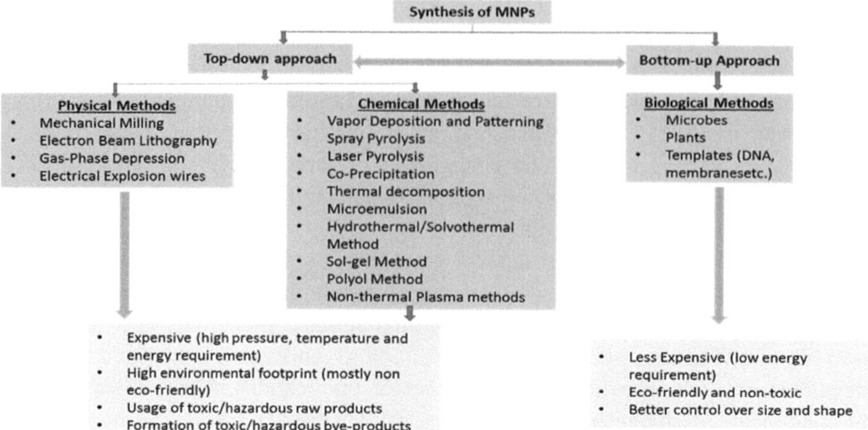

Figure 1. Different synthetic methods of nanoparticle.

For the production of magnetic nanoparticles, various chemical, physical, and biological approaches have been used. Flow injection, micro-emulsion methods, co-precipitation, hydrothermal reactions, sol-gel synthesis, reverse micelles, and other processes are typically used in chemical synthesis. The fundamental challenge in these procedures is particle dispersion, clumping, and size equality. Additionally, the use of solvents and other reagents in chemical-based procedures, including hydrazine, sodium borohydride, sodium dodecyl sulfate, and potassium bitartrate, are all hazardous to the environment as they produce toxic effluent byproducts (Laurent et al. 2008). Also, the physical generation involves grinding, milling, and thermal ablation, which are highly expensive on an hourly basis due to their intensive energy requirements.

Compared to chemical and physical approaches, green chemistry nanoparticle synthesis techniques have advantages such as being safe, environmentally benign, and inexpensive. Biological synthesis of different magnetic nanoparticles is synthesized based on the utilization of living organisms, including bacteria, fungi, plants, algae, viruses, and actinomycetes, as well as green solvents like water. This results in the nanoparticles being free of harmful chemical contaminants, which leads to their widespread acceptance in the biomedical area. The primary focus of this review is biological production of iron oxide MNPs by using plants, fungi, bacteria, and algae, as well as the advantages and disadvantages of these process and its applications in different fields.

2. Different Green Synthesis Approaches for Iron-Oxide Nanoparticles

The biological generation of iron-oxide NPs involves a variety of biota, including plants, fungi, viruses, bacteria, algae, and actinomycetes. The synthesis of these nanoparticles, which are free of harmful chemical contaminants are widely used in the biomedical area, depending on the use of water, a universal green solvent.

The first two fundamental processes in natural synthesis are bioreduction and biosorption. Metal cations are chemically converted into stable elemental forms through the process of bioreduction, and biosorption is the process of attaching metal ions to the interfaces of species, including peptides and cell walls, to make more stable complexes. Furthermore, biobased synthesis discards further procedures like encapsulating or attaching the bioactive molecules to their surface to produce stable particles. Additionally, compared to physicochemical techniques, these synthesis processes require less time.

2.1 Using Plant Extracts

Plants are typically regarded as a readily available, sustainable, safe, and inexpensive source of substance for the production of different types of NPs. The biosynthesis process makes use of various parts of plants, including leaves, seeds, roots, flowers, fruits, petals, seed husks, peels, and the entire plant since they are very rich in different biomolecules such as proteins, amino acids, carbohydrates, flavonoids, saponins, terpenoids, and nitrogenous compounds which acts as a stabilizer, capping agents, reducer, and redox mediators for the synthesis of NPs (Mittal et al. 2013). In order to adjust the size and morphology of the desired NPs, the source of the extract can be varied due to the diversity of plants. In addition to its economic benefits, the extract of a plant leaf utilization for the synthesis of NP can be expanded and employed for large-scale production. The metal oxide and metal nanoparticles synthesized from a plant extract are typically stable and do not exhibit any significant changes even after a month. According to current data, the metal oxide and metal nanoparticles made from extracts of plants are significantly more stable than those made in a conventional manner and do not exhibit any significant changes even after a month, and typically have unique geometries, such as spheres, prisms, cylindrical needles, stems, dendrites, and cubes (Bibi et al. 2019). The characteristics of synthesized NPs are primarily influenced by discrete factors, such as the nature of plant extracts used, the ratio of the volume of the plant extract to the solutions of metallic salt, and different reaction conditions such as temperature, pH, time of incubation, etc. Therefore, based on these outlined variables, researchers have employed various methods of synthesizing iron-oxide magnetic nanoparticles that vary a bit from each other in the choosing of precursor materials, the method used to prepare plant extracts, the iron salt chosen, availability or lack of NaOH, various reaction conditions, and the procedure for collecting synthesized NPs. As per the recent studies, aquous extracts of flowers, stems, and leaves of *Tridax procumbens*, *Cymbopogon citratus*, *Tinospora cordifolia*, *Euphorbia milii*, *Calotropis procera*, *Datura inoxia*, *Ziziphora tenuior*, *Solanum trilobatum*, *Abutilon indicum*, *Persia americana*, *Abutilon indicum*, and *Azadirachta indica* were used to synthesize different sized magnetic iron-oxide nanoparticles.

Washing the initial plant material and extracting it is typically the first step in a controlled study on the biosynthesis of Fe oxide nanoparticles from seeds, leaves, fruits, seed coats, petals, flowers, peels, or the entire plant. The most typical process entails collecting plant pieces (fresh or dried) from the garbage or market, cleaning them with normal tap water, and then washing them with either deionized or distilled water. Next, these cleaned plant pieces are generally left to dry at RT (room temperature) or in an oven at 40 to 60°C for 1–4 hours, in the shade of natural sunlight, or by air drying (Khalil et al. 2017). The sample is next processed into a powdered form using a pestle and mortar, Willy mill, blender, or electric grinder. Double distilled water, deionized water, and water and ethanol/methanol (1:1) are added to the sample powder, and the mixture is heated over a heating mantle or water bath while being continuously stirred.

Curcuma and tea leaves were used to synthesize hematite (Fe_2O_3) nanoparticles that were successfully synthesized (Cai et al. 2003). In order to synthesize iron nanoparticles from tea leaf extract, 10 mL of the tea extract solution was added into 100 mL of $Fe(NO_3)_2.9H_2O$ (1 mM) aqueous solution under vigorous stirring for 24 hours, keeping at a temperature of 50°C. The Curcuma-based synthesis process involves the addition of Curcuma (0.1 g) to around 20 mL of deionized water while stirring for 6 hours, followed by the addition of 40 mL of $Fe(NO_3)_3.9H_2O$ (1 mM) while stirring for 24 hours at 50°C. The obtained products were then centrifuged for 20 minutes at 15,000 rpm, washed several times with DI water and ethanol, and vacuum-dried at 70°C. Murraya koenigii (curry leaves) were used in another technique of making Fe NPs since the leaf extract contains significant amounts of water-soluble compounds like alkaloids, flavonoids, polyphenols and carbazole (Babu and Prabu 2011). In summary, 3 mL of curry leaf extract was mixed with 7 mL of a $FeSO_4$ (1 mM) solution, and the mixture was agitated for 5 minutes using a magnetic stirrer. The final product was then centrifuged for 20 minutes at 15,000 rpm and 20°C, washed numerous times in distilled water, and dried in a hot air oven. Within five minutes, the color changed from a translucent yellow to a black color, signifying the synthesis of Fe NPs. Iron-oxide NP (Fe_3O_4) were also synthesized using the Tridox procumbens leaf extract in a different study that sought to examine the antibacterial activity of iron NPs (Kiruba Daniel et al. 2013). Tridox procumbens was chosen because it contains biomolecules that can function as reducing agents. The color changed from brown to black after 10 minutes when 10 mL of Tridax procumbens leaf extract was added to 10 mL of $FeCl_3$ solution due to the presence of iron-oxide nanoparticles. While the SEM revealed irregular sphere forms of Fe_3O_4 with rough surfaces, the XRD spectra revealed that the crystalline Fe_3O_4 having crystallite size within the range of 80–100 nm. Another method of synthesis that demonstrated the rapid synthetic method of Fe nanoparticles was accomplished with the addition of 5 mL of Dodonaea viscosa leaf extract to 10 mL of 10 mM $FeCl_3$ solution, and the instantaneous change in color indicated the production of Zero-valent iron (ZVI) nanoparticles (Fahmy et al. 2018). High-resolution transmission electron microscopy (HRTEM) revealed that the nanoparticles' shape was spherical and ranged in size from 50 nm to 60 nm. Due to the abundance of flavonoids, including tannins, Pendleton, santin, pinocembrin, and saponins, in Dodonaea viscosa leaf extract, which functions as capping and reducing agents, it was utilized to develop ZVI NPs.

The extracts of green tea leaf were used to make iron nanoparticles in the following way: the extract was prepared by boiling green tea (Alwald brand;

60.0 gL^{-1}) for about 1 hour, allowing it to cool, and then filtering it using the vacuum filtration method (Lunge et al. 2014). Furthermore, a 0.10-M $FeCl_2.4H_2O$ solution was prepared by mixing 19.9 g of $FeCl_2.4H_2O$ with 1 L of DI water. After that, 60 gL^{-1} of green tea was mixed with a 0.10 M $FeCl_2.4H_2O$ solution to produce a 2:3 volume ratio solution. The pH was then increased to 6 by adding 1 M aqueous solution of NaOH. The formation of intense black precipitate showed the existence of green tea leaf Fe NPs that were ready to use. This process produced nanoparticles that were isolated from the iron solution by evaporating water and then allowed to dry overnight in fumehood.

Zero-valent iron nanoparticles (ZVINPs) were synthesized using additional techniques using Syzygium aromaticum (clove) (Ojha et al.). With regular stirring at 50–60°C, freshly prepared 0.001M $FeCl_3$ solution and clove extract were mixed in a 1:1 proportion. In order to stabilize Fe NPs, 1% chitosan and 1% PVA were also added. Changing from a high acidic to a low acidic pH was accompanied by the production of Fe NPs (from a pH value of 4.22 to 1.88). Scanning electron microscope (SEM) analysis confirmed the produced Syzygium aromaticum Fe nanoparticles to have distributed spheres with a diameter of about 100 nm.

The method for producing nanoparticles from Lactobacillus casei and Lactobacillus fermentum extract has been slightly altered to include 10-minute centrifugation at 2,500 rpm, followed by 24-hour drying of the nanoparticle pellet in an oven at 40°C. The synthesized IONPs were filtered using 0.2 m syringe filters and Whatman filter paper in accordance with the procedure used by Fatemi et al. (2018). The NPs-containing filtrate was subsequently concentrated with a rotary balloon (20–30 rpm), a steam bath with temperature around 60°C, and a circular at 15–10°C. The nanoparticle was finally dried under sterile environment in a laminar air flow. Due to their non-pathogenic nature and ability to produce a variety of enzymes, LAB (lactic acid bacteria) are typically regarded as an attractive microbiological source of different NPs. The LAB also have a high negative electrokinetic potential and are facultative anaerobic bacteria. As a result, under both oxidizing and reducing circumstances, the LAB became attracted to various metal ions, which helps the formation of metal oxide and metal nanoparticles. In addition, because it is a Gram-positive bacteria, lipoteichoic and teichoic acid, peptidoglycan, proteins, and polysaccharides make up the cell wall, which may work as an active site for the bioreduction and biosorption of metal ions.

2.2 *Biosynthesis of Iron-oxide NPs Using Fungi*

Fungi are considered to be suitable for the extracellular production of iron-oxide NPs due to their high biomass formation ability, usage of affordable raw resources for development, ease of scaling up, simplicity of downstream stages, economic viability, and low toxicity of residue. The interactions of microbes and metals have been extensively investigated, and this ability of microbes to accumulate and absorb metals has been utilized in a variety of biological systems, such as bioleaching, heavy metal elimination, and environmental remediation. It is envisaged that a variety of extracellularly generated nanoparticles are produced by this enzymatic reduction of metal ions. Nitrate reductase is one of these enzymes that is present in fungus.

However, the general mechanisms of utilizing fungus to produce metal nanoparticles are not fully understood. The biological material and reaction conditions employed in the fungi's MNP production are thought to be important factors. The type of microbe utilized, growth parameters, and sample preparation methodology are all considered biological materials. While surveying the literature, one would find a large deal of variation in the initial biomaterial used for biosynthesis. To produce the nanoparticles, some techniques utilized fungus biomass, fungus culture cell-free filtrate, or fungus biomass homogenate, while others used fungus cell filtrate. The technique that was most frequently used entailed first cultivating the fungi in the proper growth medium and then centrifuging the mycelia to separate them. The mycelia produced in this way is combined with a metal salt precursor and incubated for several days at 28°C (4–5 days). A suspension of produced nanoparticles is developed after additional filtration of this solution (Kaul et al. 2012). In a different technique, the mycelia are isolated by filtration after the fungus is cultivated in a suitable medium. The fungal biomaterial is dispersed in sterile deionized water and cultured on a rotary shaker for some days after being thoroughly washed with sterile distilled water.

After incubation, CFF (cell-free filtrate) is obtained by removing mycelium using Whatman filter paper 1. Then, the resultant CFF is mixed with a salt mixture of iron precursor consisting of ferrous sulfate ($FeSO_4.7H_2O$) and ferric chloride ($FeCl_3.6H_2O$) in a 1:2 M ratio (Chatterjee et al. 2020). Abdeen et al. employed a different method that integrates physical and microbiological processes to make iron-oxide NPs (Abdeen et al. 2016). In this procedure, a cyclomixer was used to form a fine, homogeneous solution of fungal mycelium in deionized water. $FeCl_3$ or $FeSO_4$ salt was added to this homogenate metal precursor mixture of concentration 2,000 ppm. The final mixture is then incubated under static conditions for almost six days. The obtained IONPs were separated by centrifugation and then thoroughly cleaned with absolute ethanol. With the aid of supercritical pressure reactor operating at 850 psi pressure and 300°C, the nanoparticles were further dried.

Another remarkable variant was the production of iron-oxide NPs using fungal cell filtrate (FCF) (Fani 2018). This technique involves cultivating the chosen fungus in a suitable medium and, after the required period of incubation using a vacuum pump, filtering the culture medium with Whatman filter paper 1. In order to achieve the resultant fungal cell filtrate (FCF), the obtained filtrate is centrifuged and re-filtered. $FeCl_3$ and $FeCl_2$ salt solution (1:2 mM) are stirred in an equal volume of FCF for just five minutes. The synthesis of IONPs can be seen in an instantaneous color change.

2.3 Biosynthesis of Iron-Oxide NPs Using Algae

The term "algae" refers to photosynthetic organisms lacking root and leaf structures. Both single-celled microalgae and multicellular macroalgae, such as seaweeds, are widely employed in the field of nanotechnology to synthesize various kinds of metallic nanoparticles (copper, gold, silver, palladium, iron, and other elements). Similar to plants and other microorganisms, algal bodies are also a source of a variety of biomolecules, including carbohydrates, proteins, fat, alkaloids, macrolides, peptides, polysaccharides, terpenes, glycoproteins (composed of different functional groups including hydroxyl, carboxyl, carbonyl, and sulfonate), and enzymes which are essential for the reduction, fabrication, capping, and stabilization of NPs. As a result, this method is believed to be clean, easy, cost-effective, and safe for producing nanoparticles.

The first approach to synthesizing iron-oxide nanomaterials to be discussed here is from Sargassum muticum, brown algae (Mahdavi et al. 2013). It is a type of food available in coastal areas that may be used in the fast single-step production of iron-oxide nanoparticles. Sulfated polysaccharides were found in Sargassum muticum extract. This served as a reducing agent and stabilizer and provided a capping for the produced nanoparticles. In this method, 100 g of deionized water was boiled with one gram of brown seaweed dried from the sample and cooled at 20°C. The extract was then added to a 1:1 aqueous solution of $FeCl_3$. The color of the solution immediately changes from yellow color to dark brown color, signifying the generation of Fe_3O_4 NPs. In order to confirm the superparamagnetic nature of the formed iron nanoparticles, a new method was used to determine the magnetic properties of iron-oxide NP. This method revealed that compared to the co-precipitation chemical method, green synthesis produced nanoparticles with less magnetic properties.

Aspergillus oryzae TFR9 utilization is another method for producing Fe NPs from microorganisms (fungi) (Tarafdar and Raliya 2013). This Aspergillus oryzae TFR9 fungus is placed in 50 mL of potato dextrose liquid medium and allowed to develop for 72 hours at pH 5.8, 28°C, and on a shaker at 150 rpm media using a Whatman filter paper no. 1. The mycelia were separated from the culture using a 0.45-micron membrane filter and kept after being recovered in 50 mL Milli-Q deionized water for 12 hours at 28°C on a shaker at 150 rpm.

2.4 Biosynthesis of Iron-Oxide NPs Using Bacteria

Due to the widespread availability, rapid rate of doubling, ability to grow in challenging environments, and cheap and easy cultivation media, prokaryotic systems have also been intensively investigated as having a significant role in the field of nanotechnology. This process has been regarded as the best method of producing nanoparticles with a variety of forms, sizes, structural frameworks, and various chemical and physical properties by the reduction of metal ions. By employing metal salts as reaction precursors, scientists have taken advantage of the fact that microorganisms accumulate and detoxicate metals through the activity of the reductase enzyme. Metal ion reduction is an essential step in the synthetic process. This process is influenced by a number of variables, including the growth media, salt concentration, strain type, environmental parameters, such as pH and temperature, and functional groups on the cell wall (which are necessary for biomineralization). As this eco-friendly synthetic method is controlled by the enzymatic system for metal ions reduction by NADH reductase, the biosynthesis method depends on optimal conditions. Among these parameters, pH, temperature, and the time of incubation for the microbial culture are the most important ones. The pH and high temperature totally deactivate the enzymes and inhibit their activity, which slows the formation of nanoparticles.

Along with enzymes, organic molecules, such as proteins, peptides, cofactors, and organic compounds, play stabilizing, capping, and reducing roles in the synthesis process. A series of publications on the production of nanoparticles suggest that extracellular polysaccharides, supernatant, and cytoplasmic extracts can all be used as reduction activating molecules. In general, bacteria have investigated two separate pathways, intracellular and extracellular, for the synthesis of NPs. Ion transport occurs within microbial cells, which is the basis for intracellular biosynthesis. The main participant in this process is the cell wall because the enzymes contain in the cell wall are responsible for the reduction of metal, whereas the extracellular process relies on the accumulation of metal ions on the microbial cell surface through the electrostatic force of interaction between the ions and soluble enzymes secreted by microbes and negatively charged cell wall enzymes. According to numerous reports, the bacterial strains that have been used for both extracellular and intracellular synthesis of iron-oxide nanomaterials include *Lactobacillus fermentum, Lactobacillus casei, Bacillus licheniformis, Microbacterium hominis, Thermophilicbacteria TOR-39, Geobacter metallireducens GS-15, M. gryphiswaldense, M. magnetotacticum (MS-1), Magnetospirillum magnetotacticum, Magnetotactic bacterium MV-1, Aquaspirillum magnetotacticum, Thiobacillus thioparus, Bacillus subtilis, Thermoanaerobacter* sp., *Actinobacter* sp., etc. Selecting the appropriate bacteria is usually the first step in synthesizing iron-oxide nanoparticles (from soil and lyophilized culture).

The sample is first grown on a medium such as BH1 agar, Blood agar, or nutrient agar, after which it is incubated under controlled conditions for 24–48 hours at 35–37°C, or in the case of blood agar medium, under controlled conditions of carbon dioxide. Once sufficient growth has been achieved, a single colony is shifted to a nutrient medium, such as MRS broth, nutrient broth, or Luria broth, after which it is incubated and centrifuged to remove the bacterial biomass. The most important phase in the synthesis of IONPs begins after centrifugation and involves taking either the supernatant or cell pellet. The methodology included an additional step that involved centrifuging at 3,000 rpm for 10 minutes and then rinsing with a relatively short incubation in liquid N_2 at a temperature 196°C for 5 minutes provides the cells a quick heat shock by steam bathing the cell pellet for 15 minutes at 37°C. Phosphate buffer saline was used for dissolving the pellet during the first step of the protocol. Next, centrifugation at 12,000 rpm for 30 minutes is performed, and the supernatant is collected as cytoplasmic extract. In the protocol developed by Kaul et al. (2012), pellets are dissolved in sterile deionized water (Kaul et al. 2012). The desired volume of precursor salts (Fe_2O_3, FeSO4, and $FeCl_3.6H_2O$) is mixed with the supernatant and pellet suspension. By changing the pH to the desired range, the reaction conditions are optimized (5–9). Finally, depending on the type of bacteria utilized, the mixed solution is incubated under various temperature conditions. Brick red to dark brown, colorless to black, golden yellow to turbid brown, and dark green to blackish brown are some of the colors that change after incubation to show the presence of nanoparticles.

Table 1. Biosynthesis of iron oxide NPs from plant extract.

Sl. no.	Name of plants	Biomaterial used	Iron precursor used	Size (nm)	References
1.	Sageretia thea	Fine powdered fresh leaves	Iron sulfate hepta hydrate	29 nm	(Khalil et al. 2017)
2.	Lagenaria siceraria	Air dried leaves	$FeCl_3.6H_2O$ (0.01 M)	30–100 nm	(Kanagasubbulakshmi and Kadirvelu 2017)
3.	Daphne mezereum	Dried leaves	$FeCl_3.6H_2O$ (0.1 M)	6.5–14.9 nm	(Beheshtkhoo et al. 2018)
4.	Ruellia tuberose	Leaf extract	$FeSO_4$ (1M), 1g NaOH	20–80 nm	(Vasantharaj et al. 2019)
5.	Avecinnia marina	Flowers	Aqueous ferric chloride	10–40 nm	(Karpagavinayagam and Vedhi 2019)

Table 2. Biosynthesis of iron-oxide NPs using fungi.

Sl. no	Fungal strain	Biomaterial used	Iron precursor	Size (nm)	References
1.	Fusarium oxysporum	Fungal biomass	$K_3[Fe(CN)_6]$ and $K_4[Fe(CN)_6]$ (2:1 mM)	20–50 nm	(Bharde et al. 2006)
2.	Verticillium sp	Fungal biomass	$K_3[Fe(CN)_6]$ and $K_4[Fe(CN)_6]$ (2:1 mM)	100–400 nm	(Bharde et al. 2006)
3.	Curvularia lunata	Fungal biomass	Fe_2O_3 (1,000 ppm)	20.7 nm	(Kaul et al. 2012)
4.	Chaetomium globosum	Fungal biomass	Fe_2O_3 (1,000 ppm)	25 nm	(Kaul et al. 2012)
5.	A. fumigatus	Fungal biomass	Fe_2O_3 (1,000 ppm)	42.4 nm	(Kaul et al. 2012)
6.	A. wentii	Fungal biomass	Fe_2O_3 (1,000 ppm)	46 nm	(Kaul et al. 2012)

Table 3. Biosynthesis of iron-oxide NPs using algae.

Sl. No	Algae	Biomaterial used	Iron precursor	Size (nm)	References
1.	Sargassum muticum (brown seaweed)	Freeze-dried powder	$FeCl_3.6H_2O$ (0.1 M)	18 ± 4 nm	(Mahdavi et al. 2013)
2.	Kappaphycus alvarezii (red seaweed)	Dried, grinded powder	$FeCl_2.4H_2O$, $FeCl_3.6H_2O$, NaOH (1 M)	14.7 nm	(Yew et al. 2016)
3.	Chaetomorphaantennina (green algae)	Cell-free extract	$FeCl_2.4H_2O$, NaOH	9–10 nm	(Priya et al. 2021)

Table 4. Biosynthesis of iron-oxide NPs using bacteria.

Sl. No	Bacteria	Iron precursor	Size (nm)	References
1.	Bacillus subtilis	Fe_2O_3 (2 mM)	60–80 nm	(Sundaram et al. 2012)
2.	Alcaligens faecalis	a. Fe_2O_3 b. $FeSO_4$ (1,000 ppm)	a. 12 nm b. 43.6 nm	(Kaul et al. 2012)
3.	Lactobacillus fermentum	Iron sulfate (10−3M)	10–15 nm	(Fani et al. 2018)

3. Mechanism of Biosynthesis of Iron-oxide NPs by Plants, Algae, Fungi, and Bacteria

Metal oxide nanoparticles can be synthesized by utilizing biological organisms, such as plants, algae, fungi, and bacteria. These biological sources consist of a variety of biomolecules such as alkaloids, tannins, saponins, proteins, amino acids, flavonoids, steroids, carbohydrates, enzymes (iron reductases and nitrate reductases), cell wall components (laminarian and alginate), and EPS (exopolysaccharides), which assist the bioreduction, stabilization, and capping of the NPs. Although the exact mechanism by which these important biomolecules reduce metal ions is yet not known, it has been inferred from scientific reports that these reducing agents surfaces (which contain functional groups like -COOH, -CHO, -C-O-C, -CO, -OH, and -CC) play a significant role in the metal ions reduction.

Typically, bioreduction occurs in four different phases: the activation phase, the nucleation phase, the growth phase, and the termination phase (Figure 2). In the first stage, known as the activation phase, the metal ions from the precursor salt solution are reduced with the aid of biomolecules contained in the extract. Next, during the nucleation phase, the crystal grows on the metal nuclei while biometabolites simultaneously reduce the metal nuclei. Reducing substances, such as flavonoid quercetin, are adsorbed on the metal NPs' surface, which act as capping and chelating substances. During the growth phase, all metal ions are fully reduced to zero-valent oxidation from a monovalent or bivalent oxidation state. In this step of growth progression, metallic nanoparticles group together to develop a variety of morphologies. The final stable morphology of NPs is obtained in the termination phase, and they are covered by biomolecules to improve the steric repulsion among them, which is expected to reduce the issue of agglomeration. Typically, iron-oxide NPs made by physical and chemical methods are highly reactive and easily get oxidized by ambient oxygen to form aggregates.

Additionally, nanoparticles often agglomerate to reduce the surface energy that arises from the high value of surface area to volume ratio. In order to prevent the agglomeration of the synthesized nanoparticles, a variety of stabilizers are added during the physical and chemical synthesis process. In biological synthesis methods, no such stabilizing chemicals are needed because the biomolecules in the extract serve as capping agents and stabilize the generated nanoparticles. For the green synthesis method to generate nanoparticles with the desired size, shape, crystallinity, purity, stability, and morphology, it is essential to control a number of reaction parameters, including stirring, pH, extract volume, and temperature. The choice of the starting

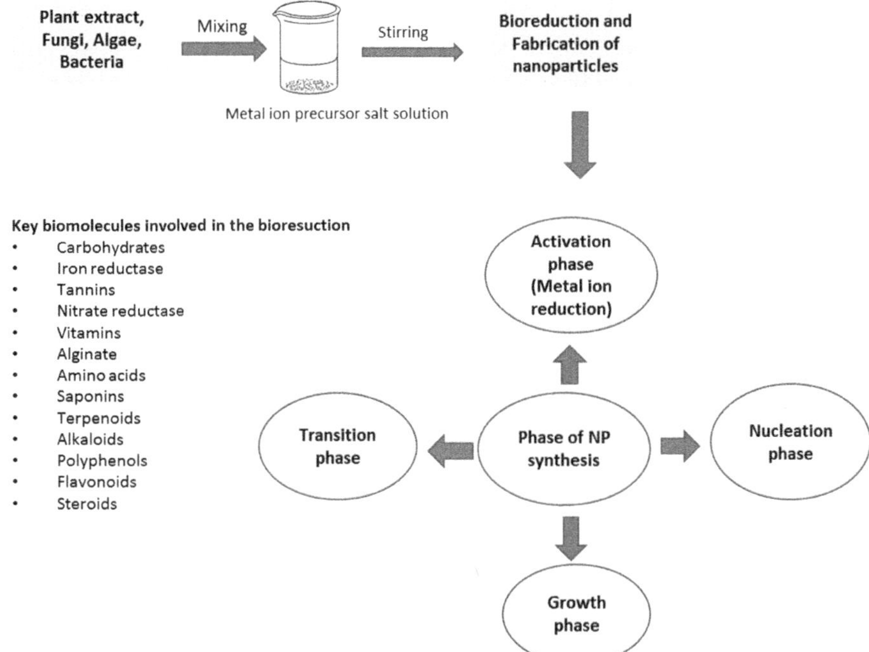

Figure 2. Schematic representation of the mechanism of biosynthesis of nanoparticles.

material is crucial for nanoparticle synthesis mediated by plants. Each component of the plant has a different concentration of reducing substances like terpenoids, flavones, sugars, amides, ketones, alkaloids, and anthracenes. Due to its high content of phytochemicals that aid in the bioreduction of metal ions, the leaf extract is thought to be the best source for the creation of metal oxide nanoparticles. Fungi are thought to be a superior source for the synthesis of metal oxide nanoparticles when compared to bacteria.

4. Applications of Reusable Iron-oxide Nanomaterials

Iron oxide magnetic nanomaterials have been employed in a variety of applications, such as magnetic inks and seals, magnetic recording media, catalysts, ferrofluids, contrast agents for magnetic resonance imaging, and medicinal substances for the treatment of cancer (Azharuddin et al. 2008). Nanomaterials with specified sizes, shapes, surface properties, and magnetic properties are essential for these applications. The particles used in data storage applications must have a stable, switchable magnetic state unaffected by temperature changes. The particles should be consistently tiny, have high coercivity and high remanence, and be resistant to corrosion, friction, and temperature changes for best recording performance. The application of magnetite nanoparticles in magnetic hyperthermia, targeted drug delivery systems, magnetic resonance imaging contrast agents, and immunoassays has drawn much interest. Particles that display superparamagnetic behavior at room temperature are required for these applications.

4.1 Biomedicine

Due to their various physical and chemical characteristics, ease of preparation, stability, and biocompatibility, MNPs are currently widely used in a variety of biological applications. External magnetic fields can affect magnetic behavior of MNPs. Additionally, MNPs have the ability to change the magnetic fields around them, which improves magnetic resonance imaging (MRI). Different types of force and torque are developed and generated by the externally applied magnetic field at dipoles, which causes energy to be transferred, rotated, and dissipated. Such effects have a wide range of uses, including cell separation or biomarker, magnetic drug delivery, magneto-mechanical activation of cell surface receptors, bacterial theranostics, biomedical imaging, hyperthermia, and drug release triggering. Depending on how they will be used in various applications, different materials will exhibit different physical and magnetic properties in forming MNPs. However, one must take into account their possible toxicity or biocompatibility when performing biomedical research (Chen et al. 2017).

4.2 Cancer Therapies

Recently, new methods for cancer theranostics have arisen in the disciplines of molecular biology and nanobiotechnology. These systems primarily entail the synthesis of modified NPs with several functions to overcome the drawbacks of currently available cancer therapeutics and diagnostics (Chen et al. 2017). For the early identification and imaging of cancer development, NPs can be utilized to fabricate nanoscale imaging probes. Additionally, NPs are being developed as delivery systems for anticancer drugs, genes, or proteins using the electron paramagnetic resonance (EPR) effect to reach the targeted tumor areas successfully. MNPs are, therefore, excellent candidates for MRI, biosensing, hyperthermia therapy for cancer, and targeted drug administration due to their special physicochemical and super-magnetic properties. A new cancer therapy called magnetic hyperthermia uses an external AMF (alternating magnetic field) to heat MNP suspensions inside the body (Noh et al. 2017). Because of lower pH and less thermotolerance in the cancerous microenvironment, cancerous cells are more sensitive to hyperthermia cells than normal cells. This method is seen to be a promising one for the treatment of cancer since it produces localized hysteric heat that is focused and causes less heating of surrounding tissues. The ability of magnetic NP-based hyperthermia therapy to penetrate deep tissues and kill cancer cells specifically while protecting healthy tissues from damage is its key benefit.

4.3 Environmental Remediation

The degradation and contamination of soil, water, and the environment are evolving into a major environmental issue as a result of the increasing rate of release of hazardous and harmful compounds and chemicals from different human activities. Many different types of organic pollutants, including pharmaceuticals, pesticides, industrial wastes, polychlorinated biphenyls (PCBs), and polycyclic aromatic hydrocarbons (PAHs), are consistently present in the atmosphere (Roy et al. 2021). In sewage effluents, drinking water, groundwater, and ocean, several forms of organic contaminants can be found. When such contaminants enter into the food chain, they can cause major health issues for humans. Therefore, it is essential to develop an effective technology to improve the water quality. Compared to traditional treatments, nanotechnology has recently emerged as one of the more advantageous and reliable approaches. It is worth noting that MNPs have been shown to be effective at purifying water by removing organic species, degrading dye, and specifically targeting bacteria.

It has become increasingly common to use iron-oxide nanoparticles as a catalyst to reduce pollution caused by organic dyes. Methylene blue, an industrial dye, is a contaminant in water bodies and cannot be degraded in visible light. Iron-oxide nanoparticles were used by Ahamed et al. to successfully degrade the methylene blue dye (Ahmad et al. 2020). Both sunlight and UV lamps can serve as the source of light for photocatalytic degradation. It has been reported that blue 4 dye can be degraded by Fe_2O_3 nanoparticles prepared using an environmentally friendly method. Water and carbon dioxide, which are both safe chemical compounds, are produced as a result of photocatalytic degradation utilizing Fe_3O_4 (Bibi et al. 2019). Wu et al. found that by fabricating core-shell heterostructures of Fe_2O_3/ZnO, Rhodamine blue dye can be completely degraded within 105 minutes (Wu et al. 2012). However, the size of the nanoparticle also plays a significant role in degradation. According to Khedr et al., when the particle sizes were from 100 nm to 135 nm, the degradation was extremely slow. However, when the particle sizes were about 35 nm, the degradation was significantly faster (Khedr et al. 2009).

In recent years, research has increasingly focused on magnetic iron-oxide/TiO_2 composite photocatalysts. For instance, Wang and colleagues have described the synthesis of core-shell Fe_3O_4@SiO_2@TiO_2 microspheres using a wet chemical method. The microspheres have both photocatalytic and ferromagnetic characteristics. Under the influence of UV radiation, the TiO_2 nanoparticles on the microsphere surfaces degraded organic dyes. The ferromagnetic Fe_3O_4 core of the microspheres also made it simple to separate them from the mixture after the photocatalytic process. The photocatalysts were reused multiple times, and after six cycles, the degradation percentage of methyl orange was still 91% (Wang et al. 2012).

Chalasani and Vasudevan demonstrated photocatalytic Fe_3O_4@TiO_2 core-shell magnetic nanoparticles by attaching β-cyclodextrin (CMCD) cavities into the TiO_2 shell and photocatalytically destructed endocrine-disrupting compounds, dibutyl phthalate and bisphenol A (BPA), contained in water. The particles, which had an average diameter of 12 nm, were magnetic and separated magnetically from the dispersion medium before being reused. Liquid chromatography was used to determine the concentration of BPA solution, and after that, the solution was exposed to UV radiation for 60 minutes. The magnet-separated CMCD-Fe_3O_4@TiO_2 nanoparticles can be employed again to photodegrade freshly prepared BPA solutions. For the photodegradation of BPA, the recycling photocatalytic performance of CMCD-Fe_3O_4@TiO_2 was outstanding and stable, having 90% efficacy of the BPA after ten cycles (Meng et al. 2015).

For the magnetic separation of pollutants from wastewater, Fe_3O_4@amino acid is one of the best examples (Jin et al. 2014). The surface of Fe_3O_4 NPs was modified and functionalized with arginine, lysine, and poly-L-lysine to produce Fe_3O_4@Arinine, Fe_3O_4@Lysine, and Fe_3O_4@Poly-L-lysine. For Gram-positive and Gram-negative bacteria like *Escherichia coli* and *Bacillus subtilis*, respectively, the functionalized Fe_3O_4@AA has shown better capturing capacity. Using all three types of Fe_3O_4@AA, about 97% of the bacteria were eliminated and collected. Iron oxide has a magnetic property that makes it cost-effective and simple to separate from aqueous solutions because they agglomerate quickly when an external magnetic field is applied.

Currently, it has been found that magneto-catalysis is the most efficient approach for removing dyes and chronic contaminants in stimulus-response systems. The fundamental mechanism is the production of free radicals by magnetoelectric induction of catalytic decomposition of organic pollutants, which then reacts with parent molecules to transform them into less risky compounds. The ability of the magnetic NPs has been examined by using an AMF (alternating magnetic field) to break down organic contaminants like RhB (Rhodamine B).

In order to catalytically degrade Rhodamine B (RhB) and other medicinal chemicals, Pane's group developed cobalt ferrite-bismuth ferrite (CFO-BFO) core-shell nanoparticles with magnetoelectric properties. They developed a magnetoelectric device without using catalytic molecules to clean water using oxidation processes under wireless magnetic fields and a combination of magnetostrictive CFO and multiferroic BFO. They discovered that synthetic nanoparticles (NPs) produced by magnetoelectric induction produce superoxide and hydroxyl radicals, which catalytically degrade Rhodamine B with 97% removal efficacy and a combination of pharmaceutical pollutants with 85% removal efficiency.

In the past, much research has revealed that MNPs with catalytic and photocatalytic characteristics are used to degrade chemical pollutants, such as pesticides and antibiotics, through oxidation and reduction processes (Hodges et al. 2018).

5. Catalyst Recyclability and Stability

The reusability and stability of the catalysts are important factors for commercial applications. Several studies have already been reported concerning the employment of magnetic nanoparticles for the treatment of wastewater and the degradation or removal of pollutants since they can be easily recovered from the reaction mixture using an external magnet. According to the research cited in this review paper, magnetic nanoparticles can effectively remove a wide range of harmful inorganic and organic pollutants from sewage water. Studies have shown that these methods can achieve more than 95% removal efficiency in a short period, demonstrating their high efficiency, cost-effectiveness, and speed of results. However, it is important to note that, when compared to other types of MNPs, iron magnetic nanocomposites have higher removal efficiencies. In addition to being more sustainable, the most effective methods for contaminant removal are adsorption, photocatalytic degradation, and transformation via redox reactions. MNPs are demonstrated to be more stable, have the ability to target particular contaminants, can make separation simple, and have high reusability characteristics and a high capacity for regeneration. Additionally, the results showed that removal efficacy was not impacted even after recycling for five cycles and still maintained above 95% removal effectiveness. It has also been established that using MNPs, particularly biologically synthesized MNPs, are more cost-effective than using most conventional methods for treating wastewater, such as membrane filtration, the activated sludge process of activated carbon, etc. However, it is necessary to conduct more research on the toxicity and economics of using magnetic nanomaterials to treat wastewater.

6. Conclusion and Discussion

The primary research area in the realm of nanotechnology is the synthesis of biocompatible NPs. Due to the existence of a large diversity of biomolecules that exhibit substantial metal ions reduction to produce NPs from the bulk metal, green sources have become the key performers in the biological production of iron-oxide NPs. There are several factors that make biological nanoparticle synthesis particularly profitable. First of all, this method appears to be a cheap method because only basic conditions are needed for the synthesis method, no sophisticated equipment is needed, and the biological agents used are widely accessible. The next aspect that is important to take into account is that they are devoid of harmful chemicals needed for synthesis, which makes these particles biocompatible and environmentally safe. Additionally, the majority of green synthesis methods use water as their primary solvent, providing the term "green technology" further significance. Finally, the biomolecules in the extract are self-sufficient in carrying out this action; therefore, no agglomeration is encountered. Moreover, no separate capping agents are required to stabilize the biologically produced nanoparticles, and no agglomeration is observed. The repeatability of this approach, however, is the major difficulty. The reactants are used passively and to a limited extent in this process, which gives a passive approach to the production of the nanoparticles. The process is not repeatable because the unreacted elements can result in the production of undesirable products with a variety of characteristics. Although the issue may be somewhat resolved by the speedy recovery of the product, it remains a difficulty for the researchers using this green synthesis. Because size is the primary determinant of the characteristics and, consequently, the usability of the nanoparticles, researchers must concentrate on the size-controlled production of the nanoparticles. Furthermore, an analysis of recent research shows that there are still many uncertainties regarding the variables influencing size, shape, crystalline nature, quantity of synthesized nanoparticles, and precise magnetic characteristics of nanoparticles synthesized using green methodology. To fill these gaps and develop better-controlled studies, further research in these domains is therefore absolutely required.

Researchers are very interested in magnetically structured iron-oxide nanoparticles because they can be used as sensors, photocatalysts, storage materials, and bio-transporters. Concerning the development and catalytic usage of iron-oxide nanoparticles, significant progress and accomplishment were attained. The three key components of an effective catalytic material are stability, reusability, and regeneration, along with the appropriate reagent selection and synthesis process.

In this review, various environmentally friendly processes for synthesizing iron nanoparticles from various plants and microbes were explored. The produced iron nanoparticles were employed in a variety of applications, including bioremediation, antibacterial activity, and dye removal. These activities will significantly contribute to protecting our environment from contamination.

Acknowledgment

The authors would like to acknowledge NIT Silchar, Assam, for providing financial support.

References

Abdeen Mai, Soraya Sabry, Hanan Ghozlan, Ahmed A. El-Gendy and Everett E. Carpenter. (2016). Microbial-physical synthesis of Fe and Fe_3O_4 magnetic nanoparticles using *Aspergillus niger* YESM1 and supercritical condition of ethanol. *Journal of Nanomaterials*, 2016: 1–7. doi:10.1155/2016/9174891.

Ahmad Waqas, Arif Ullah Khan, Saira Shams, Lei Qin, Qipeng Yuan, Aftab Ahmad, Yun Wei, Zia Ul Haq Khan, Sadeeq Ullah and Aziz Ur Rahman. (2020). Eco-benign approach to synthesize spherical iron oxide nanoparticles: a new insight in photocatalytic and biomedical applications. *Journal of Photochemistry and Photobiology B: Biology*, 205(April): 111821. doi:10.1016/j.jphotobiol.2020.111821.

Azharuddin, M., Tsuda, H., Wu, S. and Sasaoka, E. (2008). Catalytic decomposition of biomass tars with iron oxide catalysts. *Fuel*, 87 (4-5): 451–459. doi:10.1016/j.fuel.2007.06.021.

Babu S. Ananda and Gurumallesh Prabu, H. (2011). Synthesis of AgNPs using the extract of calotropis procera flower at room temperature. *Materials Letters*, 65(11): 1675–1677. doi:10.1016/j.matlet.2011.02.071.

Beheshtkhoo Nasrin, Mohammad Amin Jadidi Kouhbanani, Amir Savardashtaki, Ali Mohammad Amani and Saeed Taghizadeh. 2018. Green Synthesis of iron oxide nanoparticles by aqueous leaf extract of daphne mezereum as a novel dye removing material. *Applied Physics A*, 124(5): 363. doi:10.1007/s00339-018-1782-3.

Bharali Linkon, Juri Kalita, Siddhartha S. Dhar and Shaemningwar Moyon, N. (2022). Excellent photocatalytic activity of a novel hydroxyapatite based composite (ZnFe2O4/HAp-Sn2+) towards degradation of ofloxacin and norfloxacin antibiotics.

Bharde Atul, Debabrata Rautaray, Vipul Bansal, Absar Ahmad, Indranil Sarkar, Seikh Mohammad Yusuf, Milan Sanyal and Murali Sastry. (2006). Extracellular biosynthesis of magnetite using fungi. *Small*, 2(1): 135–141. doi:10.1002/smll.200500180.

Bibi Ismat, Nosheen Nazar, Sadia Ata, Misbah Sultan, Abid Ali, Ansar Abbas, Kashif Jilani et al. (2019). Green synthesis of iron oxide nanoparticles using pomegranate seeds extract and photocatalytic activity evaluation for the degradation of textile dye. *Journal of Materials Research and Technology*, 8(6): 6115–6124. doi:10.1016/j.jmrt.2019.10.006.

Cai Yizhong, Mei Sun and Harold Corke. (2003). Antioxidant Activity of betalains from plants of the amaranthaceae. *Journal of Agricultural and Food Chemistry*, 51(8): 2288–2294. doi:10.1021/jf030045u.

Chatterjee Shreosi, Shouvik Mahanty, Papita Das, Punarbasu Chaudhuri and Surajit Das. (2020). Biofabrication of iron oxide nanoparticles using manglicolous fungus Aspergillus Niger BSC-1 and removal of Cr(VI) from aqueous solution. *Chemical Engineering Journal*, 385(April): 123790. doi:10.1016/j.cej.2019.123790.

Chen Hongmin, Weizhong Zhang, Guizhi Zhu, Jin Xie and Xiaoyuan Chen. (2017). Rethinking cancer nanotheranostics. *Nature Reviews Materials*, 2(7): 17024. doi:10.1038/natrevmats.2017.24.

Fahmy Heba Mohamed, Fatma Mahmoud Mohamed, Mariam Hisham Marzouq, Amira Bahaa El-Din Mustafa, Asmaa M. Alsoudi, Omnia Ashoor Ali, Maha A. Mohamed and Faten Ahmed Mahmoud.

(2018). Review of green methods of iron nanoparticles synthesis and applications. *BioNanoScience*, 8(2): 491–503. doi:10.1007/s12668-018-0516-5.

Fani Mina, Fereshteh Ghandehari and Malahat Rezayi. (2018). Biosynthesis of iron oxide nanoparticles by cytoplasmic extract of bacteria.

Hao Rui, Ruijun Xing, Zhichuan Xu, Yanglong Hou, Song Gao and Shouheng Sun. (2010). Synthesis, functionalization, and biomedical applications of multifunctional magnetic nanoparticles. *Advanced Materials*, 22(25): 2729–2742. doi:10.1002/adma.201000260.

Hodges Brenna C., Ezra L. Cates and Jae-Hong Kim. (2018). Challenges and prospects of advanced oxidation water treatment processes using catalytic nanomaterials. *Nature Nanotechnology*, 13(8): 642–650. doi:10.1038/s41565-018-0216-x.

Jin Yinjia, Fei Liu, Chao Shan, Meiping Tong and Yanglong Hou. (2014). Efficient bacterial capture with amino acid modified magnetic nanoparticles. *Water Research*, 50(March): 124–134. doi:10.1016/j.watres.2013.11.045.

Kalita Juri, Linkon Bharali and Siddhartha S. Dhar. (2022a). Zn-doped hydroxyapatite@g-C_3N_4: A novel efficient visible-light-driven photocatalyst for degradation of pharmaceutical pollutants. *New Journal of Chemistry*, 46(42): 20182–20192. doi:10.1039/D2NJ04087E.

Kalita Juri, Bishal Das and Siddhartha S. Dhar. (2022b). Synergistic effect of iron and copper in hydroxyapatite nanorods for fenton-like oxidation of organic dye. *Colloids and Surfaces A: Physicochemical and Engineering Aspects*, 643(June): 128750. doi:10.1016/j.colsurfa.2022.128750.

Kanagasubbulakshmi, S. and Kadirvelu, K. (2017). Green synthesis of iron oxide nanoparticles using lagenaria siceraria and evaluation of its antimicrobial activity. *Defence Life Science Journal*, 2(4): 422. doi:10.14429/dlsj.2.12277.

Karpagavinayagam, P. and Vedhi, C. (2019). Green synthesis of iron oxide nanoparticles using avicennia marina flower extract. *Vacuum*, 160(February): 286–292. doi:10.1016/j.vacuum.2018.11.043.

Kaul, R. K., Kumar, P., Burman, U., Joshi, P., Agrawal, A., Raliya, R. and Tarafdar, J.C. (2012). Magnesium and iron nanoparticles production using microorganisms and various salts. *Materials Science-Poland*, 30(3): 254–258. doi:10.2478/s13536-012-0028-x.

Khalil Ali Talha, Muhammad Ovais, Ikram Ullah, Muhammad Ali, Zabta Khan Shinwari and Malik Maaza. (2017). Biosynthesis of iron oxide (Fe_2O_3) nanoparticles via aqueous extracts of *Sageretia Thea* (Osbeck.) and their pharmacognostic properties. *Green Chemistry Letters and Reviews*, 10(4): 186–201. doi:10.1080/17518253.2017.1339831.

Khedr, M. H., Abdel Halim, K. S. and Soliman, N. K. (2009). Synthesis and photocatalytic activity of nano-sized iron oxides. *Materials Letters*, 63(6-7): 598–601. doi:10.1016/j.matlet.2008.11.050.

Kiruba Daniel, S. C. G., Vinothini, G., Subramanian, N., Nehru, K. and Sivakumar, M. (2013). Biosynthesis of Cu, ZVI, and Ag nanoparticles using dodonaea viscosa extract for antibacterial activity against human pathogens. *Journal of Nanoparticle Research*, 15(1): 1319. doi:10.1007/s11051-012-1319-1.

Laurent Sophie, Delphine Forge, Marc Port, Alain Roch, Caroline Robic, Luce Vander Elst and Robert N. Muller. (2008). Magnetic iron oxide nanoparticles: Synthesis, stabilization, vectorization, physicochemical characterizations, and biological applications. *Chemical Reviews*, 108(6): 2064–2110. doi:10.1021/cr068445e.

Lunge Sneha, Shripal Singh and Amalendu Sinha. (2014). Magnetic iron oxide (Fe3O4) nanoparticles from Tea waste for arsenic removal. *Journal of Magnetism and Magnetic Materials*, 356(April): 21–31. doi:10.1016/j.jmmm.2013.12.008.

Mahdavi Mahnaz, Farideh Namvar, Mansor Ahmad and Rosfarizan Mohamad. (2013). Green biosynthesis and characterization of magnetic iron oxide (Fe_3O_4) nanoparticles using seaweed (sargassum muticum) aqueous extract. *Molecules*, 18(5): 5954–5964. doi:10.3390/molecules18055954.

Meng Jingxin, Pengchao Zhang, Feilong Zhang, Hongliang Liu, Junbing Fan, Xueli Liu, Gao Yang, Lei Jiang and Shutao Wang. (2015). A self-cleaning TiO_2 nanosisal-like coating toward disposing nanobiochips of cancer detection. *ACS Nano*, 9(9): 9284–9291. doi:10.1021/acsnano.5b04230.

Mittal Amit Kumar, Yusuf Chisti and Uttam Chand Banerjee. (2013). Synthesis of metallic nanoparticles using plant extracts. *Biotechnology Advances*, 31(2): 346–356. doi:10.1016/j.biotechadv.2013.01.003.

Noh Seung-hyun, Seung Ho Moon, Tae-Hyun Shin, Yongjun Lim and Jinwoo Cheon. (2017). Recent advances of magneto-thermal capabilities of nanoparticles: From design principles to biomedical applications. *Nano Today*, 13(April): 61–76. doi:10.1016/j.nantod.2017.02.006.

Ojha Akshaya Kumar, Jogeswari Rout, Shikha Behera and Nayak, P. L. (2013). Green synthesis and characterization. International Journal of Pharmaceutical Research & Allied Sciences, 2: 31–35.

Panigrahi Sudipa, Subrata Kundu, Sujit Ghosh, Sudip Nath and Tarasankar Pal. (2004). General method of synthesis for metal nanoparticles. *Journal of Nanoparticle Research*, 6(4): 411–414. doi:10.1007/s11051-004-6575-2.

Peigneux Ana, Jose D. Puentes-Pardo, Alejandro B. Rodríguez-Navarro, Maxwell T. Hincke and Concepción Jimenez-Lopez. (2020). Development and characterization of magnetic eggshell membranes for lead removal from wastewater. *Ecotoxicology and Environmental Safety*, 192(April): 110307. doi:10.1016/j.ecoenv.2020.110307.

Priya Naveen, Kamaljit Kaur and Amanpreet K. Sidhu. (2021). Green synthesis: An eco-friendly route for the synthesis of iron oxide nanoparticles. *Frontiers in Nanotechnology*, 3(June): 655062. doi:10.3389/fnano.2021.655062.

Rajkumari Kalyani, Juri Kalita, Diparjun Das and Samuel Lalthazuala Rokhum. (2017). Magnetic $Fe_3O_4@$ silica sulfuric acid nanoparticles promoted regioselective protection/deprotection of alcohols with dihydropyran under solvent-free conditions. *RSC Advances*, 7(89): 56559–56565. doi:10.1039/C7RA12458A.

Roy Saikatendu Deb, Krishna Chandra Das and Siddhartha Sankar Dhar. (2021). Conventional to green synthesis of magnetic iron oxide nanoparticles; its application as catalyst, photocatalyst and toxicity: A short review. *Inorganic Chemistry Communications*, 134(December): 109050. doi:10.1016/j.inoche.2021.109050.

Shah Monaliben, Derek Fawcett, Shashi Sharma, Suraj Tripathy and Gérrard Poinern. (2015). Green synthesis of metallic nanoparticles via biological entities. *Materials*, 8(11): 7278–7308. doi:10.3390/ma8115377.

Sundaram P. Alagu, Robin Augustine and Kannan, M. (2012). Extracellular biosynthesis of iron oxide nanoparticles by bacillus subtilis strains isolated from rhizosphere soil. *Biotechnology and Bioprocess Engineering*, 17(4): 835–840. doi:10.1007/s12257-011-0582-9.

Tarafdar Jagadish Chandra and Ramesh Raliya. (2013). Rapid, low-cost, and ecofriendly approach for iron nanoparticle synthesis using *Aspergillus Oryzae* TFR9. *Journal of Nanoparticles*, 2013(March): 1–4. doi:10.1155/2013/141274.

Tombácz Etelka, Rodica Turcu, Vlad Socoliuc and Ladislau Vékás. (2015). Magnetic iron oxide nanoparticles: Recent trends in design and synthesis of magnetoresponsive nanosystems. *Biochemical and Biophysical Research Communications*, 468(3): 442–453. doi:10.1016/j.bbrc.2015.08.030.

Vasantharaj Seerangaraj, Selvam Sathiyavimal, Palanisamy Senthilkumar, Felix LewisOscar and Arivalagan Pugazhendhi. (2019). Biosynthesis of iron oxide nanoparticles using leaf extract of ruellia tuberosa: antimicrobial properties and their applications in photocatalytic degradation. *Journal of Photochemistry and Photobiology B: Biology*, 192(March): 74–82. doi:10.1016/j.jphotobiol.2018.12.025.

Wang Zhenghua, Ling Shen and Shiyu Zhu. (2012). Synthesis of core-shell @@ microspheres and their application as recyclable photocatalysts. *International Journal of Photoenergy*, 2012: 1–6. doi:10.1155/2012/202519.

Wu Wei, Shaofeng Zhang, Xiangheng Xiao, Juan Zhou, Feng Ren, Lingling Sun and Changzhong Jiang. (2012). Controllable synthesis, magnetic properties, and enhanced photocatalytic activity of spindlelike mesoporous α-Fe_2O_3/ZnO core–shell heterostructures. *ACS Applied Materials & Interfaces*, 4(7): 3602–3609. doi:10.1021/am300669a.

Yew Yen Pin, Kamyar Shameli, Mikio Miyake, Noriyuki Kuwano, Nurul Bahiyah Bt Ahmad Khairudin, Shaza Eva Bt Mohamad and Kar Xin Lee. (2016). Green synthesis of magnetite (Fe_3O_4) Nanoparticles using seaweed (kappaphycus alvarezii) extract. *Nanoscale Research Letters*, 11(1): 276. doi:10.1186/s11671-016-1498-2.

CHAPTER 13

Green Synthesized Biocompatible Nanomaterial for Environmental Application

Aishee Ghosh, Utsa Saha, Somagni Roy, Snehagni Roy
*and Suresh K. Verma**

1. Introduction

The rapid progression of science and technology has sparked significant interest in the research community to delve into the field of nanotechnology. Nanotechnology involves manipulating matter at the nanoscale level, resulting in unique particle properties like high surface energy, specific surface area, and quantum confinement (*Introduction to Nanomaterials and Nanotechnology*, n.d.; Patel et al. 2021). Because of these distinctive physicochemical properties, nanoparticles have countless applications in various industries.

However, the physical methods used for nanoparticle synthesis, such as laser ablation and inert gas condensation, require a considerable amount of time to achieve thermal stability and consume significant energy, raising the environmental temperature around the source material (Attarad et al. 2016, Syafiuddin et al. 2017). Furthermore, these physical methods occupy large spaces, making them unsuitable for nanoparticle generation. In contrast, chemical synthesis of nanoparticles uses harsh reducing agents and organic solvents, causing toxicity and environmental issues (Virkutyte and Varma 2011). In fact, modern technological advances in chemical processes have led to increased levels of pollutants, which are beyond the self-cleaning capacities of the environment (Yunus et al. 2012). As a result, detecting and treating existing contaminants while preventing the generation of new pollutants

School of Biotechnology, KIIT University, Bhubaneswar, Odisha, 751024, India.
* Corresponding author: sureshverma22@gmail.com

has become a challenging task in the 21st century. Even though remediation of contaminants using existing technologies has not been found to be effective and efficient in cleaning the environment, nanotechnology dedicated to environmental clean-up has evolved to sophisticated and efficient levels (Pandey and Fulekar 2012). Such advancements allow contaminated areas to be "engineered" back to their original integrity, thus resetting the conditions for restoring a balanced environment.

Furthermore, the optical and electronic properties of nanoparticles, which depend on their size and shape, provide scientists with an intriguing opportunity to explore their role in making products and processes compatible with the environment. As a result, biological synthesis pathways, which involve using living organisms, such as bacteria, fungi, plants, and algae to produce nanoparticles, are preferred over physical and chemical synthesis methods (Jeevanandam et al. 2016, Purohit et al. 2019). Besides being a green and sustainable approach, biological synthesis offers several advantages, such as eco-friendliness, cost-effectiveness, scalability, and the ability to produce nanoparticles of various sizes, shapes, and compositions (Parveen et al. 2016, Singh et al. 2016). More on this has been discussed in the chapter.

2. Approaches in the Synthesis of NPs

There are two main techniques for nanomaterial synthesis. They include the "top-down" approach and the "bottom-up" approach (see Figure 1).

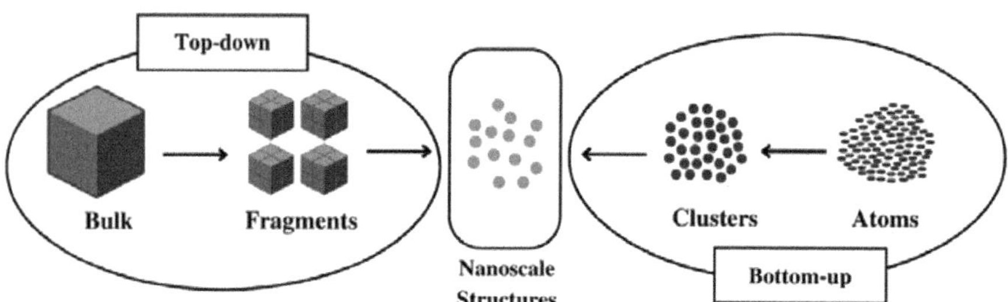

Figure 1. The two main approaches taken up for the production of nanostructures.

2.1 Top-Down Approach

The top-down approach in nanoparticle synthesis involves breaking down a bulk material into smaller particles to create nanoparticles. This method usually involves mechanical or physical processes such as milling, lithography, or laser ablation (Jiang et al. 2022). In this approach, the starting material is usually larger in size, and it is reduced to smaller nanoparticles through a series of processes. This approach is suitable for creating well-defined and uniform nanoparticles with controlled properties (Abid et al. 2022). The top-down approach is widely used in industry to produce nanoparticles with high throughput and scalability. However, it requires specialized equipment and can be challenging to achieve consistent results.

2.2 Bottom-Up Approach

The bottom-up approach in nanoparticle synthesis involves building small particles into larger nanoparticles (Abid et al. 2022). This method typically involves chemical or biological processes such as precipitation, sol-gel synthesis, or biomimetic synthesis. In this approach, nanoparticles are formed by controlling the growth and assembly of molecules, atoms, or ions. The starting materials are usually smaller and assemble into larger nanoparticles through a series of chemical reactions or self-assembly. The bottom-up approach can produce highly uniform and complex nanoparticles with precise size, shape, and composition. However, it can be challenging to scale up and control reproducibility (Jiang et al. 2022).

3. Methods of Nanoparticle Generation

Day after day, new techniques are devised for synthesizing nanoparticles. In general, these techniques can be divided into three major categories: physical methods, chemical methods and biological methods (as depicted in Figure 2).

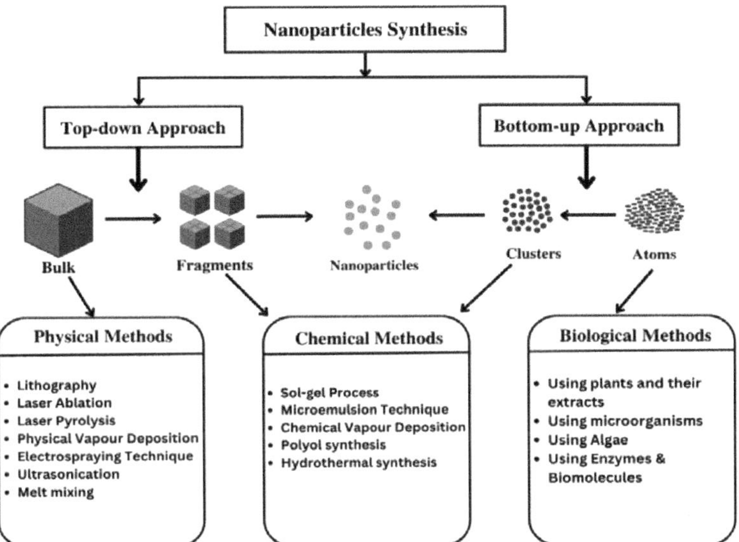

Figure 2. Methods depicting synthesis of nanoparticles (Devatha and Thalla 2018).

3.1 Physical Methods

Physical methods involve using physical forces to generate nanoparticles, such as mechanical milling, laser ablation, inert gas condensation, physical vapor deposition, and spray pyrolysis. In mechanical milling, the material is ground using a ball mill or other types of milling equipment to produce nanoparticles with varying shapes and dimensionalities (Xing et al. 2013). In contrast, inert gas condensation uses inert gases (such as He or Ar) and a substrate holder chilled with liquid nitrogen to produce nanoparticles. Ward and colleagues adopted this technique to produce Mn nanoparticles (Ward et al. 2006). In physical vapor deposition, an environmentally friendly vacuum deposition method is made up of three basic steps: (1) vaporizing the material from a solid source, (2) transporting the vaporized material, and (3) nucleation and growth to produce thin films and nanoparticles (Okuyama and Lenggoro 2003). Laser ablation involves using a laser beam to vaporize a target material, which is then condensed to form nanoparticles (Ullmann et al. 2002). Spray pyrolysis involves spraying a precursor solution into a high-temperature reactor, which decomposes to form nanoparticles (Okuyama and Lenggoro 2003). This suggests that physical technique uses expensive machinery, high temperature and pressure, and a sizable area for putting up machines, suggesting they are not economical. However, physical methods produce uniform monodisperse nanoparticles free of solvent contamination.

3.2 Chemical Methods

Chemical methods involve using chemical reactions to produce nanoparticles, such as sol-gel synthesis, co-precipitation, polyol synthesis, chemical vapor deposition, and microemulsion. In sol-gel synthesis, a precursor solution is hydrolyzed and condensed to form nanoparticles (Zhang and Gao 2004). The co-precipitation method involves mixing metal salts with a base under controlled conditions, leading to the formation of nanoparticles (Lagrow et al. 2019). This method produces nanoparticles with narrow size distributions. In polyol synthesis, nanoparticles are produced with controlled sizes and shapes. Metal ions are reduced in a polyol solvent at a controlled temperature, forming nuclei that grow into nanoparticles (Park et al. 2007). The process is influenced by the concentration of the metal precursor, capping agents, and reaction conditions (Lin et al. 2014). Microemulsion involves using a microemulsion system to prepare nanoparticles. This method involves using a surfactant and oil mixture that stabilizes water droplets containing the precursor material (Chugh et al. 2021). The reaction takes place inside the droplets, forming nanoparticles with narrow size distributions and controlled shapes. The chemical vapor deposition method involves the reaction of precursor gases in a heated chamber, leading to the formation of nanoparticles on a substrate (Kwok and Chiu 2005). The substrate acts as a template for the growth of the nanoparticles.

This suggests that compared to physical methods, chemical methods of producing nanoparticles offer certain advantages, such as the ability to produce nanoparticles with controlled sizes and shapes and the possibility of modifying their surfaces for various applications. However, chemical methods require potentially hazardous chemicals and may generate toxic waste.

3.3 Biological Methods

Recently, the utilization of biological systems has emerged as a novel method for the synthesis of nanoparticles owing to the growing need to develop environmentally benign nanoparticle synthesis processes that do not use toxic chemicals in the synthesis protocol. The potential of living organisms in nanoparticle synthesis varies from simple bacterial cells to fungi and plants (Mohanpuria et al. 2008). For instance, bacteria can be used for producing gold, silver, cadmium, zinc, magnetite, and iron nanoparticles; yeasts for silver, lead, and cadmium nanoparticles; fungi for gold, silver, and cadmium nanoparticles; algae for silver and gold nanoparticles; and plants for silver, gold, palladium, zinc oxide, platinum, and magnetite nanoparticles (Bhattacharya and Gupta 2005, Senapati 2005). More about this has been explained in the next section under green synthesis of nanoparticles.

4. Green Synthesis of Nanoparticles

As previously noted, nanoparticles can be produced in a variety of methods (physical, chemical, or biological routes), but synthesizing nanoparticles by physical and chemical pathways raises toxicity issues as well as environmental concerns (Ealias and Saravanakumar 2017). Many of the physical methods require a lot of heat, raising the environmental temperature in addition to being time-consuming. The chemical procedures, on the other hand, use highly toxic reductants and stabilizing agents, which can have detrimental effects on our already suffering environment as well as animal life forms (Singh et al. 2018). Therefore, an alternative technology was required, which led to the development of green synthesis of nanoparticles, also known as green nanotechnology. Green nanotechnology is the synthesis of nanoparticles by biological organisms or materials like bacteria, fungi, yeast, plants, and enzymes with the help of biotechnological procedures (refer to Figure 3). The resulting nanoparticles are free of hazardous chemicals and ecologically friendly (El-Rafie et al. 2013, Husen and Siddiqi 2014, Patel et al. 2015, Wadhwani et al. 2016). In contrast with other synthesis pathways, green synthesis of nanoparticles is a single-step, eco-friendly approach that uses less energy to produce and is cost-effective (Pal et al. 2018). This approach has gained popularity in recent years and has created nanoparticles with a range of uses.

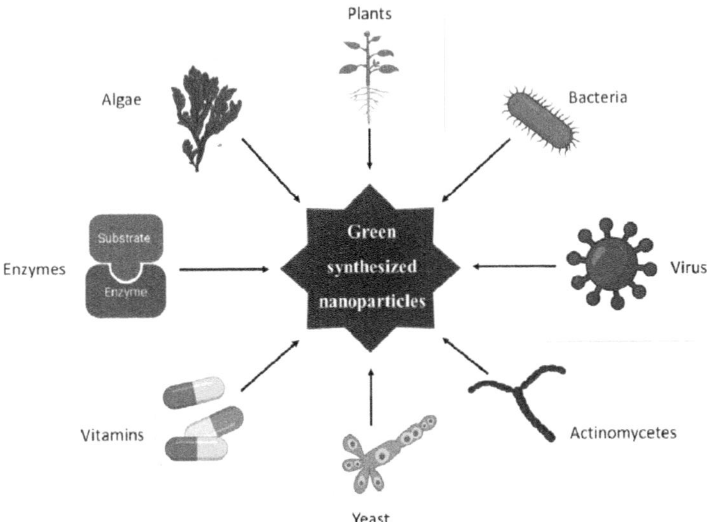

Figure 3. Various sources for nanoparticle synthesis.

4.1 Bacteria-Mediated Nanoparticle Synthesis

The main reason bacteria can synthesize nanoparticles is attributed to their defense mechanism. The resistance of bacterial cells toward reactive ions in the environment is responsible for nanoparticle synthesis. High ionic concentration is usually harmful to bacterial cells. In order to avoid cell death, their cellular machinery helps convert reactive ions to stable atoms. That is how stable nanoparticles are generated through bacterial synthesis (Singh 2022). Bacterial synthesis of nanoparticles has been adopted worldwide owing to their ease of manipulation (Thakkar et al. 2010). A study from 2012 using *Pseudomonas stutzeri* AG259 strain gave first-hand evidence of bacterial synthesis of silver nanoparticles (Prabhu and Poulose 2012). There are many examples of bacterial strains that have been extensively used to produce silver and gold nanoparticles by bio-reduction method, like *Escherichia coli* (Singh et al. 2010), *Lactobacillus casei* (Xu et al. 2018), *Bacillus cereus* (Ganesh Babu and Gunasekaran 2009) and *Bacillus megaterium* D01 (Wen et al. 2009), *Desulfovibrio desulfuricans* (Deplanche and Macaskie 2008), and *Bacillus subtilis* 168 (Alsamhary 2020).

4.2 Fungi-Mediated Nanoparticle Synthesis

Nanoparticle synthesis using fungi is more popular for several reasons, such as ease of handling, quick growth, ease of maintenance in lab settings, high tolerance, capable of metal bioaccumulation, and economic feasibility (Yadav et al. 2015). The fungal system has a strong binding capacity as well as intracellular metal uptake ability (Bhainsa and D'Souza 2006a, Ingle et al. 2008, Naveen et al. 2010). Verticillium sp., for example, demonstrated excellent reduction of silver ions, leading to the formation of silver nanoparticles beneath the surface of fungal cells (Priyabrata et al. 2001). Fungi release a large amount of extracellular enzymes essential for nanoparticle formation and greater yield (Alani et al. 2012, Birla et al. 2009). In comparison to other bacteria and plants, fungal mycelia can endure high-pressure flow, agitation, and other conditions in a bioreactor (Prakash and Soni 2012). Furthermore, because the nanoparticles precipitated outside the cell are without biological components, they can be employed directly for a wide range of applications (Narayanan and Sakthivel 2010).

4.3 Actinomycetes-Mediated Nanoparticle Synthesis

Actinomycetes are a group of Gram-positive bacteria that are abundant in soil, and they are known for their ability to produce a wide range of secondary metabolites with various biological activities (Kumar et al. 2008). Though they are least studied, in recent years, it has been discovered that some actinomycetes species can also synthesize nanoparticles.

The actinomycetes-mediated synthesis of nanoparticles involves the reduction of metal ions to nanoparticles using actinomycetes biomass as a reducing and stabilizing agent. The biomass contains various enzymes, proteins, and other biomolecules that can reduce metal ions and stabilize the nanoparticles (Ahmad et al. 2019). The process takes place on the surface of mycelia along with cytoplasmic membranes under mild conditions, such as room temperature and atmospheric pressure, making it an environmentally friendly method.

Various actinomycetes species have been reported to synthesize nanoparticles, including streptomyces, nocardiopsis, actinoplanes, micromonospora, and frankia. These actinomycetes can produce nanoparticles of different shapes and sizes, such as spherical, rod-shaped, and triangular.

Actinomycetes-mediated nanoparticle generation has been applied in various fields, including medicine, biotechnology, and environmental science. For example, actinomycetes-mediated silver nanoparticles have been used as antimicrobial agents against various pathogenic microorganisms, including bacteria, fungi, and viruses (Singh et al. 2016). Actinomycetes-mediated gold nanoparticles have been used in cancer therapy, drug delivery, and bio-imaging (Singh et al. 2016). Actinomycetes-mediated magnetite nanoparticles have been used for environmental remediation, such as removing heavy metals from contaminated water (Kapoor et al. 2021).

4.4 Yeast-Mediated Nanoparticle Synthesis

Yeast-mediated nanoparticle generation is a green synthesis method that utilizes yeast cells' reducing and stabilizing properties to produce nanoparticles. The process is similar to the actinomycetes-mediated nanoparticle generation, but it uses yeast cells instead of actinomycetes biomass.

Yeast has a high metabolic rate and can produce various biomolecules, including enzymes, proteins, and organic compounds, that can serve as the reducing and stabilizing agents in nanoparticle synthesis (Yusof et al. 2019). The process of yeast-mediated nanoparticle generation involves growing yeast cells in a suitable environment and adding metal salts to the culture. The reducing and stabilizing agents within the yeast cells then reduce the metal ions to form nanoparticles. For instance, when *Candida glabrata* is exposed to Cd2+ ions, it forms CdS quantum dots intracellularly via the activation of phytochelation (Bae and Chen 2004). Additionally, yeast is mostly studied for bio-synthesizing semiconductor nanoparticles, such as CdS quantum dots formed intracellularly in *Schizosaccharomyces pombe* yeast cells, which exhibit ideal diode characteristics (Kowshik et al. 2002).

4.5 Algae-Mediated Nanoparticle Synthesis

Among several microbial systems, algae have demonstrated promising multi-purpose potential for nanoparticle's green synthesis with the added advantages of wastewater bioremediation, bioenergy generation, and mass-scale manufacture of marketable products such as pigments and therapeutic goods. The bioactive substances found in green algae, including proteins, lipids, carbohydrates, carotenoids, vitamins, and other secondary metabolites, are thought to have a variety of biological functions (Oza et al. 2012, Rajeshkumar et al. 2012). In wastewater treatment processes, algae-based biosystems can be employed in a variety of ways, such as the elimination of harmful bacteria, the decrease of biological and chemical oxygen requirements, and the elimination of nitrogen, phosphorus, and other nutrients/pollutants (Agarwal et al. 2019). *Chlorella vulgaris*, *Scenedesmus platydiscus*, *Scenedesmus quadricauda*, *Scenedesmus capricornutum*, and other freshwater algae species have demonstrated accumulation and degradation of polycyclic aromatic hydrocarbons. Algae-based NPs (AuNPs, AgNPs, etc.) exhibit a variety of features, such as antibacterial, photocatalytic activities, and antifouling properties, making them appropriate for usage in several fields, such as forensics, optical biosensing, and bioremediation, among many others (Chaudhary et al. 2020). As algae-derived green synthesized nanoparticles implement biomolecules as reducing agents and do not require expensive chemical reductants like hydrazine and sodium borohydride, these NPs offer a cost-effective and environmentally friendly alternative to chemical and physical synthesis. Due to their small size and substantial surface area, silver nanoparticles (AgNPs) have the ability to interact strongly with microbial surfaces and serve as potential antibacterial agents (Durán et al. 2016). Particularly, the penetrating properties of algae-based nanoparticles through extra-polysaccharide (EPS) and cell membranes make them suitable for application as antibiofilm agents that act against some multidrug-resistant bacteria (Saravanan et al. 2021). The production of algal NPs has been revealed to be significantly influenced by the operational circumstances comprising a number of variables, including temperature, the pH of the medium, light, the length of the incubation period, biomass, precursor concentration, and more (Hamida et al. 2020).

These factors influence the stability, nucleation, and production of NPs. Although still in its infancy, algae-based biosynthesis of nanoparticles shows promise for future environmental applications. Although micro algal systems play a significant role in the synthesis of nanoparticles, there is a scarcity of information regarding them, which limits their potential as a green synthesis option (Saravanan et al. 2021).

4.6 *Plant-Mediated Nanoparticle Synthesis*

The process of green synthesis of nanoparticles (NPs) is a safe, biocompatible, and eco-friendly procedure compared to other chemical synthesis processes in existence. When it involves using the plant as a whole or its specific parts, this synthesis is referred to as plant-derived NP synthesis, resulting in NPs bearing a specific shape, size, and composition. Due to its simplicity of application and the wide variety of plants, plant-based NP green synthesis is now regarded as the gold standard among the green biological approaches currently in existence. The process of producing NPs from plant extract is a simple method involving the plant extract being mixed with the metal salt solution at a specific temperature. The large variety of phytochemicals in their extract serves as an organic stabilizer and/or reducing agent for the metal ions produced in the salt solution and converts them to their elemental state. It is widely acknowledged that plant-derived NPs have a lower risk of having negative side effects on people than chemically produced NPs do. They also have a high biological potential and can be used in a variety of fields, including areas like agriculture, food science and technology, nanomedicine, human health protection, etc.

Plant-based silver nanoparticles (AgNPs) are among the easiest to prepare since almost all plants have the potential to be exploited to prepare AgNPs. A silver metal ion solution and a reducing biological agent are employed in producing these AgNPs. Reducing and stabilizing Ag ions with a combination of biomolecules, such as amino acids, polysaccharides, proteins, vitamins, saponins, phenolics, terpenes, and/or alkaloids, is the simplest and least expensive way to make AgNPs (Tolaymat et al. 2010). Gold nanoparticles (AuNPs) have drawn much interest due to their simple manufacture. According to many research (Khan et al. 2020, Perveen et al. 2021), biomolecules like proteins, flavonoids, and phenols play a significant influence in the reduction of metal ions and the coating of gold nanoparticles in plant extracts. Compared to Au and Ag, copper (Cu) is a more affordable metal and has been used to make CuNPs by reducing aqueous Cu ions with various plant extracts (Cherian et al. 2020, Letchumanan et al. 2021). Other metals used for NP preparation with plant extracts include zinc (Zn), nickel (Ni), and manganese (Mn). Compared to NPs generated chemically, plant-derived green NPs are less likely to have serious adverse effects on people. Thus, they offer a wide range of possible uses. Due to their antiproliferative, antiparasitic, antibacterial, and anti-inflammatory properties, they are useful in nanomedicine. Precision farming includes target-specific biomolecule delivery, improved nutrient uptake, controlled release of agrochemicals, detection of plant diseases, etc. In the bioengineering sectors, they are used as biocatalysts, biosensors, photocatalysts, etc., as well as find applications in the food and cosmetics industries (Hano and Abbasi 2022).

An intriguing and newly discovered component of nanotechnology that significantly affects the environment and advances the long-term sustainability and advancement of nanoscience lies in using plants to manufacture green nanoparticles. Many special substances found in plants aid in the synthesis process, thus quickening the synthesis kinetic. This environmentally friendly method of NP synthesis is gaining ground and is anticipated to grow dramatically in the next few years.

4.7 Enzyme-Mediated Nanoparticle Synthesis

Enzymes are biological catalysts that can catalyze the reduction of metal ions to nanoparticles (Ovais et al. 2018). Enzymes such as laccase, peroxidase, and alkaline phosphatase have successfully synthesized nanoparticles. The mechanism of enzyme-mediated synthesis involves the oxidation of substrate molecules and the reduction of metal ions to nanoparticles (Sanket and Das 2021). Enzymes are particularly useful in synthesizing nanoparticles as they can produce nanoparticles at mild reaction conditions and with high yields.

For example, laccase, a copper-containing enzyme, has been used to synthesize copper nanoparticles. The synthesis of copper nanoparticles using laccase involves the reduction of copper ions to nanoparticles by the enzyme (Gour and Jain 2019). The nanoparticles produced were uniform in size and shape and had good stability (Rudakiya and Gupte 2019).

4.8 Vitamin-Mediated Nanoparticle Synthesis

Vitamins are organic compounds that play important roles in biological processes. They can also be useful in the synthesis of nanoparticles. Vitamins such as ascorbic

Table 1. Green nanoparticle synthesis from various biological sources.

Nanoparticle (NP)	Origin/Source	Morphology	Size (nm)	Precursor	Application	References
Bacteria						
Silver	*Bacillus cereus*	Spherical	20–40	$AgNO_3$	Antibacterial activity against *E. coli, Pseudomonas aeruginosa, Staphylococcus aureus, Salmonella typhi*, and *Klebsiella pneumonia* bacteria	(Sunkar and Nachiyar 2012)
	Pseudomonas proteolytica, Bacillus cecembensis	Spherical	6–13	$AgNO_3$	Antibacterial activity against *A. kerguelensis, A. gangotriensis, B. indicus, P. antarctica, P. proteolytica,* and *E. coli*	(Shivaji et al. 2011)
	Lactobacillus casei	Spherical	20–50	$AgNO_3$	Drug delivery, cancer treatments, bio-labeling	(Korbekandi et al. 2012)
Gold	*Rhodopseudomonas capsulate*	Triangular	10–20	$HAuCl_4$	Cancer hyperthermia	(Shiying He et al. 2007)
Fungi						
Silver	*Rhizopus nigricans*	Round	35–40	$AgNO_3$	Bactericidal, catalytic	(Ravindra and Rajasab 2014)
	Verticillium	Spherical	21–25	$AgNO_3$	Catalysis	(Mukherjee et al. 2001)
	Aspergillus fumigates	Spherical	5–25	$AgNO_3$	Coating for solar energy absorption and intercalation material for electrical batteries	(Bhainsa and D'Souza 2006b)
	Penicillium brecompactum	Crystalline spherical	23–105	$AgNO_3$	Antimicrobial agent	(Soliman et al. 2021)

Plants						
Gold and silver	*Aloe barbadensis Miller* (Aloe vera)	Spherical, Triangular	10–30	–	Cancer hyperthermia, optical coatings	(Chandran et al. 2006)
	Camellia sinensis (black tea leaf extracts)	Spherical, prism	20	–	Catalysts, sensors	(Mondal et al. 2011)
Gold, silver and silver_x0002_gold alloys	*Azadirachta indica* (neem)	Spherical, Triangular, Hexagonal	5–35 and 50–100	–	Remediation of toxic metals	(Shankar et al. 2004)
Gold	*Cymbopogon fexuosus* (lemongrass)	Spherical, Triangular	200–500	–	Infrared-absorbing optical coatings	(Shankar et al. 2005)
Yeast						
Silver	MKY3	Hexagonal	2–5	–	Coatings for solar energy absorption and intercalation material for electrical batteries	(Kowshik et al. 2002)
Gold, Silver	*Saccharimyces cerevisae* broth	Spherical	4–15	–	Catalysis	(Mourato et al. 2011)
Gold	*Yarrowia lipolytica*	Triangular	15	$HAuCl_4$	–	(Agnihotri et al. 2009)
Actinomycetes						
CdS	*Schizosaccharomyces pombe*	Wurtzite-Hexagonal	1–1.5	–	Semiconductors and fabrication	(Kowshik et al. 2002)
Gold	*Thermomonospora* sp.	Spherical	8	$HAuCl_4$	–	(Ahmad et al. 2003)

Table 1 contd. ...

...Table 1 contd.

Nanoparticle (NP)	Origin/Source	Morphology	Size (nm)	Precursor	Application	References
Algae						
Silver	*Caulerpa racemosa*	Spherical, Triangular	5–25	$AgNO_3$	Antimicrobial agent	(Kathiraven et al. 2015)
	Sargassum tenerrimum	Spherical	11–20	$AgNO_3$	Antimicrobial agent	(Kumar et al. 2012)
Vitamins						
Silver	Vitamin B2	Wire, rod-shaped	6.1 ± 0.1	$AgNO_3$	–	(Nadagouda and Varma 2008)
Palladium	Vitamin B2	Wire, rod-shaped	4.1 ± 0.1	$PdCl_2$	–	(Nadagouda and Varma 2008)
Enzymes						
Silver	Keratinase	Spherical	5–30	$AgNO_3$	Antibacterial activity	(Lateef et al. 2015b)
	Laccase	Spherical	50–100	$AgNO_3$	Antibacterial activity	(Lateef et al. 2015a)

acid (vitamin C) (Malassis et al. 2016) and riboflavin (vitamin B2) have been used for the green synthesis of nanoparticles. The mechanism of vitamin-mediated synthesis involves the reduction of metal ions to nanoparticles by the vitamin. Vitamins are particularly useful in the synthesis of nanoparticles as they are readily available, biocompatible, and non-toxic.

For example, ascorbic acid has been used to synthesize silver nanoparticles (Suriati et al. 2014). The synthesis of silver nanoparticles using ascorbic acid involves the reduction of silver ions to nanoparticles by the vitamin. The nanoparticles produced were uniform in size and shape and had good stability.

5. Characterization of NPs

Different analytical methods have been developed to evaluate the properties of nanomaterials produced through green synthesis. These techniques can be used to measure the size, shape, solubility, porosity, and fractal dimension of the product. Additionally, they can be used to determine the metal content, diameter, crystallinity, optical properties, functional groups, and thermal stability of the nanomaterials (Kshtriya et al. 2021). These techniques can be categorized into two types:

(i) Spectroscopic characterization techniques, which provide information on chemical composition and physical characteristics like size, crystallinity, color, aggregation behavior, and polydispersity of the nanomaterial;

(ii) Microscopic characterization techniques mainly provide information on the size and morphology of the nanomaterials.

5.1 Spectroscopic Techniques Used for Characterization of Green-Synthesized Nanoparticles

5.1.1 UV Visible Spectroscopy

The NPs must be accurately and fully described to ensure reproducibility in manufacturing nanoparticles (NP), biological activity, and safety. In order to achieve this, the synthesized NPs are accurately characterized using a wide variety of physicochemical techniques, and one such commonly used technique is ultraviolet-visible spectroscopy. UV-visible spectroscopy (ultraviolet-visible spectroscopy) is a technique that follows the principle of absorption or transmission of discrete wavelengths of light (UV or visible) by the sample, resulting in the formation of a distinct spectrum. The amount of discrete light is measured by comparing it to a reference sample or blank. Ultraviolet-visible (UV/Vis) spectroscopy uses an electromagnetic absorption range of 200–400 nm (ultraviolet spectroscopy) and 400–800 (visible spectroscopy).

In nanotechnology studies, UV/Vis spectroscopy has gained importance in deciphering information about the plasmonic resonance of nanoparticles. The absorbance data obtained after a UV/Vis spectroscopy analysis confirms the formation of a nanoparticle and the plasmonic resistance of the nanoparticle under study. This technique can also decipher the average size of a metal nanoparticle (e.g., AuNP), which is about 6% accurate compared to transmission electron microscope (TEM) data. Information about the concentration of the nanoparticle in the sample solution and the amount of non-spherical nanoparticles can also be obtained, which assists in aggregation studies (Amendola and Meneghetti 2009). The analysis of the extent to which the nanoparticles have been stabilized by these biologically active agents, resulting in their successful synthesis, is where characterization by UV-Vis spectroscopy comes into play.

5.1.2 Fourier-Transform Infrared Spectroscopy (FT-IR)

FTIR analysis utilizes infrared light to scan samples and identify organic, inorganic, and polymeric materials. Changes in the absorption bands' characteristic pattern indicate a modification in the material composition (Titus et al. 2018). FTIR is a valuable tool for identifying and characterizing unknown materials, detecting contaminants, identifying additives, and identifying decomposition and oxidation. Additionally, FTIR helps identify biomolecules responsible for capping, reduction, and stabilization (Kshtriya et al. 2021).

5.1.3 XRD Spectroscopy

Due to their high surface area-to-volume ratio, the properties of nanoparticles undergo significant changes as their size varies. Therefore, characterizing nanoparticles is an essential step in understanding their properties at different molecular levels. In the field of nanotechnology, X-ray diffraction (XRD) spectroscopy is a powerful analytical technique used to determine various characteristics of materials at the nanoscale, including their crystal structure, crystalline phase, crystalline defect, and lattice parameters (F and E 2019). This technique is particularly well-suited for analyzing powdered samples that are freshly prepared from the respective colloidal solutions of the samples. The XRD equipment is used to compare the position and intensity of peaks in a sample with diffraction patterns from various databases, which helps to quantify the composition of nanoparticles (F and E 2019).

In XRD spectroscopy, a beam of X-rays is directed at a sample, and the resulting diffraction pattern is analyzed to determine the crystal structure of the material. The X-rays are diffracted by the atoms in the crystal lattice, and the resulting pattern of diffraction is unique to the crystal structure of the material. By analyzing the diffraction pattern, researchers can determine the orientation, size, and shape of the crystalline domains within a material (Bunaciu et al. n.d.). Hence, nanoparticles characterized through XRD require a source, a detector, and a sample.

In addition to its use in the study of nanomaterials, XRD spectroscopy is also used in the characterization of thin films and coatings, as well as in the study of phase transformations and defects in materials (van de Krol et al. 1999). Its high resolution and sensitivity make it an indispensable tool in the field of nanotechnology.

5.2 Microscopic Techniques Used for Characterization of Green Synthesized Nanoparticles

5.2.1 Scanning Electron Microscopy (SEM)

Scanning electron microscopy is a method to obtain high-resolution surface imaging. It uses electrons to produce clear images. It has greater magnification power as well as depth of field. Through SEM, the surface topography and composition of any green-synthesized nanoparticle can be obtained. SEM examination can give us the size of the obtained nanoparticles in nanometres. The purity and monodispersity or polydispersity can be confirmed by this technique (Dipankar and Murugan 2012, Thamer et al. 2018).

5.2.2 *Transmission Electron Microscopy (TEM)*

The most versatile microscopic technique used to characterize green-synthesized nanoparticles is Transmission Electron Microscopy (TEM). TEM is used to analyze the two-dimensional morphology of nanomaterials and metallic NPs in various fields, such as agriculture, drug delivery, and biological and biomedical sample analysis (Kshtriya et al. 2021). TEM can also be used to analyze the crystalline or non-crystalline properties of NPs by utilizing the selected area electron diffraction (SEAD) technique, which is a part of an advanced version of TEM known as high-resolution transmission electron microscopy (HRTEM), for more precise characterization of nanoparticles.

5.2.3 *Atomic Force Microscopy (AFM)*

Atomic Force Microscopy (AFM) is another microscopy technique that captures high-resolution images of nanomaterials, their surface, morphology, and texture. AFM is a very precise scanning probe microscopy technique that has better resolution than normal optical microscopy techniques. Over the last three decades, AFM has been extensively used for the precise characterization of nanoparticles, bio-nanomaterials, peptide-assembly, viruses, bacteria, and different nanostructures, making it an important technique for nanotechnology. Unlike SEM or TEM, AFM uses laser pulses as the light source to measure the force by which the sharp tip of a cantilever moves across the surface of the sample (Kshtriya et al. 2021). This setup determines the surface topography of materials on the atomic scale vertically and the nanometre scale horizontally, producing an image.

6. Factors Affecting the Green Synthesis of NPs

6.1 *pH and Conductivity*

When using green technology to synthesize nanoparticles, pH is a significant component. The size and texture of the produced nanoparticle are influenced by the pH of the solution medium, according to research (Armendariz et al. 2004, Gardea-Torresdey et al. n.d.). Consequently, by adjusting the pH of the solution media, green-synthesized nanoparticle size can be adjusted.

A study (Prakash and Soni 2012) showed the impact of pH on the size and structure of the produced silver nanoparticle. The generation of NPs is greatly influenced by the reaction's pH, which controls how nucleation centers are formed. The number of nucleation centers increases along with the pH, which enhances the creation of metallic NPs through green synthesis. It has been demonstrated that pH has a significant effect in determining the size and shape of the NPs. At various pH levels, the production of Au NPs from *A. sativa* was investigated by Armendariz et al. (2004). The size of the NPs was significantly greater at lower pH (pH 2), which led researchers to hypothesize that since Au NPs do not create as many nucleation centers at lower pH, they aggregate to produce larger-sized NPs (Armendariz et al. 2004)

6.2 Temperature

Worldwide, a substantial amount of research is being done to examine how NPs regulate their temperature. One of the most important variables that affects the size and morphology of the NPs, as well as their rate of synthesis, is temperature.

Temperature affects the size, synthesis, and shapes of different NPs (such as triangles, octahedral platelets, spheres, and rod-like structures). The nucleation center formation and the pace of reactions both increase with temperature (Sneha et al. 2010). The influence of temperature on the shape of the Au NPs produced from Piper betel leaf extract was examined by the research team led by Sneha et al. (2010). Triangular-shaped NPs were found to be produced at 20°C, whereas octahedral-shaped NPs were produced at temperatures between 30°C and 40°C. Moreover, the sizes and shapes of the NPs were more uniform and spherical at higher temperatures (50–60°C). The majority of the time, ambient temperature or temperatures lower than 100°C are needed for the green technology-based synthesis of nanoparticles where the type of nanoparticle that forms depends on the reaction medium's temperature (Rai et al. 2006).

6.3 Pressure

For nanoparticle production, pressure is crucial. The size and shape of the produced nanoparticles are influenced by the pressure that is applied to the reaction media (Abhilash and Pandey 2012). At ambient pressure circumstances, it has been observed that the rate of metal ion reduction by biological agents is significantly faster (Tran et al. 2013)

6.4 Time

One of the key variables that affects the morphology of NPs, alongside temperature and pH, is reaction time. The amount of time the reaction medium is incubated has a significant impact on the type and quality of nanoparticles that can be manufactured utilizing green technology (Darroudi et al. 2011).

In a similar way, the synthesis procedure, exposure to light, storage conditions, and other factors all had a significant impact on how the features of the generated nanoparticles changed over time (Kuchibhatla et al. 2012, Mudunkotuwa et al. 2012). Particles may aggregate due to long-term storage; they may contract or expand during that time, they may have a shelf life, and so forth, all of which impact their potential (Baer 2011). The research examined and clarified how reaction time affected the size, reduction, and stability of Au NPs produced from oil palm extract (Ahmad et al. 2016). It was observed that when reaction time was increased, the rate of reduction for AgNP generation increased.

7. Stability and Reusability of Green-Synthesized Nanoparticles

The biogenic synthesis of nanoparticles (NPs) involves the use of biomolecules from plants or microbes to reduce metal ions formed in a salt solution. The reduction of a metal ion to a base metal can be carried out at room temperature and pressure quite quickly and conveniently. Several commercial goods and manufacturing processes call for the stability of nanoparticles in concentrated aqueous solutions. The most frequent method for stabilizing nanoparticles is the adsorption of a dispersion layer onto the particle surface. The stabilization of suspensions with high nanoparticle concentrations depends on the creation of a dispersion layer (adlayer) of proper thickness. While narrow adlayers cause particle aggregation, thick adlayers produce an enormous excluded volume around the particles. The maximum concentration of nanoparticles in the suspension is decreased by both effects. Traditional dispersants, however, do not enable systematic control of the adlayer thickness on the particle surface. In a study (Sadeghi and Gholamhoseinpoor 2015), the stability of bio-produced AgNPs using *Ziziphora tenuior* was assessed using a zeta potentiometer at various pH levels. The stability of the produced AgNPs in a wide pH range, from 6 to 12, was identified where the value of AgNP's zeta potential rose as pH increased. Ag-NPs that have been created by the use of leaf extract have strong negative zeta potential values, making them stable over a wide pH range. Another study (Gavamukulya et al. 2020) evaluated the stability of green-synthesized AgNPs, utilizing *Annona muricata* fruit extracts at varied pH and temperature levels. Addressing pH stability, it was apparent that the AgNPs maintained their characteristic absorption maxima between 410 nm and roughly 420 nm at all pH settings examined. This implies that AgNPs can be stable under various pH circumstances without losing their potency, which is essential. According to the findings regarding temperature stability, AgNPs maintained their distinctive absorption maximum of 420–430 nm, which is within the range of AgNPs at all temperatures examined. This is crucial because it suggests that green-synthesized AgNPs can be stable at different temperatures and levels of heating without compromising their efficacy (Gavamukulya et al. 2020).

The reusability of NPs, frequent recycling, and adsorbent stability are vitally important for commercial success and economic considerations. In terms of regeneration and reusability, the application prospective of nanoparticles in pollutant removal cannot be disregarded because these elements are significant in the cost-benefit analysis of the nano-based elements. According to reports, green-synthesized NPs can be recycled and used repeatedly for a variety of applications, like as antibacterial agents (M et al. 2012) or as catalysts in chemical reactions. According to a study, the heterogeneous, green-synthesized TiO_2 NPs being used as catalysts can be easily extracted by centrifugation once the reaction finishes. The TiO_2 NPs that had been filtered were cleaned, dried, and reused for a subsequent reaction run with the standard reaction conditions. The outcomes showed that the catalytic adequacy was not considerably impacted. When the retrieved NPs were employed four more times, the conversion and selectivity remained unchanged. The confirmation was more compact when TEM analysis of the recoverable catalyst revealed that there had been barely any size change conferring its reusability properties.

8. Environmental Remediation Applications

Engineered nanomaterials from biological sources are used in nano-remediation, which is less expensive and more efficient than the majority of conventional approaches. In addition to being economical, using nanomaterials for environmental remediation is gaining popularity. Green nanotechnology can create new methods or goods for the production of renewable energy, the creation of nanoscale sensors, and the monitoring of pollutants.

8.1 Antimicrobial Activity

One of the most urgent challenges of recent years is antibiotic resistance, as the prevalence of bacteria with multidrug resistance has increased the incidence of infectious diseases, which are now the leading cause of death worldwide (Havelaar et al. 2013). The fast development of the bacterial genome has resulted in bacteria developing resistance to antimicrobial substances. As a result, the creation of a novel, all-natural antibacterial agent is required due to the rising threat posed by foodborne microorganisms that are multidrug resistant. Biogenic NPs have thus demonstrated positive results in the treatment of multidrug-resistant bacteria and may represent a possible option in the struggle against such resistant pathophysiology in the search for a novel medication (Nadeem et al. 2018).

There have been several postulated mechanisms for NP antibacterial activity, including breakdown of cell walls, cell membrane disintegration, enormous free radical production, particular (targeted) and/or specific activities against proteins, essential enzyme inhibition, DNA fragmentation, disruption in electron transport, and loss of cellular fluids (Havelaar et al. 2013). A wide variety of human pathogenic strains were successfully combated by AgNPs made from the leaf extract of *Clerodendrum inerme* (Khan et al. 2020). A range of human pathogenic bacteria was successfully inhibited by green AgNPs made from a leaf extract of *Carissa carandas*, particularly *Shigella flexineri* (Gram-negative), which causes shigellosis, being more likely to be inhibited (Singh et al. 2021). The growth of *Escherichia coli*, a Gram-negative bacteria, can also be inhibited by bimetallic nanostructures covered with reduced graphene oxide produced from a stevia leaf extract, such as Pd-Ag nanostructures (Mallikarjuna et al. 2021). As a result of the synergistic action of physiologically active phytochemicals from the plant *C. inerme* extract, AuNPs generated from this also show very similar inhibitory capacity (Havelaar et al. 2013). Antibiofilm activity of AuNPs, made from a seed extract, against some bacteria was observed, which was most likely as a result of intracellular ROS generation (Perveen et al. 2021). Direct cell contact, which damaged bacterial cell integrity, caused ZnONPs, produced from a *C. auriculata* leaf extract, to demonstrate antibacterial activity (Prasad et al. 2020). A substantial amount of antimicrobial activity, including antibiofilm capabilities, can be seen in other metallic NPs, such as CuONPs made from *Cymbopogon citratus* (Cherian et al. 2020). This demonstrates that not only does the type of NPs created affect antimicrobial action but also the makeup of the coated phytochemicals on their surfaces, which is regulated by the plant extract utilized to make NPs.

8.2 Catalytic Activity

Nanoparticles have emerged as promising materials for catalytic environmental remediation due to their unique physicochemical properties. The high surface area-to-volume ratio of nanoparticles provides a large number of active sites for catalytic reactions, which enhances their efficiency compared to conventional catalysts. Metallic NPs accelerate the rate of reaction by boosting reactant adsorption on their surface, lowering activation energy barriers (Gangula et al. 2011, Sarina et al. 2014). Additionally, the size and shape of nanoparticles can be tuned to optimize their catalytic activity and selectivity for specific reactions.

One such example of the catalytic activity of nanoparticles in environmental remediation is the degradation of organic pollutants in wastewater. Nanoparticles such as titanium dioxide (TiO_2), iron oxide (Fe_2O_3), and zinc oxide (ZnO) have been used as photocatalysts for the degradation of organic compounds (Nagaveni et al. 2004, Xin et al. 2007). When these nanoparticles are irradiated with UV light, they generate electron-hole pairs that can react with organic pollutants in water to break them down into harmless products.

Nanoparticles have also been used as catalysts for the reduction of harmful gases such as carbon monoxide (CO) and nitrogen oxides (NOx) (Campelo et al. 2009, Joo et al. 2010). For example, palladium (Pd) nanoparticles have been shown to be effective catalysts for the reduction of CO and NOx in vehicle exhaust (Bonet et al. 2002).

Overall, nanoparticles have shown great potential as catalysts for environmental remediation due to their unique properties and tunability. However, further research is needed to understand the mechanisms behind their catalytic activity fully and to optimize their performance for specific applications.

8.3 Pollutant Dye Removal

Ionic dyes make up most of the organic pollutants. There is a huge demand for dyes as they are used in almost every industry, including textile, paper, plastic, leather, food, and pharmaceuticals. In the textile and leather industry, the maximum amount of dyes are used up. About 60% of dyes are required during the tanning and coloring of the fabric. After processing the fabric, a huge amount of dyes goes to waste directly into our atmosphere, polluting soil, water, and life forms that come in contact with it (Padhi 2012). These pollutants are the main cause of ecological pollution. They cause the water bodies to turn turbid, which prevents sunrays penetration, blocking photochemical synthesis. This affects aquatic and marine life severely. Therefore, managing pollutant effluent dyes from the environment is critical (Jyoti and Singh 2016).

Metal oxide nanoparticles, such as ZnO, CuO, SnO, and TiO_2, are sometimes preferred for the photocatalytic activity of synthetic dyes. These nano-photocatalysts are advantageous due to their high surface area-to-mass ratio, which improves organic pollutant adsorption. Because of the enormous number of surface reactive sites available on the nanoparticle surfaces, the surface energy of the nanoparticles increases. This increases the rate of pollutant elimination at low concentrations.

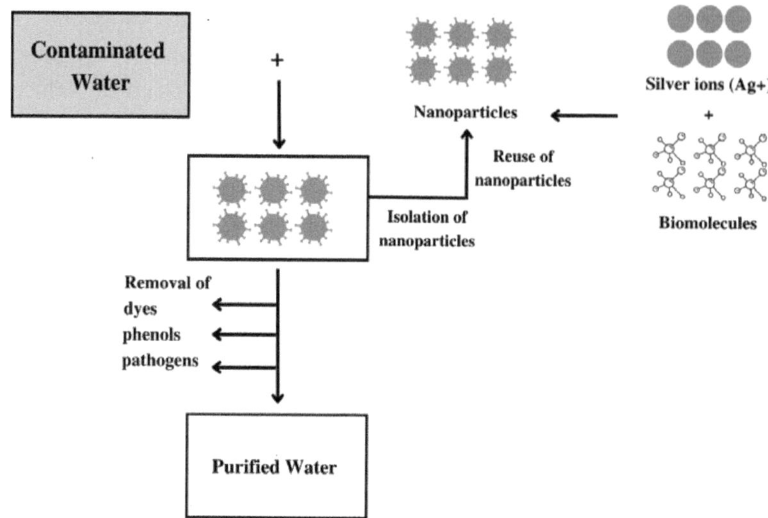

Figure 4. Removal of water contaminants by using green synthesized nanoparticles from the effluent stream (Khatoon et al. 2017).

Hence, fewer nanocatalysts will be required to remediate dirty water relative to the bulk material (Dror et al. 2005, Tsuda and Konduru 2016). Like metal oxide nanoparticles, metal nanoparticles exhibit improved photocatalytic degradation of several pollutant dyes; for example, silver nanoparticles prepared from *Z. armatum* leaf extract were used for pollutant dye degradation (Jyoti and Singh 2016) (refer to Figure 4).

8.4 Heavy Metal Ion Sensing

Heavy metal contamination is an everlasting problem that has persisted throughout scientific history, giving rise to worldwide water shortage and drinking water contamination. Green nanotechnology is in hopes of bringing a solution to this persistent issue (Palit and Hussain 2018). Heavy metals like Ni, Cu, Pb, Cr, Hg, and Cd are known to pollute the air, soil, and water. Heavy metals arise from a variety of sources like leather, dye, paper, plastic, and coal industries (Zhang et al. 2012). Mining waste, natural gas, and vehicular emissions are also leading causes of heavy metal pollution. Some of the heavy metals show acute toxicity even at very minute amounts (Que et al. 2008). Therefore, bioremediation of these hazardous metals has become crucial for the biological and aquatic environment (Aragay et al. 2011). The traditional techniques based on instrument systems are extremely sensitive for multi-element analysis, but the experimental setups required to conduct such assessments are extremely costly, time-consuming, skill-requiring, and non-portable (Singh et al. 2018). Metallic nanoparticles have been used to detect heavy metal ions in polluted water systems due to their variable size and distance-dependent optical characteristics (Annadhasan et al. 2014). The benefits of utilizing metal nanoparticles as colorimetric sensors for heavy metal ions in environmental samples are due to their simplicity, low cost, and great sensitivity, even at sub-ppm levels. Studies show that Ag nanoparticles synthesized from plant extract can be employed for colorimetric sensing of various heavy metal ions like zinc, cadmium, chromium, calcium, mercury (Zn^{2+}, Cd^{2+}, Cr^{3+}, Ca^{2+}, and Hg^{2+}), and in water. Moreover, other studies showed that AgNPs extracted from green tea and pepper seed were selectively sensitive toward zinc, lead, and mercury ions (Zn^{2+}, Pb^{2+}, and Hg^{2+}) (Karthiga and Anthony 2013).

8.5 *Wastewater Treatment*

Wastewater treatment is a process that involves the removal of pollutants and contaminants from water before it is released back into the environment. Environmental remediation using NPs in wastewater treatment is a relatively new area of research, but it shows a lot of potential for improving the efficiency and effectiveness of wastewater treatment. The most commonly used nanoparticles in environmental remediation are metallic nanoparticles, such as silver, copper, and iron nanoparticles, and metal oxide nanoparticles, such as titanium dioxide and zinc oxide NPs.

Nanoparticles can be used to eliminate pollutants from wastewater through a variety of mechanisms, such as adsorption, catalysis, and photocatalysis (Anjum et al. 2019). Adsorption involves the physical attraction and binding of pollutants to the surface of nanoparticles, which can be increased by manipulating their surface area and charge. Catalysis involves the use of NPs to promote chemical reactions that break down pollutants into less harmful substances, and this process can be optimized by adjusting the size and shape of the nanoparticles. Photocatalysis is a light-activated form of catalysis that generates reactive oxygen species to break down pollutants. NPs have a range of applications in wastewater treatment, such as removing organic and inorganic pollutants, disinfecting pathogens, and eliminating heavy metals (Ghasemzadeh et al. 2014). For instance, metal oxide NPs can remove organic pollutants via adsorption or photocatalysis, while metallic NPs can remove inorganic pollutants through adsorption or catalysis. Silver NPs can disinfect wastewater by eliminating pathogens such as bacteria and viruses.

Nanoparticles have shown great potential for improving the efficiency and effectiveness of wastewater treatment. They can remove a wide range of pollutants through a variety of mechanisms, and their properties can be tailored to optimize their performance. While there are still some challenges to overcome, such as the potential toxicity of nanoparticles, environmental remediation using NPs in wastewater treatment is a promising area of research that could help to address some of the world's most pressing environmental challenges.

9. Conclusion and Future Prospective

The field of green chemistry has been dedicated to creating 'green' nanomaterials that reduce environmental and health risks during their production and disposal. However, being labeled as 'green' does not necessarily mean they are sustainable, so the focus should shift toward "sustainable synthesis." Although biocompatible nanomaterials created through green methods have shown promise in various environmental applications, there are still risks associated with exposure and the need for more sustainable synthesis and analysis techniques. To ensure the long-term sustainability of nanomaterial production, a sustainable synthesis that considers technical, environmental, economic, and social factors is necessary. Biocompatible nanomaterials produced through sustainable methods offer significant advantages over traditional methods, such as reducing environmental impact, minimizing harm to living organisms and the environment, and eliminating toxic reagents and hazardous by-products. Continuing research in this area is crucial to overcoming challenges, improving methods and processes, and making sustainable synthesis the standard in nanomaterial production. Integrating sustainable synthesis into scientific research and mass production is essential to provide a viable solution to environmental challenges while ensuring long-term sustainability. Green nanotechnology is a promising field for research and academic integrity and developing novel green route strategies based on biogenic synthesis offers many opportunities. However, before applying nanomaterials to desorbed pollutants, it is crucial to evaluate the long-term toxicity of green-synthesized nanomaterials to humans, as they may exhibit unexpected behavior if released into the environment. Thus, industries like environmental remediation, food, cosmetics, and pharmaceuticals will require a greener approach based on bio-derived materials or nanomaterials. In fact, for effective pollutant removal and NP magnetic separability, efforts must be made to design the NP shape and saturation magnetization value by optimizing the synthesis procedure. Moreover, biomaterials derived from marine plants and algae occurring in certain places are areas that still remain hugely unexplored. Therefore, while green-synthesized biocompatible nanomaterials offer potential solutions to environmental challenges, sustainable synthesis is the ultimate goal for the future of nanomaterial production. Addressing current challenges and future opportunities while considering all necessary factors is necessary to ensure long-term sustainability. By doing so, we can provide a promising future for green nanotechnology while protecting human health and the environment.

References

Abhilash and Pandey, B. D. (2012). Synthesis of zinc-based nanomaterials: A biological perspective. *IET Nanobiotechnology*, 6(4): 144–148. https://doi.org/10.1049/iet-nbt.2011.0051.

Abid, N., Khan, A. M., Shujait, S., Chaudhary, K., Ikram, M., Imran, M., Haider, J., Khan, M., Khan, Q. and Maqbool, M. (2022). Synthesis of nanomaterials using various top-down and bottom-up approaches, influencing factors, advantages, and disadvantages: A review. *Advances in Colloid and Interface Science*, 300. https://doi.org/10.1016/j.cis.2021.102597.

Agarwal, P., Gupta, R. and Agarwal, N. (2019). Advances in synthesis and applications of microalgal nanoparticles for wastewater treatment. *Journal of Nanotechnology*, 2019. https://doi.org/10.1155/2019/7392713.

Agnihotri, M., Joshi, S., Kumar, A. R., Zinjarde, S. and Kulkarni, S. (2009). Biosynthesis of gold nanoparticles by the tropical marine yeast Yarrowia lipolytica NCIM 3589. *Materials Letters*, 63(15): 1231–1234. https://doi.org/10.1016/j.matlet.2009.02.042.

Ahmad, A., Senapati, S., Khan, M. I., Kumar, R., Ramani, R., Srinivas, V. and Sastry, M. (2003). Intracellular synthesis of gold nanoparticles by a novel alkalotolerant actinomycete, Rhodococcus species. *Nanotechnology*, 14(7): 824–828. https://doi.org/10.1088/0957-4484/14/7/323.

Ahmad, F., Ashraf, N., Ashraf, T., Zhou, R. bin and Yin, D. C. (2019). Biological synthesis of metallic nanoparticles (MNPs) by plants and microbes: their cellular uptake, biocompatibility, and biomedical applications. *Applied Microbiology and Biotechnology*, 103(7): 2913–2935. https://doi.org/10.1007/s00253-019-09675-5.

Ahmad, T., Irfan, M., Bustam, M. A. and Bhattacharjee, S. (2016). Effect of reaction time on green synthesis of gold nanoparticles by using aqueous extract of elaise guineensis (Oil Palm Leaves). *Procedia Engineering*, 148: 467–472. https://doi.org/10.1016/j.proeng.2016.06.465.

Alani, F., Moo-Young, M. and Anderson, W. (2012). Biosynthesis of silver nanoparticles by a new strain of Streptomyces sp. compared with Aspergillus fumigatus. *World Journal of Microbiology and Biotechnology*, 28(3): 1081–1086. https://doi.org/10.1007/s11274-011-0906-0.

Alsamhary, K. I. (2020). Eco-friendly synthesis of silver nanoparticles by Bacillus subtilis and their antibacterial activity. *Saudi Journal of Biological Sciences*, 27(8): 2185–2191. https://doi.org/10.1016/j.sjbs.2020.04.026.

Amendola, V. and Meneghetti, M. (2009). Size evaluation of gold nanoparticles by UV-vis spectroscopy. *Journal of Physical Chemistry C*, 113(11): 4277–4285. https://doi.org/10.1021/jp8082425.

Anjum, M., Miandad, R., Waqas, M., Gehany, F. and Barakat, M. A. (2019). Remediation of wastewater using various nano-materials. *Arabian Journal of Chemistry*, 12(8): 4897–4919. https://doi.org/10.1016/j.arabjc.2016.10.004.

Annadhasan, M., Muthukumarasamyvel, T., Sankar Babu, V. R. and Rajendiran, N. (2014). Green synthesized silver and gold nanoparticles for colorimetric detection of Hg2+, Pb2+, and Mn2+ in aqueous medium. *ACS Sustainable Chemistry and Engineering*, 2(4): 887–896. https://doi.org/10.1021/sc400500z.

Aragay, G., Pons, J. and Merkoçi, A. (2011). Recent trends in macro-, micro-, and nanomaterial-based tools and strategies for heavy-metal detection. *Chemical Reviews*, 111(5): 3433–3458. https://doi.org/10.1021/cr100383r.

Armendariz, V., Herrera, I., Peralta-Videa, J. R., Jose-Yacaman, M., Troiani, H., Santiago, P. and Gardea-Torresdey, J. L. (2004). Size controlled gold nanoparticle formation by Avena sativa biomass: Use of plants in nanobiotechnology. *Journal of Nanoparticle Research*, 6(4): 377–382. https://doi.org/10.1007/s11051-004-0741-4.

Attarad, A., Hira, Z., Muhammad, Z., Ihsan ul, H., Abdul Rehman, P., Joham, S. A. and Altaf, H. (2016). Synthesis, characterization, applications, and challenges of iron oxide nanoparticles. *Nanotechnology, Science and Applications*, 9: 49–67.

Baer, D. R. (2011). Surface characterization of nanoparticles: Critical needs and significant challenges. *Journal of Surface Analysis (Online)*, 17(3): 163–169. https://doi.org/10.1384/jsa.17.163.

Bae, W. and Chen, X. (2004). Proteomic study for the cellular responses to Cd2+ in Schizosaccharomyces pombe through amino acid-coded mass tagging and liquid chromatography tandem mass spectrometry. *Molecular and Cellular Proteomics*, 3(6): 596–607. https://doi.org/10.1074/mcp.M300122-MCP200.

Bhainsa, K. C. and D'Souza, S. F. (2006a). Extracellular biosynthesis of silver nanoparticles using the fungus Aspergillus fumigatus. *Colloids and Surfaces B: Biointerfaces*, 47(2): 160–164. https://doi.org/10.1016/j.colsurfb.2005.11.026.

Bhainsa, K. C. and D'Souza, S. F. (2006b). Extracellular biosynthesis of silver nanoparticles using the fungus Aspergillus fumigatus. *Colloids and Surfaces B: Biointerfaces*, 47(2): 160–164. https://doi.org/10.1016/j.colsurfb.2005.11.026.

Bhattacharya, D. and Gupta, R. K. (2005). Nanotechnology and potential of microorganisms. *Critical Reviews in Biotechnology*, 25(4): 199–204. https://doi.org/10.1080/07388550500361994.

Birla, S. S., Tiwari, V. v., Gade, A. K., Ingle, A. P., Yadav, A. P. and Rai, M. K. (2009). Fabrication of silver nanoparticles by Phoma glomerata and its combined effect against Escherichia coli, Pseudomonas aeruginosa and Staphylococcus aureus. *Letters in Applied Microbiology*, 48(2): 173–179. https://doi.org/10.1111/j.1472-765X.2008.02510.x.

Bonet, F., Grugeon, S., Urbina, R. H., Tekaia-Elhsissen, K. and Tarascon, J. M. (2002). *In situ* deposition of silver and palladium nanoparticles prepared by the polyol process, and their performance as catalytic converters of automobile exhaust gases. *Solid State Sciences*, 4(5): 665–670. https://doi.org/10.1016/S1293-2558(02)01311-0.

Bunaciu, A. A., UdriŞTioiu, E. G. and Aboul-Enein, H. Y. (n.d.). *X-ray diffraction: Instrumentation and applications.*

Campelo, J. M., Luna, D., Luque, R., Marinas, J. M. and Romero, A. A. (2009). Sustainable preparation of supported metal nanoparticles and their applications in catalysis. *ChemSusChem,* 2(1): 18–45. https://doi.org/10.1002/cssc.200800227.

Chandran, S. P., Chaudhary, M., Pasricha, R., Ahmad, A. and Sastry, M. (2006). Synthesis of gold nanotriangles and silver nanoparticles using Aloe vera plant extract. *Biotechnology Progress,* 22(2): 577–583. https://doi.org/10.1021/bp0501423.

Chaudhary, R., Nawaz, K., Khan, A. K., Hano, C., Abbasi, B. H. and Anjum, S. (2020). An overview of the algae-mediated biosynthesis of nanoparticles and their biomedical applications. *Biomolecules,* 10(11): 1–35. https://doi.org/10.3390/biom10111498.

Cherian, T., Ali, K., Saquib, Q., Faisal, M., Wahab, R., and Musarrat, J. (2020). Cymbopogon citratus functionalized green synthesis of cuo-nanoparticles: Novel prospects as antibacterial and antibiofilm agents. *Biomolecules, 10*(2). https://doi.org/10.3390/biom10020169.

Chugh, D., Viswamalya, V. S. and Das, B. (2021). Green synthesis of silver nanoparticles with algae and the importance of capping agents in the process. *Journal of Genetic Engineering and Biotechnology,* 19(1). https://doi.org/10.1186/s43141-021-00228-w.

Darroudi, M., Ahmad, M. bin, Zamiri, R., Zak, A. K., Abdullah, A. H. and Ibrahim, N. A. (2011). Time-dependent effect in green synthesis of silver nanoparticles. *International Journal of Nanomedicine,* 6(1): 677–681. https://doi.org/10.2147/IJN.S17669.

Deplanche, K. and Macaskie, L. E. (2008). Biorecovery of gold by Escherichia coli and Desulfovibrio desulfuricans. *Biotechnology and Bioengineering,* 99(5): 1055–1064. https://doi.org/10.1002/bit.21688.

Devatha, C. P. and Thalla, A. K. (2018). Green synthesis of nanomaterials. *Synthesis of Inorganic Nanomaterials: Advances and Key Technologies,* 169–184. https://doi.org/10.1016/B978-0-08-101975-7.00007-5.

Dipankar, C. and Murugan, S. (2012). The green synthesis, characterization and evaluation of the biological activities of silver nanoparticles synthesized from Iresine herbstii leaf aqueous extracts. *Colloids and Surfaces B: Biointerfaces,* 98: 112–119. https://doi.org/10.1016/j.colsurfb.2012.04.006.

Dror, I., Baram, D. and Berkowitz, B. (2005). Use of nanosized catalysts for transformation of chloro-organic pollutants. *Environmental Science and Technology,* 39(5): 1283–1290. https://doi.org/10.1021/es0490222.

Durán, N., Durán, M., de Jesus, M. B., Seabra, A. B., Fávaro, W. J. and Nakazato, G. (2016). Silver nanoparticles: A new view on mechanistic aspects on antimicrobial activity. *Nanomedicine: Nanotechnology, Biology, and Medicine,* 12(3): 789–799. https://doi.org/10.1016/j.nano.2015.11.016.

Ealias, A. M. and Saravanakumar, M. P. (2017). A review on the classification, characterisation, synthesis of nanoparticles and their application. *IOP Conference Series: Materials Science and Engineering,* 263(3). https://doi.org/10.1088/1757-899X/263/3/032019.

El-Rafie, H. M., El-Rafie, M. H. and Zahran, M. K. (2013). Green synthesis of silver nanoparticles using polysaccharides extracted from marine macro algae. *Carbohydrate Polymers,* 96(2): 403–410. https://doi.org/10.1016/j.carbpol.2013.03.071.

F, H. C. and E, S. R. (2019). Tutorial on powder x-ray diffraction for characterizing nanoscale materials. *ACS Nano,* 13: 7359–7365.

Ganesh Babu, M. M. and Gunasekaran, P. (2009). Production and structural characterization of crystalline silver nanoparticles from Bacillus cereus isolate. *Colloids and Surfaces B: Biointerfaces,* 74(1): 191–195. https://doi.org/10.1016/j.colsurfb.2009.07.016.

Gangula, A., Podila, R., M, R., Karanam, L., Janardhana, C. and Rao, A. M. (2011). Catalytic reduction of 4-nitrophenol using biogenic gold and silver nanoparticles derived from breynia rhamnoides. *Langmuir,* 27(24): 15268–15274. https://doi.org/10.1021/la2034559.

Gardea-Torresdey, J. L., Tiemann, K. J., Gamez, G., Dokken, K. and N. E. P. (n.d.). *Recovery of gold (III) by alfalfa biomass and binding characterization using X-ray microfluorescence.*

Gavamukulya, Y., Maina, E. N., Meroka, A. M., Madivoli, E. S., El-Shemy, H. A., Wamunyokoli, F. and Magoma, G. (2020). Green synthesis and characterization of highly stable silver nanoparticles from ethanolic extracts of fruits of Annona muricata. *Journal of Inorganic and Organometallic Polymers and Materials,* 30(4): 1231–1242. https://doi.org/10.1007/s10904-019-01262-5.

Ghasemzadeh, G., Momenpour, M., Omidi, F., Hosseini, M. R., Ahani, M. and Barzegari, A. (2014). Applications of nanomaterials in water treatment and environmental remediation. *Frontiers of Environmental Science and Engineering*, 8(4): 471–482. https://doi.org/10.1007/s11783-014-0654-0.

Gour, A. and Jain, N. K. (2019). Advances in green synthesis of nanoparticles. *Artificial Cells, Nanomedicine and Biotechnology*, 47(1): 844–851. https://doi.org/10.1080/21691401.2019.1577878.

Hamida, R. S., Ali, M. A., Redhwan, A. and Bin-Meferij, M. M. (2020). Cyanobacteria—A promising platform in green nanotechnology: A review on nanoparticles fabrication and their prospective applications. *International Journal of Nanomedicine*, 15: 6033–6066. https://doi.org/10.2147/IJN.S256134.

Hano, C. and Abbasi, B. H. (2022). Plant-based green synthesis of nanoparticles: Production, characterization and applications. *Biomolecules*, 12(1). https://doi.org/10.3390/biom12010031.

Havelaar, A. H., Cawthorne, A., Angulo, F., Bellinger, D., Corrigan, T., Cravioto, A., Gibb, H., Hald, T., Ehiri, J., Kirk, M., Lake, R., Praet, N., Speybroeck, N., de Silva, N., Stein, C., Torgerson, P. and Kuchenmüller, T. (2013). WHO initiative to estimate the global burden of foodborne diseases. *The Lancet*, 381: S59. https://doi.org/10.1016/s0140-6736(13)61313-6.

Husen, A. and Siddiqi, K. S. (2014). Plants and microbes assisted selenium nanoparticles: Characterization and application. *Journal of Nanobiotechnology*, 12(1). https://doi.org/10.1186/s12951-014-0028-6.

Ingle, A., Gade, A., Pierrat, S., Sonnichsen, C. and Rai, M. (2008). Mycosynthesis of silver nanoparticles using the fungus fusarium acuminatum and its activity against some human pathogenic bacteria. *Current Nanoscience*, 4(2): 141–144. https://doi.org/10.2174/157341308784340804.

Pokropivny, V., Lohmus, R., Hussainova, I., Pokropivny, A. and Vlassov, S. (2007). *Introduction to Nanomaterials and Nanotechnology* (pp. 45–100). Ukraine: Tartu University Press.

Jeevanandam, J., Chan, Y. S. and Danquah, M. K. (2016). Biosynthesis of metal and metal oxide nanoparticles. *ChemBioEng Reviews*, 3(2): 55–67. https://doi.org/10.1002/cben.201500018.

Jiang, Z., Li, L., Huang, H., He, W. and Ming, W. (2022). Progress in laser ablation and biological synthesis processes: "top-down" and "bottom-up" approaches for the green synthesis of Au/Ag nanoparticles. *International Journal of Molecular Sciences*, 23(23). https://doi.org/10.3390/ijms232314658.

Joo, S. H., Park, J. Y., Renzas, J. R., Butcher, D. R., Huang, W. and Somorjai, G. A. (2010). Size effect of ruthenium nanoparticles in catalytic carbon monoxide oxidation. *Nano Letters*, 10(7): 2709–2713. https://doi.org/10.1021/nl101700j.

Jyoti, K. and Singh, A. (2016). Green synthesis of nanostructured silver particles and their catalytic application in dye degradation. *Journal of Genetic Engineering and Biotechnology*, 14(2): 311–317. https://doi.org/10.1016/j.jgeb.2016.09.005.

Kapoor, R. T., Salvadori, M. R., Rafatullah, M., Siddiqui, M. R., Khan, M. A. and Alshareef, S. A. (2021). Exploration of microbial factories for synthesis of nanoparticles – A sustainable approach for bioremediation of environmental contaminants. *Frontiers in Microbiology*, 12. https://doi.org/10.3389/fmicb.2021.658294.

Karthiga, D. and Anthony, S. P. (2013). Selective colorimetric sensing of toxic metal cations by green synthesized silver nanoparticles over a wide pH range. *RSC Advances*, 3(37): 16765–16774. https://doi.org/10.1039/c3ra42308e.

Kathiraven, T., Sundaramanickam, A., Shanmugam, N. and Balasubramanian, T. (2015). Green synthesis of silver nanoparticles using marine algae Caulerpa racemosa and their antibacterial activity against some human pathogens. *Applied Nanoscience (Switzerland)*, 5(4): 499–504. https://doi.org/10.1007/s13204-014-0341-2.

Khan, S. A., Shahid, S. and Lee, C. S. (2020). Green synthesis of gold and silver nanoparticles using leaf extract of clerodendrum inerme; characterization, antimicrobial, and antioxidant activities. *Biomolecules*, 10(6). https://doi.org/10.3390/biom10060835.

Khatoon, N., Mazumder, J. A. and Sardar, M. (2017). Biotechnological applications of green synthesized silver nanoparticles. *Journal of Nanosciences: Current Research*, 02(01). https://doi.org/10.4172/2572-0813.1000107.

Korbekandi, H., Iravani, S. and Abbasi, S. (2012). Optimization of biological synthesis of silver nanoparticles using Lactobacillus casei subsp. casei. *Journal of Chemical Technology and Biotechnology*, 87(7): 932–937. https://doi.org/10.1002/jctb.3702.

Kowshik, M., Deshmukh, N., Vogel, W., Urban, J., Kulkarni, S. K. and Paknikar, K. M. (2002). Microbial synthesis of semiconductor CdS nanoparticles, their characterization, and their use in the fabrication of an ideal diode. *Biotechnology and Bioengineering*, 78(5):, 583–588. https://doi.org/10.1002/bit.10233.

Kowshik, M., Vogel, W., Urban, J., Kulkarni, S. K. and Paknikar, K. M. (2002). Microbial synthesis of semiconductor PbS nanocrystallites. *Advanced Materials*, 14(11): 815–818. https://doi.org/10.1002/1521-4095(20020605)14:11<815::AID-ADMA815>3.0.CO;2-K.

Kshtriya, V., Koshti, B. and Gour, N. (2021). Green synthesized nanoparticles: Classification, synthesis, characterization, and applications. *Comprehensive Analytical Chemistry*, 94: 173–222. https://doi.org/10.1016/bs.coac.2020.12.009.

Kuchibhatla, S. V. N. T., Karakoti, A. S., Baer, D. R., Samudrala, S., Engelhard, M. H., Amonette, J. E., Thevuthasan, S. and Seal, S. (2012). Influence of aging and environment on nanoparticle chemistry: Implication to confinement effects in nanoceria. *Journal of Physical Chemistry C*, 116(26): 14108–14114. https://doi.org/10.1021/jp300725s.

Kumar, P., Senthamil Selvi, S., Lakshmi Prabha, A., Prem Kumar, K., Ganeshkumar, R. S. and Govindaraju, M. (2012). Synthesis of silver nanoparticles from Sargassum tenerrimum and screening phytochemicals for its antibacterial activity. *Nano Biomedicine and Engineering*, 4(1): 12–16. https://doi.org/10.5101/nbe.v4i1.p12-16.

Kumar, S. A., Peter, Y. A. and Nadeau, J. L. (2008). Facile biosynthesis, separation and conjugation of gold nanoparticles to doxorubicin. *Nanotechnology*, 19(49). https://doi.org/10.1088/0957-4484/19/49/495101.

Kwok, K. and Chiu, W. K. S. (2005). Growth of carbon nanotubes by open-air laser-induced chemical vapor deposition. *Carbon*, 43(2): 437–446. https://doi.org/10.1016/j.carbon.2004.10.005.

Lagrow, A. P., Besenhard, M. O., Hodzic, A., Sergides, A., Bogart, L. K., Gavriilidis, A. and Thanh, N. T. K. (2019). Unravelling the growth mechanism of the co-precipitation of iron oxide nanoparticles with the aid of synchrotron X-Ray diffraction in solution. *Nanoscale*, 11(14): 6620–6628. https://doi.org/10.1039/c9nr00531e.

Lateef, A. and Adeeyo, A. O. (2015a). Green synthesis and antibacterial activities of silver nanoparticles using extracellular laccase of *Lentinus edodes*. *Notulae Scientia Biologicae*, 7(4): 405–411. https://doi.org/10.15835/nsb749643.

Lateef, A., Adelere, I. A., Gueguim-Kana, E. B., Asafa, T. B. and Beukes, L. S. (2015b). Green synthesis of silver nanoparticles using keratinase obtained from a strain of Bacillus safensis LAU 13. *International Nano Letters*, 5(1): 29–35. https://doi.org/10.1007/s40089-014-0133-4.

Letchumanan, D., Sok, S. P. M., Ibrahim, S., Nagoor, N. H. and Arshad, N. M. (2021). Plant-based biosynthesis of copper/copper oxide nanoparticles: An update on their applications in biomedicine, mechanisms, and toxicity. *Biomolecules*, 11(4). https://doi.org/10.3390/biom11040564.

Lin, J. Y., Hsueh, Y. L. and Huang, J. J. (2014). The concentration effect of capping agent for synthesis of silver nanowire by using the polyol method. *Journal of Solid State Chemistry*, 214: 2–6. https://doi.org/10.1016/j.jssc.2013.12.017.

Malassis, L., Dreyfus, R., Murphy, R. J., Hough, L. A., Donnio, B. and Murray, C. B. (2016). One-step green synthesis of gold and silver nanoparticles with ascorbic acid and their versatile surface post-functionalization. *RSC Advances*, 6(39): 33092–33100. https://doi.org/10.1039/c6ra00194g.

Mallikarjuna, K., Nasif, O., Alharbi, S. A., Chinni, S. v., Reddy, L. V., Reddy, M. R. V. and Sreeramanan, S. (2021). Phytogenic synthesis of Pd-Ag/rGO nanostructures using stevia leaf extract for photocatalytic H2 production and antibacterial studies. *Biomolecules*, 11(2): 1–15. https://doi.org/10.3390/biom11020190.

M, H., K, F., A, A., D, de A., I, de L., T, R., V, S., P, W. and M., M. (2012). Antibacterial properties of nanoparticles. *Trends in Biotechnology*, 30(10): 499–511.

Mohanpuria, P., Rana, N. K. and Yadav, S. K. (2008). Biosynthesis of nanoparticles: Technological concepts and future applications. *Journal of Nanoparticle Research*, 10(3): 507–517. https://doi.org/10.1007/s11051-007-9275-x.

Mohd Yusof, H., Mohamad, R., Zaidan, U. H. and Abdul Rahman, N. A. (2019). Microbial synthesis of zinc oxide nanoparticles and their potential application as an antimicrobial agent and a feed supplement in animal industry: A review. *Journal of Animal Science and Biotechnology*, 10(1). https://doi.org/10.1186/s40104-019-0368-z.

Mondal, S., Roy, N., Laskar, R. A., Sk, I., Basu, S., Mandal, D. and Begum, N. A. (2011). Biogenic synthesis of Ag, Au and bimetallic Au/Ag alloy nanoparticles using aqueous extract of mahogany (Swietenia mahogani JACQ.) leaves. *Colloids and Surfaces B: Biointerfaces*, 82(2): 497–504. https://doi.org/10.1016/j.colsurfb.2010.10.007.

Mourato, A., Gadanho, M., Lino, A. R. and Tenreiro, R. (2011). Biosynthesis of crystalline silver and gold nanoparticles by extremophilic yeasts. *Bioinorganic Chemistry and Applications*, 2011. https://doi.org/10.1155/2011/546074.

Mudunkotuwa, I. A., Pettibone, J. M. and Grassian, V. H. (2012). Environmental implications of nanoparticle aging in the processing and fate of copper-based nanomaterials. *Environmental Science and Technology*, 46(13): 7001–7010. https://doi.org/10.1021/es203851d.

Mukherjee, P., Ahmad, A., Mandal, D., Senapati, S., Sainkar, S. R., Khan, M. I., Parishcha, R., Ajaykumar, P. v., Alam, M., Kumar, R. and Sastry, M. (2001). Fungus-mediated synthesis of silver nanoparticles and their immobilization in the mycelial matrix: a novel biological approach to nanoparticle synthesis. *Nano Letters*, 1(10): 515–519. https://doi.org/10.1021/nl0155274.

Nadagouda, M. N. and Varma, R. S. (2008). Green synthesis of Ag and Pd nanospheres, nanowires, and nanorods using vitamin B2: Catalytic polymerisation of aniline and pyrrole. *Journal of Nanomaterials*, 2008(1). https://doi.org/10.1155/2008/782358.

Nadeem, M., Tungmunnithum, D., Hano, C., Abbasi, B. H., Hashmi, S. S., Ahmad, W. and Zahir, A. (2018). The current trends in the green syntheses of titanium oxide nanoparticles and their applications. *Green Chemistry Letters and Reviews*, 11(4): 492–502. https://doi.org/10.1080/1751 8253.2018.1538430.

Nagaveni, K., Sivalingam, G., Hegde, M. S. and Madras, G. (2004). Solar photocatalytic degradation of dyes: High activity of combustion synthesized nano TiO_2. *Applied Catalysis B: Environmental*, 48(2): 83–93. https://doi.org/10.1016/j.apcatb.2003.09.013.

Narayanan, K. B., and Sakthivel, N. (2010). Biological synthesis of metal nanoparticles by microbes. *Advances in Colloid and Interface Science*, 156(1–2), 1–13. https://doi.org/10.1016/j.cis.2010.02.001

Naveen, H. K., Kumar, G. and Rao, B. K. (2010). Extracellular biosynthesis of silver nanoparticles using the filamentous fungus Penicillium sp. *Archives of Applied Science Research*, 2(6): 161–167. www.scholarsresearchlibrary.com.

Okuyama, K. and Lenggoro, W. W. (2003). Preparation of nanoparticles via spray route. *Chemical Engineering Science*, 58(3-6): 537–547. https://doi.org/10.1016/S0009-2509(02)00578-X.

Ovais, M., Khalil, A. T., Ayaz, M., Ahmad, I., Nethi, S. K. and Mukherjee, S. (2018). Biosynthesis of metal nanoparticles via microbial enzymes: A mechanistic approach. *International Journal of Molecular Sciences*, 19(12). https://doi.org/10.3390/ijms19124100.

Oza, G., Pandey, S., Mewada, A., Kalita, G., Sharon, M., Phata, J. and Ambernath, W. (2012). Facile biosynthesis of gold nanoparticles exploiting optimum pH and temperature of fresh water algae Chlorella pyrenoidusa. *Pelagia Research Library*, 3(3): 1405–1412. www.pelagiaresearchlibrary.com.

Padhi, B. (2012). Pollution due to synthetic dyes toxicity and carcinogenicity studies and remediation. *International Journal of Environmental Sciences*, 3(3): 940–955. https://doi.org/10.6088/ijes.2012030133002.

Pal, G., Rai, P. and Pandey, A. (2018). Green synthesis of nanoparticles: A greener approach for a cleaner future. *Green Synthesis, Characterization and Applications of Nanoparticles*, 1–26. https://doi.org/10.1016/B978-0-08-102579-6.00001-0.

Palit, S. and Hussain, C. M. (2018). Recent advances in green nanotechnology and the vision for the future. *The Macabresque: Human Violation and Hate in Genocide, Mass Atrocity and Enemy-Making*, 1–21. https://doi.org/10.1002/9781119418900.ch1.

Pandey, B. and Fulekar, M. H. (2012). Nanotechnology: Remediation technologies to clean up the environmental pollutants. *Res. J. Chem. Sci.*, 2(2): 90–96. http://www.isca.in/rjcs/Archives/vol2/i2/15.pdf.

Park, B. K., Jeong, S., Kim, D., Moon, J., Lim, S. and Kim, J. S. (2007). Synthesis and size control of monodisperse copper nanoparticles by polyol method. *Journal of Colloid and Interface Science*, 311(2): 417–424. https://doi.org/10.1016/j.jcis.2007.03.039.

Parveen, K., Banse, V. and Ledwani, L. (2016). Green synthesis of nanoparticles: Their advantages and disadvantages. *AIP Conference Proceedings*, 1724. https://doi.org/10.1063/1.4945168.

Patel, J. K., Patel, A. and Bhatia, D. (2021). Introduction to nanomaterials and nanotechnology. *Emerging Technologies for Nanoparticle Manufacturing*, 3–23. https://doi.org/10.1007/978-3-030-50703-9_1.

Patel, V., Berthold, D., Puranik, P. and Gantar, M. (2015). Screening of cyanobacteria and microalgae for their ability to synthesize silver nanoparticles with antibacterial activity. *Biotechnology Reports*, 5(1): 112–119. https://doi.org/10.1016/j.btre.2014.12.001.

Perveen, K., Husain, F. M., Qais, F. A., Khan, A., Razak, S., Afsar, T., Alam, P., Almajwal, A. M. and Abulmeaty, M. M. A. (2021). Microwave-assisted rapid green synthesis of gold nanoparticles using seed extract of trachyspermum ammi: Ros mediated biofilm inhibition and anticancer activity. *Biomolecules*, 11(2): 1–16. https://doi.org/10.3390/biom11020197.

Prabhu, S. and Poulose, E. K. (2012). Silver nanoparticles: mechanism of antimicrobial action, synthesis, medical applications, and toxicity effects. *International Nano Letters*, 2(1). https://doi.org/10.1186/2228-5326-2-32.

Prakash, S. and Soni, N. (2012). Synthesis of gold nanoparticles by the fungus Aspergillus niger and its efficacy against mosquito larvae. *Reports in Parasitology*, 1. https://doi.org/10.2147/rip.s29033.

Prasad, K. S., Prasad, S. K., Ansari, M. A., Alzohairy, M. A., Alomary, M. N., Alyahya, S., Srinivasa, C., Murali, M., Ankegowda, V. M. and Shivamallu, C. (2020). Tumoricidal and bactericidal properties of znonps synthesized using cassia auriculata leaf extract. *Biomolecules*, 10(7): 1–14. https://doi.org/10.3390/biom10070982.

Priyabrata, M., Absar, A., Deendayal, M., Satyajyoti, S., R., S. S., I., K. M., R., R., Renu, P., V., A. P., Mansoor, A., Murali, S. and Rajiv, K. (2001). Bioreduction of AuCl4−Ions by the Fungus, Verticillium sp. and surface trapping of the gold nanoparticles formed. *Angewandte Chemie International Edition*, 40(19): 3585–3588.

Purohit, J., Chattopadhyay, A. and Singh, N. K. (2019). Green synthesis of microbial nanoparticle: approaches to application. *Nanotechnology in the Life Sciences*, 35–60. https://doi.org/10.1007/978-3-030-16534-5_3.

Que, E. L., Domaille, D. W. and Chang, C. J. (2008). Metals in neurobiology: Probing their chemistry and biology with molecular imaging. *Chemical Reviews*, 108(5): 1517–1549. https://doi.org/10.1021/cr078203u.

Rai, A., Singh, A., Ahmad, A. and Sastry, M. (2006). Role of halide ions and temperature on the morphology of biologically synthesized gold nanotriangles. *Langmuir*, 22(2), 736–741. https://doi.org/10.1021/la052055q.

Rajeshkumar, S., Kannan, C. and Annadurai, G. (2012). Green synthesis of silver nanoparticles using marine brown Algae turbinaria conoides and its antibacterial activity. *International Journal of Pharma and Bio Sciences*, 3(4): 502–510.

Ravindra, B. K. and Rajasab, A. H. (2014). A comparative study on biosynthesis of silver nanoparticles using four different fungal species. *International Journal of Pharmacy and Pharmaceutical Sciences*, 6(1): 372–376.

Rudakiya, D. M. and Gupte, A. (2019). Laccases in nano-biotechnology: Recent trends and advanced applications. *Recent Advances in Biotechnology*, 39–62.

Sadeghi, B. and Gholamhoseinpoor, F. (2015). A study on the stability and green synthesis of silver nanoparticles using Ziziphora tenuior (Zt) extract at room temperature. *Spectrochimica Acta - Part A: Molecular and Biomolecular Spectroscopy*, 134: 310–315. https://doi.org/10.1016/j.saa.2014.06.046.

Sanket, S. and Das, S. K. (2021). Role of enzymes in synthesis of nanoparticles. *Bioprospecting of Enzymes in Industry, Healthcare and Sustainable Environment*, 139–153. https://doi.org/10.1007/978-981-33-4195-1_7.

Saravanan, M., Barabadi, H., Vahidi, H., Webster, T. J., Medina-Cruz, D., Mostafavi, E., Vernet-Crua, A., Cholula-Diaz, J. L. and Periakaruppan, P. (2021). Emerging theranostic silver and gold nanobiomaterials for breast cancer: Present status and future prospects. *Handbook on Nanobiomaterials for Therapeutics and Diagnostic Applications*, 439–456. https://doi.org/10.1016/B978-0-12-821013-0.00004-0.

Sarina, S., Bai, S., Huang, Y., Chen, C., Jia, J., Jaatinen, E., Ayoko, G. A., Bao, Z. and Zhu, H. (2014). Visible light enhanced oxidant free dehydrogenation of aromatic alcohols using Au-Pd alloy nanoparticle catalysts. *Green Chemistry*, 16(1): 331–341. https://doi.org/10.1039/c3gc41866a.

Senapati, S. (2005). Biosynthesis and Immobilization of Nanoparticles and their Applications. *Ph.D. Thesis, July*, 1–171.

Shankar, S. S., Rai, A., Ahmad, A. and Sastry, M. (2004). Rapid synthesis of Au, Ag, and bimetallic Au core-Ag shell nanoparticles using Neem (Azadirachta indica) leaf broth. *Journal of Colloid and Interface Science*, 275(2): 496–502. https://doi.org/10.1016/j.jcis.2004.03.003.

Shankar, S. S., Rai, A., Ahmad, A. and Sastry, M. (2005). Controlling the optical properties of lemongrass extract synthesized gold nanotriangles and potential application in infrared-absorbing optical coatings. *Chemistry of Materials*, 17(3): 566–572. https://doi.org/10.1021/cm048292g.

Shivaji, S., Madhu, S. and Singh, S. (2011). Extracellular synthesis of antibacterial silver nanoparticles using psychrophilic bacteria. *Process Biochemistry*, 46(9): 1800–1807. https://doi.org/10.1016/j.procbio.2011.06.008.

Shiying He, Zhirui Guo, Yu Zhang, Song Zhang, Jing Wang and Gu, N. (2007). Biosynthesis of gold nanoparticles using the bacteria Rhodopseudomonas capsulata. *Materials Letters*, 61(18): 3984–3987.

Singh, J., Dutta, T., Kim, K. H., Rawat, M., Samddar, P. and Kumar, P. (2018). "Green" synthesis of metals and their oxide nanoparticles: Applications for environmental remediation. *Journal of Nanobiotechnology*, 16(1). https://doi.org/10.1186/s12951-018-0408-4.

Singh, N. B. (2022). Green synthesis of nanomaterials. *Handbook of Microbial Nanotechnology*, 225–254. https://doi.org/10.1016/B978-0-12-823426-6.00007-3.

Singh, P., Kim, Y. J., Zhang, D. and Yang, D. C. (2016). Biological synthesis of nanoparticles from plants and microorganisms. *Trends in Biotechnology*, 34(7): 588–599. https://doi.org/10.1016/j.tibtech.2016.02.006.

Singh, R., Hano, C., Nath, G. and Sharma, B. (2021). Green biosynthesis of silver nanoparticles using leaf extract of carissa carandas 1. And their antioxidant and antimicrobial activity against human pathogenic bacteria. *Biomolecules*, 11(2): 1–11. https://doi.org/10.3390/biom11020299.

Singh, V. A., Patil, R., Anand, A., Milani, P. and Gade, W. N. (2010). Biological synthesis of copper oxide nano particles using Escherichia coli. *Current Nanoscience*, 6(4): 365–369. https://doi.org/10.2174/157341310791659062.

Sneha, K., Sathishkumar, M., Kim, S. and Yun, Y. S. (2010). Counter ions and temperature incorporated tailoring of biogenic gold nanoparticles. *Process Biochemistry*, 45(9): 1450–1458. https://doi.org/10.1016/j.procbio.2010.05.019.

Soliman, A. M., Abdel-Latif, W., Shehata, I. H., Fouda, A., Abdo, A. M. and Ahmed, Y. M. (2021). Green approach to overcome the resistance pattern of candida spp. using biosynthesized silver nanoparticles fabricated by Penicillium chrysogenum F9. *Biological Trace Element Research*, 199(2): 800–811. https://doi.org/10.1007/s12011-020-02188-7.

Sunkar, S. and Nachiyar, C. V. (2012). Biogenesis of antibacterial silver nanoparticles using the endophytic bacterium Bacillus cereus isolated from Garcinia xanthochymus. *Asian Pacific Journal of Tropical Biomedicine*, 2(12): 953–959. https://doi.org/10.1016/S2221-1691(13)60006-4.

Suriati, G., Mariatti, M. and Azizan, A. (2014). Synthesis of silver nanoparticles by chemical reduction method: Effect of reducing agent and surfactant concentration. *International Journal of Automotive and Mechanical Engineering*, 10(1): 1920–1927. https://doi.org/10.15282/ijame.10.2014.9.0160.

Syafiuddin, A., Salmiati, Salim, M. R., Beng Hong Kueh, A., Hadibarata, T. and Nur, H. (2017). A review of silver nanoparticles: research trends, global consumption, synthesis, properties, and future challenges. *Journal of the Chinese Chemical Society*, 64(7): 732–756. https://doi.org/10.1002/jccs.201700067.

Thakkar, K. N., Mhatre, S. S. and Parikh, R. Y. (2010). Biological synthesis of metallic nanoparticles. *Nanomedicine: Nanotechnology, Biology, and Medicine*, 6(2): 257–262. https://doi.org/10.1016/j.nano.2009.07.002.

Thamer, N. A., Muftin, N. Q. and Al-Rubae, S. H. N. (2018). Optimization properties and characterization of green synthesis of copper oxide nanoparticles using aqueous extract of cordia myxa L. Leaves. *Asian Journal of Chemistry*, 30(7): 1559–1563. https://doi.org/10.14233/ajchem.2018.21242.

Titus, D., James Jebaseelan Samuel, E. and Roopan, S. M. (2018). Nanoparticle characterization techniques. *Green Synthesis, Characterization and Applications of Nanoparticles*, 303–319. https://doi.org/10.1016/B978-0-08-102579-6.00012-5.

Tolaymat, T. M., el Badawy, A. M., Genaidy, A., Scheckel, K. G., Luxton, T. P. and Suidan, M. (2010). An evidence-based environmental perspective of manufactured silver nanoparticle in syntheses and applications: A systematic review and critical appraisal of peer-reviewed scientific papers. *Science of the Total Environment*, 408(5): 999–1006. https://doi.org/10.1016/j.scitotenv.2009.11.003.

Tran, Q. H., Nguyen, V. Q. and Le, A. T. (2013). Silver nanoparticles: Synthesis, properties, toxicology, applications and perspectives. *Advances in Natural Sciences: Nanoscience and Nanotechnology*, 4(3). https://doi.org/10.1088/2043-6262/4/3/033001.

Tsuda, A. and Konduru, N. V. (2016). The role of natural processes and surface energy of inhaled engineered nanoparticles on aggregation and corona formation. *NanoImpact*, 2: 38–44. https://doi.org/10.1016/j.impact.2016.06.002.

Ullmann, M., Friedlander, S. K. and Schmidt-Ott, A. (2002). Nanoparticle formation by laser ablation. *Journal of Nanoparticle Research*, 4(6): 499–509. https://doi.org/10.1023/A:1022840924336.

van de Krol, R., Goossens, A. and Meulenkamp, E. A. (1999). *In situ* x-ray diffraction of lithium intercalation in nanostructured and thin film anatase TiO_2. *Journal of the Electrochemical Society*, 146(9): 3150–3154. https://doi.org/10.1149/1.1392447.

Virkutyte, J. and Varma, R. S. (2011). Green synthesis of metal nanoparticles: Biodegradable polymers and enzymes in stabilization and surface functionalization. *Chemical Science*, 2(5): 837–846. https://doi.org/10.1039/c0sc00338g.

Wadhwani, S. A., Shedbalkar, U. U., Singh, R. and Chopade, B. A. (2016). Biogenic selenium nanoparticles: current status and future prospects. *Applied Microbiology and Biotechnology*, 100(6): 2555–2566. https://doi.org/10.1007/s00253-016-7300-7.

Ward, M. B., Brydson, R. and Cochrane, R. F. (2006). Mn nanoparticles produced by inert gas condensation. *Journal of Physics: Conference Series*, 26(1): 296–299. https://doi.org/10.1088/1742-6596/26/1/071.

Wen, L., Lin, Z., Gu, P., Zhou, J., Yao, B., Chen, G. and Fu, J. (2009). Extracellular biosynthesis of monodispersed gold nanoparticles by a SAM capping route. *Journal of Nanoparticle Research*, 11(2): 279–288. https://doi.org/10.1007/s11051-008-9378-z.

Xin, B., Ren, Z., Wang, P., Liu, J., Jing, L. and Fu, H. (2007). Study on the mechanisms of photoinduced carriers separation and recombination for Fe 3+ -TiO_2 photocatalysts. *Applied Surface Science*, 253(9): 4390–4395. https://doi.org/10.1016/j.apsusc.2006.09.049.

Xing, T., Sunarso, J., Yang, W., Yin, Y., Glushenkov, A. M., Li, L. H., Howlett, P. C. and Chen, Y. (2013). Ball milling: A green mechanochemical approach for synthesis of nitrogen doped carbon nanoparticles. *Nanoscale*, 5(17): 7970–7976. https://doi.org/10.1039/c3nr02328a.

Xu, C., Guo, Y., Qiao, L., Ma, L., Cheng, Y. and Roman, A. (2018). Biogenic synthesis of novel functionalized selenium nanoparticles by Lactobacillus casei ATCC 393 and its protective effects on intestinal barrier dysfunction caused by Enterotoxigenic Escherichia coli K88. *Frontiers in Microbiology*, 9(JUN). https://doi.org/10.3389/fmicb.2018.01129.

Yadav, A., Kon, K., Kratosova, G., Duran, N., Ingle, A. P. and Rai, M. (2015). Fungi as an efficient mycosystem for the synthesis of metal nanoparticles: progress and key aspects of research. *Biotechnology Letters*, 37(11): 2099–2120. https://doi.org/10.1007/s10529-015-1901-6.

Yunus, I. S., Harwin, Kurniawan, A., Adityawarman, D. and Indarto, A. (2012). Nanotechnologies in water and air pollution treatment. *Environmental Technology Reviews*, 1(1): 136–148. https://doi.org/10.1080/21622515.2012.733966.

Zhang, J. and Gao, L. (2004). Synthesis and characterization of nanocrystalline tin oxide by sol-gel method. *Journal of Solid State Chemistry*, 177(4-5): 1425–1430. https://doi.org/10.1016/j.jssc.2003.11.024.

Zhang, M., Liu, Y. Q. and Ye, B. C. (2012). Colorimetric assay for parallel detection of Cd2+, Ni 2+ and Co2+ using peptide-modified gold nanoparticles. *Analyst*, 137(3): 601–607. https://doi.org/10.1039/c1an15909g.

Index

About the Editors

Dr. Sathish-Kumar Kamaraj is a Senior Research Professor at IPN-CICATA, Unidad Altamira, Mexico. He earned his Doctorate in Nanoscience and Nanotechnology (2010–2014) from CINVESTAV-IPN, Mexico, supported by SEP-Mexico. He received Best Student Competition Award (2013) and the Best Ph.D. Thesis Award from the Mexican Hydrogen Society. His research focuses on sustainable energy, environment, and health solutions, resulting in multiple patents and industry collaborations. He also mentors the University of Cambridge's ReachSci Program.

Dr. Arun Thirumurugan is an Assistant Professor at the University of ATACAMA, Vallenar, Chile. He completed his Ph.D. at the National Institute of Technology, Tiruchirappalli, India (2010–2015). He was a postdoctoral fellow at the Institute of Physics, Bhubaneswar, India (2015–2017), and a FONDECYT postdoctoral fellow at the University of Chile, Santiago (2017–2020). His research interests include synthesizing and surface modification of nanomaterials for energy and environmental applications.

Dr. Shanmuga Sundar Dhanabalan leads the 'Wearable and Connected Sensors' team at RMIT University, Melbourne, Australia. His research covers materials, sensors, flexible devices, wearables, optics, and photonics. He won the ARC Industry Fellowship 2024 and is the chief investigator for three active grants in wearable and flexible electronics. Collaborations have generated over $4.1M. He has delivered over 20 invited talks, serves as a guest editor for major journals, and is a Senior Member of IEEE.

Dr. Suresh K. Verma is a scientist at KIIT, focusing on materials science and biotechnology. He completed postdoctoral studies at Karolinska Institute and Uppsala University in Sweden. A member of the EU Horizon 2020 Graphene Flagship project, he received the Carl Trygger fellowship. He serves as an associate editor and reviewer for major journals and joined the Royal Society of Chemistry in 2023. He coordinates the Zebrafish research and Advanced Imaging facilities at KIIT.

Dr. Shanavas Shajahan is with the Pilot Scale Manufacturing Lab and RIC-2D at Khalifa University, UAE. His work focuses on producing graphene and related materials for commercialization, collaborating with industries in Asia, Europe, and North America. His research impacts energy, environment, and aerospace, including developing 2D materials for the Rashid Rover – 2, Emirates Lunar Program. He is an MRS and Graphene Council member and a scientific reviewer for major publishers.